Agricultural Environment and Intelligent Plant Protection Equipment

Agricultural Environment and Intelligent Plant Protection Equipment

Editors

Xiongkui He
Fuzeng Yang
Baijing Qiu

MDPI

Basel • Beijing • Wuhan • Barcelona • Belgrade • Novi Sad • Cluj • Manchester

Editors
Xiongkui He
China Agricultural University
Beijing
China

Fuzeng Yang
Northwest A&F University
Xianyang
China

Baijing Qiu
Jiangsu University
Zhenjiang
China

Editorial Office
MDPI AG
Grosspeteranlage 5
4052 Basel, Switzerland

This is a reprint of articles from the Special Issue published online in the open access journal *Agronomy* (ISSN 2073-4395) (available at: https://www.mdpi.com/journal/agronomy/special_issues/plant_protection_equipment).

For citation purposes, cite each article independently as indicated on the article page online and as indicated below:

Lastname, A.A.; Lastname, B.B. Article Title. *Journal Name* **Year**, *Volume Number*, Page Range.

ISBN 978-3-7258-1583-8 (Hbk)
ISBN 978-3-7258-1584-5 (PDF)
doi.org/10.3390/books978-3-7258-1584-5

Cover image courtesy of Xiongkui He

Contents

agronomy

Editorial

Agricultural Environment and Intelligent Plant Protection Equipment

Xiongkui He [1,*], Fuzeng Yang [2] and Baijing Qiu [3]

1 College of Science, China Agricultural University, Beijing 100193, China
2 College of Mechanical and Electronic Engineering, Northwest A&F University, Xianyang 712100, China; yfz0701@163.com
3 Key Laboratory of Plant Protection Engineering, Ministry of Agriculture and Rural Affairs, Jiangsu University, Zhenjiang 212013, China; qbj@ujs.edu.cn
* Correspondence: xiongkui@cau.edu.cn; Tel.: +86-6273-1446

Intelligent plant protection equipment utilizes advanced sensor technology and data analysis algorithms to achieve real-time monitoring and precise management of crop growth status, pest and disease situations, and environmental parameters. By employing sensors for data collection, image recognition technology, and automated operations, intelligent plant protection equipment can accurately identify pests and diseases, locate infected areas, and achieve precise pesticide spraying during operations, thereby minimizing pesticide usage and enhancing crop yield and quality. Additionally, through remote monitoring and management functions, users can oversee equipment remotely and schedule its operations, thereby improving agricultural production efficiency, reducing labor costs, and mitigating environmental impact.

Pesticide spraying remains the primary method for pest and disease control in orchards. Orchard spraying technology and equipment, as vital components of orchard pest and disease management, play a crucial role in reducing pesticide usage, improving pesticide efficiency, and minimizing pesticide pollution to the environment. Li et al. [1] conducted a study on the effects of airflow distribution patterns on the droplet deposition coverage and penetration efficiency of multi-unit air-assisted orchard sprayers. Their experimental results indicated that increasing airflow velocity can enhance the droplet deposition coverage on the backside of pear and cherry tree leaves, though it does not significantly affect spray penetration. The article also investigated the relationship between airflow velocity, direction, and droplet deposition indicators through Partial Least Squares (PLS) analysis. The study revealed that enhancing airflow velocity in the forward and horizontal directions can significantly improve the droplet deposition coverage on the outer side of pear trees; however, airflow indicators did not significantly affect droplet deposition coverage on cherry trees. Determining the orchard spraying deposition rate is crucial for optimizing application performance and pesticide efficiency. However, the large canopy and dense foliage of orchard trees, coupled with mutual shading, pose challenges for quantitatively assessing spraying efficiency. Wang et al. [2] examined spraying deposition quantity and distribution in different heights and layers of the fruit tree canopy using orchard sprayers. Their experimental findings showed that when using a trailed orchard sprayer, the upper and lower layers of leaves in the canopy have larger leaves, while the top and bottom have smaller leaves. When using a fixed orchard sprayer, the deposition quantity of spray liquid decreases with increasing canopy height. Additionally, within the canopy, the deposition quantity of spray liquid is lower, whereas it is higher in external areas. Sun et al. [3] employed an analysis method based on CFD orthogonal experiments to study the effects of four key structural parameters—shrinkage pipe attenuation angle, diffusion pipe divergence angle, venturi diameter, and venturi length—on an online pesticide mixing device and the pesticide dissolution and mixing performance. This study provides theoretical references for the production of prototype online pesticide mixing devices based on pipeline spraying.

Citation: He, X.; Yang, F.; Qiu, B. Agricultural Environment and Intelligent Plant Protection Equipment. *Agronomy* **2024**, *14*, 937. https://doi.org/10.3390/agronomy14050937

Received: 8 April 2024
Accepted: 25 April 2024
Published: 30 April 2024

To meet the demand for precise monitoring of field pest and disease information in modern agriculture, Wang et al. [4] proposed an intelligent plant protection monitoring system. This system comprises wireless cameras, temperature and humidity sensors, smart information terminals, and probes, facilitating the collection of plant images and meteorological information. Additionally, the system features positioning and communication functions, enabling the real-time recording of field environmental information and wireless data transmission to the monitoring center. In order to enhance the path-tracking control accuracy and driving stability of orchard trailed spraying robots, Ren et al. [5] proposed a navigation path-tracking control algorithm based on Double Deep Q Network (Double DQN). Through simulation tests, this method demonstrated high accuracy and stability in both straight and "U"-shaped path-tracking scenarios. To tackle the issue of pesticide depletion during field spraying tasks when the spraying robot is incomplete, Qin et al. [6] developed a spraying–dosing robot group and proposed a collaborative navigation system based on an orchard map. Firstly, a three-dimensional orchard point cloud map was constructed, and navigation path points were set on the projected map. Secondly, a master–slave command-based collaborative navigation strategy was developed, where the spraying robot acts as the master and the dosing robot as the slave. Finally, the spraying robot and the dosing robot performed collaborative navigation on the constructed map using the pure pursuit algorithm and the D-A control algorithm, respectively. To verify the collaborative navigation system, field tests were conducted on individual communication and navigation controls. The communication experiment results showed a packet loss rate of less than 5%, meeting the communication requirements. The navigation control experiment results indicated that the maximum absolute lateral error for the spraying robot was 24.9 cm, and for the dosing robot it was 29.7 cm.

Unmanned Aerial Vehicles (UAVs) for low-altitude, low-volume and ultra-volume crop protection operations offer several advantages, including the separation of humans and machinery and of personnel and chemicals; good adaptability to terrain; flexible maneuverability; efficient labor, water, and pesticide saving; as well as a high level of automation. Gradually, they have become significant equipment in agricultural crop protection operations. Dengeru et al. [7] conducted research on the application of insecticides sprayed by UAVs agricultural sprayers on nutmeg crops, investigating the spray deposition and drift characteristics. They found a potential method to effectively enhance insect control. This research opens up new possibilities for agricultural production and demonstrates the broad prospects of UAV technology in agriculture. Thomson et al. [8] studied the spray deposition and off-target drift characteristics of plant protection materials sprayed by UAVs to enhance spraying effectiveness and reduce off-target losses. They found that wind speed is the most important weather factor affecting the local off-target drift of pesticides. In drift of very fine droplets sampling experiments, accurately describing wind speed and direction at sampling positions is challenging due to natural variations in spray movement during the sampling target approach. Although accurately tracking wind direction to the target is difficult or even impossible, analyzing spray drift using statistical models with a large amount of field experiment data obtained by varying the wind speed and the tracking method can yield many conclusions. Comparing wind tunnel experiments with other technologies, Wang et al. [9] evaluated the spray drift of an octocopter Unmanned Aerial Spraying System (UASS). Compared to ground sprayers, UASS sprays more fine droplets at higher operating heights and faster travel speeds, hence posing higher drift risks. Li et al. [10] discussed the effectiveness and distribution of pesticide spraying using UAVs and ground sprayers. Their research revealed that cooperative spraying by UAVs and ground sprayers can improve the all-round coverage of pesticides in fruit tree canopies, especially in the upper and inner layers. By adjusting the spraying parameters, pesticide utilization and operational efficiency can be enhanced. Moreover, increasing the application proportion of ground equipment can improve the uniformity of deposits in the vertical direction of the canopy. In summary, this joint spraying strategy can reduce pesticide usage while ensuring the effective use of insecticides and protective fungicides. Mu et al. [11]

investigated the complex wheat field classification problem based on multi-scale feature fusion. They conducted ground target classification research using UAV multispectral remote sensing images in diverse wheat field scenes with various varieties. Compared to satellite remote sensing, UAVs provide high-spatial-resolution remote sensing images with rich details at low altitudes. However, different wheat varieties exhibit different characteristics, which may lead to category misjudgments during semantic segmentation, reducing the classification accuracy and affecting the classification effectiveness of ground targets.

The interaction between soil erosion and farmland vegetation restoration and its impact on soil enzyme activity and microbial nutrient limitations in the loess hilly region of China is a significant research area. Tang et al. [12] investigated these interactive effects by analyzing soil macroaggregate carbon (C), nitrogen (N), and phosphorus (P) contents, alongside microbial biomass and enzyme activity. The study revealed that soil properties such as composition, organic carbon content, and microbial biomass strongly influence soil enzyme activity and microbial nutrient limitations. Furthermore, different land use types (forest, grassland, and farmland) also display variations in their effects on soil enzyme activity and microbial nutrient limitations. Given the importance of timely disease detection to mitigate crop yield reduction due to pathogen infections, Wan et al. [13] explored the potential of hyperspectral imaging as a rapid, non-destructive sensing technology for disease identification. By analyzing spectral changes, they investigated the feasibility of using hyperspectral imaging to detect plant diseases in their early stages. In addition, Zhang et al. [14] examined the light and heat environment gradients and crop growth within open-air agriculture photovoltaic systems (OAVSs) in the eastern region of China. They measured solar radiation intensity, air temperature, and soil temperature during both summer and winter seasons. Utilizing Autodesk Ecotect Analysis 2011 software, they conducted light environment analysis, energy consumption analysis, and sunlight shading analysis to assess the effects of building structural parameters, material parameters, and environmental parameters on agricultural building environments. While the research findings offer valuable insights for standardizing agricultural production and the photovoltaic agriculture industries, there are limitations to consider. The study primarily focused on the effects of continuously laid photovoltaic panels on the light and heat environment within the OAVSs at specific spans and heights. Future research should expand to explore OAVSs with different photovoltaic panel layout densities, span lengths, and heights. Moreover, there is a need for further investigation into the response mechanisms of agricultural production to environmental factors, integrating environmental analysis and agricultural production to better understand the effects of light and heat environments on crop growth.

Author Contributions: All authors listed have made a substantial, direct, and intellectual contribution to the work and approved it for publication. All authors have read and agreed to the published version of the manuscript.

Acknowledgments: We thank the authors and reviewers for their valuable contributions on this topic.

Conflicts of Interest: The authors declare that the research was conducted in the absence of any commercial or financial relationships that could be construed as potential conflicts of interest.

References

1. Li, T.; Qi, P.; Wang, Z.; Xu, S.; Huang, Z.; Han, L.; He, X. Evaluation of the Effects of Airflow Distribution Patterns on Deposit Coverage and Spray Penetration in Multi-Unit Air-Assisted Sprayer. *Agronomy* **2022**, *12*, 944. [CrossRef]
2. Wang, S.; Wang, W.; Lei, X.; Wang, S.; Li, X.; Norton, T. Canopy Segmentation Method for Determining the Spray Deposition Rate in Orchards. *Agronomy* **2022**, *12*, 1195. [CrossRef]
3. Sun, D.; Liu, W.; Li, Z.; Zhan, X.; Dai, Q.; Xue, X.; Song, S. Numerical Experiment and Optimized Design of Pipeline Spraying On-Line Pesticide Mixing Apparatus Based on CFD Orthogonal Experiment. *Agronomy* **2022**, *12*, 1059. [CrossRef]
4. Wang, S.; Qi, P.; Zhang, W.; He, X. Development and Application of an Intelligent Plant Protection Monitoring System. *Agronomy* **2022**, *12*, 1046. [CrossRef]
5. Ren, Z.; Liu, Z.; Yuan, M.; Liu, H.; Wang, W.; Qin, J.; Yang, F. Double-DQN-Based Path-Tracking Control Algorithm for Orchard Traction Spraying Robot. *Agronomy* **2022**, *12*, 2803. [CrossRef]

6. Qin, J.; Wang, W.; Mao, W.; Yuan, M.; Liu, H.; Ren, Z.; Shi, S.; Yang, F. Research on a Map-Based Cooperative Navigation System for Spraying–Dosing Robot Group. *Agronomy* **2022**, *12*, 3114. [CrossRef]
7. Dengeru, Y.; Ramasamy, K.; Allimuthu, S.; Balakrishnan, S.; Kumar, A.P.M.; Kannan, B.; Karuppasami, K.M. Study on Spray Deposition and Drift Characteristics of UAV Agricultural Sprayer for Application of Insecticide in Redgram Crop (*Cajanus cajan* L. Millsp.). *Agronomy* **2022**, *12*, 3196. [CrossRef]
8. Thomson, S.J.; Huang, Y. Comparison of Weather Acquisition Periods Influencing a Statistical Model of Aerial Pesticide Drift. *Agronomy* **2023**, *13*, 213. [CrossRef]
9. Wang, C.; Wongsuk, S.; Huang, Z.; Yu, C.; Leng, H.; Zhang, J.; Sun, W.; Zeng, A.; He, X. Comparison between Drift Test Bench and Other Techniques in Spray Drift Evaluation of an Eight-Rotor Unmanned Aerial Spraying System: The Influence of Meteorological Parameters and Nozzle Types. *Agronomy* **2023**, *13*, 270. [CrossRef]
10. Li, Y.; Han, L.; Liu, L.; Huang, Z.; Wang, C.; He, X. Design and Spray Performance Evaluation of an Air–Ground Cooperation Stereoscopic Plant Protection System for Mango Orchards. *Agronomy* **2023**, *13*, 2007. [CrossRef]
11. Mu, F.; Chu, H.; Shi, S.; Yuan, M.; Liu, Q.; Yang, F. Research on the Classification of Complex Wheat Fields Based on Multi-Scale Feature Fusion. *Agronomy* **2022**, *12*, 2658. [CrossRef]
12. Tang, F.; Yao, Y.; Song, J.; Wang, C.; Liu, Y. Interactive Influence of Soil Erosion and Cropland Revegetation on Soil Enzyme Activities and Microbial Nutrient Limitations in the Loess Hilly-Gully Region of China. *Agronomy* **2022**, *12*, 2796. [CrossRef]
13. Wan, L.; Li, H.; Li, C.; Wang, A.; Yang, Y.; Wang, P. Hyperspectral Sensing of Plant Diseases: Principle and Methods. *Agronomy* **2022**, *12*, 1451. [CrossRef]
14. Zhang, L.; Yang, Z.; Wu, X.; Wang, W.; Yang, C.; Xu, G.; Wu, C.; Bao, E. Open-Field Agrivoltaic System Impacts on Photothermal Environment and Light Environment Simulation Analysis in Eastern China. *Agronomy* **2023**, *13*, 1820. [CrossRef]

agronomy

Review

Hyperspectral Sensing of Plant Diseases: Principle and Methods

Long Wan [1,†], Hui Li [1,†], Chengsong Li [1,*], Aichen Wang [2], Yuheng Yang [3] and Pei Wang [1,2,3,4,*]

1 College of Engineering and Technology, Southwest University, Chongqing 400715, China; wl09151057@email.swu.edu.cn (L.W.); leehui@swu.edu.cn (H.L.)
2 Key Laboratory of Modern Agricultural Equipment and Technology (Jiangsu University), Ministry of Education, School of Agricultural Engineering, Jiangsu University, Zhenjiang 212013, China; acwang@ujs.edu.cn
3 Interdisciplinary Research Center for Agriculture Green Development in Yangtze River Basin, Southwest University, Chongqing 400715, China; yyh023@swu.edu.cn
4 Institute of Urban Agriculture, Chinese Academy of Agricultural Sciences, Chengdu 610213, China
* Correspondence: lcs_shz@163.com (C.L.); peiwang@swu.edu.cn (P.W.)
† These authors contributed equally to this work.

Abstract: Pathogen infection has greatly reduced crop production. As the symptoms of diseases usually appear when the plants are infected severely, rapid identification approaches are required to monitor plant diseases at early the infection stage and optimize control strategies. Hyperspectral imaging, as a fast and nondestructive sensing technology, has achieved remarkable results in plant disease identification. Various models have been developed for disease identification in different plants such as arable crops, vegetables, fruit trees, etc. In these models, important algorithms, such as the vegetation index and machine learning classification and methods have played significant roles in the detection and early warning of disease. In this paper, the principle of hyperspectral imaging technology and common spectral characteristics of plant disease symptoms are discussed. We reviewed the impact mechanism of pathogen infection on the photo response and spectrum features of the plants, the data processing tools and algorithms of the hyperspectral information of pathogen-infected plants, and the application prospect of hyperspectral imaging technology for the identification of plant diseases.

Keywords: hyperspectral imaging technology; plant disease identification; photo response; machine learning

Citation: Wan, L.; Li, H.; Li, C.; Wang, A.; Yang, Y.; Wang, P. Hyperspectral Sensing of Plant Diseases: Principle and Methods. *Agronomy* **2022**, *12*, 1451. https://doi.org/10.3390/agronomy12061451

Academic Editor: Francesco Calzarano

Received: 7 April 2022
Accepted: 13 June 2022
Published: 17 June 2022

1. Introduction

Pathogens have brought great challenges to crop production in the last two decades. The reduction of crop yield induced by plant diseases can be up to 28.1% for wheat, 40.9% for rice, 41.4% for maize, 21% for potato, and 32.4% for soybean [1]. Several reasons may contribute to the increase in incidence around the world. Firstly, the globalization of human activities is promoting a more rapid spread and wider distribution of plant pathogens [2]. Secondly, commercialized high-yield varieties are bred from genotypes that are cultivated in certain regions. These varieties may not be adapted to new regional environments and may be less resistant to the local plant pathogens compared to their native wild relatives [3]. Finally, heavy fertilizer application, the large-scale cultivation of monocultures, and reduced crop rotation impact the soil bacterial composition and damage disease-suppressive microorganisms. As a result, there might be a higher possibility for the plants to be infected by pathogens [4].

Conventionally, plant diseases are identified visually according to symptoms when plants are infected severely by pathogens. This is usually quite time-consuming and labor-intensive. Furtherly, it is also a drawback for crop protection with fungicide application, as a better control effect could occur when spraying is conducted at the early stage of pathogen infection. In addition to manual monitoring, biological and molecular detection approaches also provide accurate assessment results of plant diseases. However, a laboratory bioassay can have a high cost and poor real-time performance as well as a limited sample size, which will restrict the measurement scale [5–7]. Thus, it is of great significance to evaluate the occurrence of plant diseases and provide an early warning of pathogen infection in crops in time. To satisfy the requirements of the high-throughput detection of plant diseases for precise fungi control, non-destructive and automatic remote detection methods with high sensitivity and reliability are required.

In recent years, modern sensing technologies have been applied for the automatic detection and early warning of plant diseases [8]. For instance, the infrared thermal imaging system has been applied to identify tomato mosaic virus and wheat leaf rust [9]; chlorophyll fluorescence imaging technology was applied to monitor the photosynthetic fingerprint of citrus Huanglongbing [10]; the deep learning model significantly improved the classification accuracy and efficiency of RGB images of healthy or pathogen-infected plants. However, the machine vision technology could only identify plant diseases if the symptoms were obviously apparent on the plant leaves, when it was usually too late for fungicide application. Compared to other sensing approaches, spectral imaging technology has been applied most commonly and has presented remarkable application potential in the real-time detection and evaluation of plant diseases [11–13]. In this review, we summarize the principles, application status, technical obstacles, data processing algorithms, and development trends of using hyperspectral imaging technology for disease identification in phytopathology.

2. Literature Overview

This review investigates the relevant literature published from 2001 to 2021. The following results were obtained:

(1) A search with the keywords "hyperspectral" and "plants" resulted in 6207 publications in total. Apart from the repeated literature records, 6165 relevant literature records were obtained. The number of relevant publications in the timeframe is shown in Figure 1a.

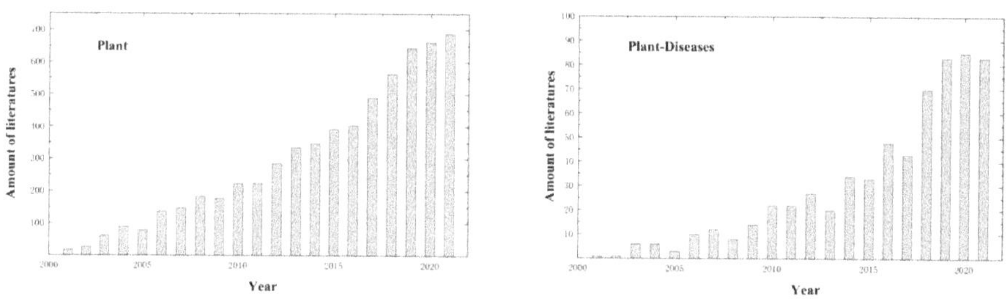

(a) (b)

Figure 1. Number of published relevant literature records each year. (**a**) Keywords "hyperspectral" and "plants"; (**b**) keywords "hyperspectral", "plants", and "disease".

(2) In a search of the subject "disease" in the results of (1), 621 literature records were obtained. The time distribution of their publication is shown in Figure 1b.

(3) The 621 literature results obtained above are analyzed in this review. Their research objects included grain crops, vegetables, fruit trees, medicinal materials, tobacco, and other crops [14–20]. Among them, it was found that wheat disease identification attracted the most attention from scientists, with 174 articles in total. Significant results were achieved in the detection of stripe rust [21], scab [22], powdery mildew [23], downy mildew [24], and several other main diseases in wheat.

To identify these diseases, various spectral data processing algorithms, including the vegetation index, spectral stoichiometry, machine learning, and deep learning were used. It should be noted that the deep learning model has become more common for spectral data processing than conventional machine learning tools in recent years. Both supervised and unsupervised classes were applied in the disease identification algorithms, which included the K-means clustering algorithm [25], support vector machine (SVM) [26], K-nearest neighbor (KNN) [27], decision tree algorithm [28], and deep learning methods, such as stacked auto-encoder (SAE) [29], deep belief network (DBN) [30], convolutional neural networks (CNNs) [31], etc. Meanwhile, the feature extraction methods of hyperspectral data have changed from extraction based on a single spatial or spectral feature to a space spectrum combination.

In general, hyperspectral imaging technology has achieved gratifying accomplishments in the field of plant disease detection, in which data processing methods, such as machine learning and deep learning have shown great potential.

3. Variation Mechanism of Spectral Information and Photo Response of Plant Diseases

3.1. Principles of Hyperspectral Imaging Technology and Sensors

Spectral analysis has been widely used for investigating the interaction between electromagnetic radiation and materials based on the wavelength and reflection intensity. It is a quantitative analysis method for photo information extraction in remote sensing. Via the analysis and interpretation of any object's spectrum information, it can reveal the variation of its surface physical characteristics or chemical features.

Hyperspectral remote sensing is also known as high spectral resolution remote sensing. According to the international remote sensing community, a spectral resolution of $10^{-1}\lambda$ is defined as multispectral, and a spectral resolution of $10^{-2}\lambda$ is defined as hyperspectral. As a comprehensive technology, hyperspectral imaging technology integrates weak signal detection, probing technology, precision optical machinery, and computing technology. It can obtain hyperspectral resolution image data with a continuous and narrow band. It is also a kind of multi-dimensional information acquisition technology that can combine imaging and spectral technologies.

Compared to non-imaging spectral technology, hyperspectral imaging data can provide additional information, such as shape, gradient, color, etc. The resolution of image data of hyperspectral imaging technology is high with a wavelength variation of $10^{-2}\lambda$ [32]. In the visible to shortwave infrared band, the spectral has a resolution level of a nanometer. There could be dozens or even hundreds of spectral bands. The spectral bands are continuous, and a complete hyperspectral resolution spectral curve can be extracted at each pixel location of the image data. Therefore, the hyperspectral data frame is formed as a three-dimensional image cube, as shown in Figure 2. The X-Y dimension expresses the spatial position information of the image. The third dimension (λ) is the spectral/wavelength dimension, which is composed of several bands in the spectral space. By making a multi-dimensional section of the spectral image cube, different types of spectral features can be obtained, such as the spectral characteristics at any point pixel, the spectral change of a spectral interval on an arbitrary spatial profile, or the spatial image of any band in the spectral dimension, etc. With the above information, we can not only identify the object according to the image feature in the spatial section but also analyze the spectral feature in the spectral dimension. As a result, it will benefit the recognition of the types, components, and contents of the substances [33].

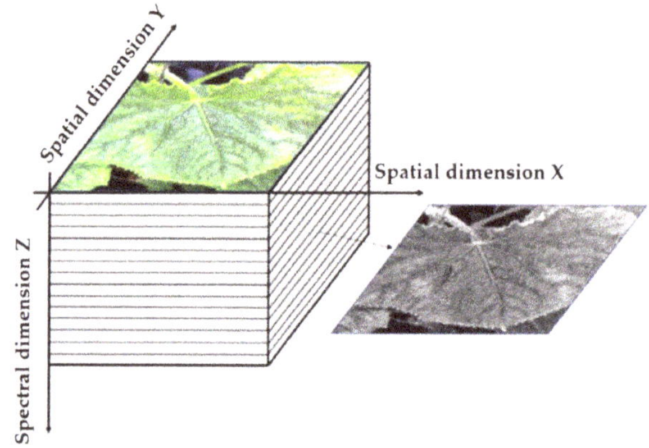

Figure 2. Structure of a hyperspectral image data cube of a cucumber leaf.

As the kernel part of the hyperspectral imaging system, spectral sensors can be classified according to their spectral resolution (the number and widths of measurement bands), spectral scales (ultraviolet, visible, near-infrared, shortwave infrared, etc.), or imaging principles (imaging or non-imaging) [34]. The hyperspectral imaging sensor can provide spectral information with a spatial resolution for the detected object, while the non-imaging sensors usually provide the average spectral information of a certain region in its field of view. In addition to RGB bands, the hyperspectral sensor can also cover the visible band of the electromagnetic spectrum from 400 nm to 700 nm, the near-infrared band from 700 nm to 1000 nm, and the shortwave infrared band from 1000 nm to 2500 nm. The possible narrow spectral resolution of a hyperspectral sensor can be less than 1 nm [35]. Multispectral sensors are similar to hyperspectral sensors, but they provide less data complexity and information content. These sensors measure the spectral information of objects in several bands, which are usually RGB and near-infrared. The reduction of the sensing expense can make multispectral sensors light and relatively cheap. In field applications, both multispectral and hyperspectral sensors are often combined with unmanned aerial vehicles for low-altitude remote sensing.

3.2. Photo Response of Pathogen-Infected Plants

Understanding the interaction between plants and light is crucial for the analysis of hyperspectral data. Many studies have shown that the modes of interaction between plant leaves and natural light can be classified into three types—transmission, absorption, and reflection, as shown in Figure 3.

Figure 3. Interaction between natural light and plant leaf surface.

Firstly, light can transmit through the leaves [36]; secondly, the light can be absorbed by chemical substances in the leaves (such as the pigments, water, sugars, lignin, and amino acids) [37,38]; thirdly, the light can be reflected by the surface or internal structure of the leaves (waxy cuticle, cell wall, etc.) [37,39].

Hyperspectral imaging technology can be used to monitor reflected and transmitted light. Hyperspectral data reveal the characteristics of the spectral absorption activity of the chemicals in the leaf. However, due to the limitation of the output of transmittance sensors and the complexity of their measurement settings, only a few systems for measuring both the reflectance and transmittance of light have been developed to detect plant diseases.

Different substances presented different reflectance and absorption values of light in specific wavebands, showing unique spectral characteristics. For example, the spectral characteristic curves of green plants are similar. They showed typical reflective spectral characteristics in some bands, which are shown in Figure 4, including the following aspects:

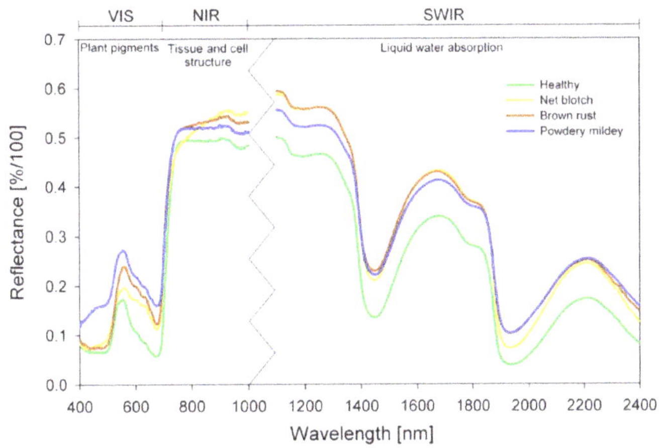

Figure 4. Spectral signatures of healthy barley leaves (green) and diseased barley leaves with net blotch (yellow), brown rust (brown), and powdery mildew (blue), respectively, at 10 days after inoculation. Sensors: "PS V10E" (400–1000 nm) and "SWIR" (1000–2500 nm) push broom scanners (Specim, Oulu, Finland). A zigzag line has been added to separate the discontinuous parts of the spectral curves. All measurements were taken with the HyperART setup according to Thomas et al. [34] (Reprinted with permission, copyright 2022 Anne-Katrin Mahlein). VIS visual, NIR—near-infrared, SWIR—shortwave infrared (parts of the electromagnetic spectrum).

(1) There is a small reflection peak near 550 nm, which corresponds to the strong reflection region of chlorophyll [40];

(2) The reflectance near 700 nm increases sharply, as chlorophyll absorption had low red reflectance and high internal leaf scattering, causing high near-infrared reflectance. The slope of the curve is closely related to the content of chlorophyll in vivo in plants. This is also known as the red edge phenomenon [41];

(3) In the band between 700 and 1200 nm, strong reflectance takes place, as the leaves almost no longer absorb light energy to avoid high-temperature burns [36];

(4) There are two moisture absorption bands with low reflectance near 1470 nm and 1940 nm, which created two troughs [42];

(5) In the spectrum range of 1300~2500 nm, the spectral reflectance characteristics are mainly related to the plant moisture content and CO_2 emissions [43].

Plant pathogenesis is a dynamic process in which a series of reactions occur continuously. In the process of plant–pathogen interactions, many physiological and biochemical reactions take place in vivo in plants. Based on the nutritional structure and ontogeny of different fungal pathogens, these changes in the plant pathogenesis in turn affect the optical properties of plants [44]. This makes it possible to detect plant diseases using the hyperspectral method and distinguish different diseases at each plant pathogenesis stage.

Conventionally, plant diseases are identified upon either symptoms (e.g., damage, wilt, gall, tumor, canker, wilt, rot, or necrotic disease) or visible pathogenic microorganisms (e.g., the spores in the genus rust, the mycelium or conidium in the genus erysipelas) [45]. Early plant–pathogen interactions are detected at a submillimeter size, which limits precise observation with visual assessment or HSI systems. The spatial resolution of a hyperspectral camera can be improved by using a hyperspectral microscope [46,47]. Using this approach, it is possible to study the microscopic and subtle resistance-related reactions and pathogenesis mechanisms in plants. Obtaining the hyperspectral imaging properties of the biological processes during plant–pathogen interactions enables the detection of plant diseases using hyperspectral imaging technology.

Complex processes control the emergence and progression of disease in each individual plant–pathogen system. Changes in the reflectance caused by plant disease could be attributed to the damage of specific chemical components of the leaf surfaces or tissues during pathogenesis, which could be chlorosis, the succession of necrotic tissue, or the appearance of typical fungal structures. Each host–pathogen interaction has specific and dynamic statuses in both spatial and temporal dimensions, which affects the spectral properties in different wavelength ranges [48–51].

4. Analysis Hyperspectral Imaging Information for Plant Disease Identification

Hyperspectral sensing is a non-contact and nondestructive detection method. Hyperspectral images contain a large amount of information that reveals environmental and chemical compound characteristics [51–53]. The challenge of hyperspectral data analysis is to extract target-related information from the huge amount of hyperspectral image data [51]. The useful information correlated with plant disease could vary during plant pathogenesis and be distributed among several regions of the measured spectrum. In recent years, a variety of data analysis methods have been applied to hyperspectral images for plant disease detection. This section reviews the most common preprocessing and analysis approaches to hyperspectral data.

4.1. Preprocessing of Hyperspectral Images

4.1.1. Image Mosaic

After the collection of crop images, it is necessary to splice several small-range images taken to obtain a complete and valuable large crop image. The mosaic of complete images is accomplished by matching pixels of the same name between images.

Fu et al. [54] reported an end-to-end deep learning method to reconstruct hyperspectral images directly from a raw mosaic image. It saves the separate demosaicing process required by other methods, which reconstruct the full-resolution RGB data from the raw mosaic image. The method reduced the computational complexity and accumulative error. Three different networks were designed based on the state-of-the-art models, including a residual network, a multiscale network, and a parallel-multiscale network. Liu et al. [55] used remote sensing images to extract the plant number information of maize seedling stages. To improve the accuracy of image mosaics, geometric reference boards were set up in different navigation belts as GPS control points to ensure subsequent mosaic work based on geographical positioning. Then the images with GPS positioning were input into Agisoft PhotoScan (Agisoft LLC, St. Petersburg, Russia) to complete the stitching process automatically. Another software, Pix4Dmapper (Pix4D S.A., Prilly, Switzerland) [56], can also automatically complete image stitching and generate high-precision normal projection images. For example, Dai et al. [57] provided a classification method of the main crops in Northern Xinjiang and He et al. [58] performed the biomass measurement of single tree trunks of Abies Minjiang using Pix4Dmapper (Pix4D S.A., Prilly, Switzerland).

4.1.2. Image Segmentation

Efficient image segmentation is important to separate the object from the background. Common image segmentation methods can be divided into three categories: region-based segmentation methods, threshold-based segmentation methods, and edge-based segmentation methods.

The selection of threshold determines the effect of image segmentation. The gray scale of the image can clearly show some characteristics of the object, so the gray histogram can be used to select the threshold of image segmentation.

Nalepa et al. [59] showed a method to effectively deal with a limited number and size of available hyperspectral ground-truth sets and apply transfer learning for building deep feature extractors. They also exploited spectral dimensionality reduction to make the technique applicable over hyperspectral data acquired using different sensors, which may capture different numbers of hyperspectral bands. The experiments were performed on several benchmarks and backed up with statistical tests. This indicates that the approach allowed for effectively training well-generalizing deep convolutional neural networks even using significantly reduced data. Cui et al. [60] proposed a novel spectral–spatial hyperspectral image classification method that exploited the spatial autocorrelation of hyperspectral images by which image segmentation was performed on the hyperspectral image to assign each pixel to a homogeneous region. Although good results were achieved, it still did not achieve the best segmentation for large field images with a complicated background, large scope, and many transformations. Therefore, it is still an urgent demand to explore innovative and efficient segmentation methods.

4.2. Vegetation Index

According to the spectral characteristics of green plants, visible and near-infrared bands can be combined to form various vegetation indices (VIS). The vegetation index method is derived from the analysis of satellite multispectral data [61]. It is also a very important tool in the process of hyperspectral data processing of medium- and near-range plant diseases. It can simply select several bands from hundreds of bands of spectral data to represent the changes in target diseases, which greatly reduces the processing cost of hyperspectral data analysis.

More than 40 VIS have been defined so far and are widely used in global and regional land cover [62,63], vegetation classification and environmental change [64,65], crop and pasture yield estimation [66,67], drought monitoring [63,68], and host–pathogen studies. Francesco et al. [69] investigated the relationships among tiger-stripe foliar symptom expression, microelements, and vegetation indices, including the NDVI, GNDVI, and WI of grapevine. The results showed that calcium could play a role in modulating the plants' response to toxic fungal metabolites, reducing the effects of an uncontrolled reaction associated with the expression of foliar symptoms in diseased vines. Therefore, increased availability of calcium and magnesium in up to pea-sized berries reduces foliar symptom expression just when they begin to appear. Rumpf et al. [70] used nine spectral vegetation indices related to physiological parameters to be features of a support vector machine with a radial basis function as a kernel. The discrimination between healthy sugar beet leaves and diseased leaves resulted in classification accuracies of up to 97%. According to Cao et al. [71], the SPAD values were predicted by calculating the spectral fractal dimension index (SFDI) from a hyperspectral curve (420 to 950 nm). The correlation between the SPAD values and hyperspectral information was further analyzed to determine the sensitive bands that correspond to different disease levels.

Table 1 shows several examples of general spectral vegetation indices used to detect and measure the severity of various plant diseases. Formulas for defining new targeted vegetation indices can be developed for specific plant diseases, such as leaf beet disease [72]. Nevertheless, vegetation index methods abandon the spectral data of most bands. They do not fully take advantage of the potential of hyperspectral data.

Table 1. Application of general spectral vegetation indices.

Vegetation Index	Formula	Application
Normalized difference vegetation index (NDVI)	$(R_{800} - R_{670})/(R_{800} + R_{670})$	Detecting vegetation coverage, related to chlorophyll content.
Green normalized difference vegetation index (GNDVI)	(NIR − GREEN)/(NIR + GREEN)	Detecting withered or aged crops and measuring the nitrogen content in leaves in the absence of red bands.
Water index (WI)	R_{900}/R_{970}	Estimating the water stress.
Ratio vegetation index (RVI)	R_{800}/R_{670}	Detecting and estimating plant biomass, related to chlorophyll content.
Plant senescence reflectance index (PSRI)	$(R_{680} - R_{500})/R_{750}$	Detecting plant senescence, related to pigment content.
Carotenoid reflectance index (CARI)	$1/R_{510} - 1/R_{550}$	Related to carotenoid content.
Anthocyanin reflectance index (ARI)	$1/R_{510} - 1/R_{700}$	Detecting yellow rust of wheat, related to anthocyanin content.
Photochemical reflectance index (PRI)	$(R_{570} - R_{531})/(R_{570} + R_{531})$	Detecting yellow rust of wheat, related to photosynthesis and carotenoid content.
Structure insensitive pigment index (SIPI)	$(R_{800} - R_{445})/R_{800} + R_{680})$	Related to the ratio of carotinoids and chlorophylla.

4.3. Machine Learning

Hyperspectral image classification tries to define each pixel sample in the image with different categories. Machine learning hyperspectral image classification methods can be divided into unsupervised learning methods and supervised learning methods according to whether prior categories of the training samples need to be introduced.

Unsupervised learning refers to "blind" classification (clustering) based on the distribution rules of spectral features of hyperspectral images without prior knowledge. The K-means cluster and the dynamic clustering analysis methods (interactive self-organizing data analysis—ISODATA) are typical unsupervised learning methods [64].

Since plants and background regions in spectral images have significant differences in the near-infrared and red bands, Zhang et al. [73] removed the non-plant background using the K-means clustering algorithm before the detection of the scabs of rice sheath blight. Compared to the traditional background removal methods based on pixel threshold segmentation, the clustering-based method can mitigate the issue of irregular edges and the "salt and pepper" phenomenon. Based on the selected spectral features, Yuan et al. [74] developed an analytical framework for disease scab detection that combines the unsupervised classification ISODATA method and adaptive two-dimensional threshold determination. A comparison with the direct pixel-based classification results suggests that the proposed method is insensitive to differences in leaf background and can effectively identify diseased tea leaves (anthracnose) and analyze the degree of infection. The accuracies for both the calibration and validation samples were satisfactory, with an overall accuracy of 96% at the pixel level. Supervised learning is based on prior knowledge, namely labeled training samples, to learn the internal relationships between the hyperspectral image pixel samples and the corresponding categories, which are used for mapping between the unlabeled pixel samples and the corresponding categories, and finally to determine the category for unlabeled pixels. Although the classification method of supervised learning requires manually labeled training samples to assist the training classification model, its disadvantages lie in a high labor cost and strong human subjective factors. Compared to the unsupervised learning method, the classification accuracy of supervised learning can be improved through the repeated testing of prior training samples. Therefore, the supervised learning method is more popular than the unsupervised learning method. Table 2 shows the common supervised learning algorithms with their advantages and disadvantages.

Table 2. Comparison of supervised learning algorithms.

Method	Advantage	Disadvantage
K-NN	(1) The theory has matured and easy application. (2) Less training time.	(1) High computational complexity and spatial complexity. (2) Sample imbalance.
NBM	(1) Performs well for small-scale data and is suitable for multi-classification. (2) Simple computation.	(1) Due to the emphasis on conditional independence, the accuracy is reduced. (2) Sensitive to the expression of input data. A priori probability needs to be calculated.
MLE	It can quickly specify the pixels to be classified into one of several classes.	When there are many types of hyperspectral data, the operation speed slows down significantly, and more training samples are needed.
Decision Tree	(1) Low computing cost. (2) Able to handle irrelevant features. (3) Suitable for datasets with many attributes and has high scalability.	(1) Easy to cause over-fitting. (2) When all kinds of samples are unbalanced, the result of information gain will be biased to the characteristics with more data values.
ELM	Has fast learning speed and good generalization ability.	The parameters of the model are randomly selected, which causes ELM instability.
SRC	Low computing cost.	Various sub-dictionaries have high atomic correlation, and similar samples may be linearly represented by different sub-dictionaries, resulting in poor classification accuracy.
SVM	(1) Low generalization error rate, low computing cost. (2) It can solve the classification problem of small-sample and high-dimensional data. Has good classification effect.	It is sensitive to parameter adjustment and function selection.

The feature extraction is an essential step in the classification process that largely affects the final classification accuracy of the algorithm in comparison and combination of thermal, fluorescence early identification of root rot disease with hyperspectral reflectance [65]. In this paper, supervised machine learning classification methods were divided into conventional machine learning classification and deep learning classification according to whether the features could be automatically learned from the original hyperspectral data.

Classification methods based on ordinary machine learning can only fit the relationship between the samples and the corresponding categories according to the inherent features of the hyperspectral images or features designed by feature engineering. According to the features, the general machine learning classification method can be classified as a spectral characteristic-based classification method or a spatial spectrum joint feature-based classification method.

Conventional machine learning classification methods only use spectral features of hyperspectral images or spectral features extracted from original data through feature engineering as the basis for classification, without utilizing the spatial information of images. The methods include K-nearest neighbor (K-NN), maximum likelihood estimate (MLE), support vector machine (SVM), Fisher's linear discrimination analysis (FLDA), naive Bayes (NB), decision tree, extreme learning machine (ELM), sparse representation-based classifier (SRC), etc.

Calderon et al. [75] applied the LDA and SVM classification methods to classify the *Verticillium dahliae* severity of olive using remote sensing at a large scale of 3000 ha. The LDA reached an overall accuracy of 59.0% and a kappa of 0.487, while the SVM obtained a higher overall accuracy of 79.2% and a similar kappa of 0.495. However, the LDA better classified trees at the initial and low severity levels, reaching accuracies of 71.4 and 75.0%, respectively, in comparison to the 14.3% and 40.6% obtained by the SVM. Based on the selected spectral bands, vegetation indices, and wavelet features from the canopy hyperspectral data of winter wheat, Huang et al. [76] established discriminate models using FLDA and SVM. The results indicated that FLDA was more suitable for differentiating stresses, with respective accuracies of 78.1% and 95.6% for powdery mildew and stripe rust. Karadag et al. [77] used an artificial neural network (ANN), naive Bayes (NB), and K-nearest neighbor (KNN) to classify the spectral features of healthy and fusarium diseased peppers. The average success rates were calculated as 100% for KNN, 88.125% for the ANN, and 82% for NB. Other relevant studies are provided in Table 3.

Table 3. Recent advances in plant disease detection based on machine learning.

Reference	Pathosystem	Scale	Methods	Detection	Early Detection	Precision
Calderon et al. (2015) [75]	Olive—Verticillium wilt	Canopy	LDA, SVM		√	Overall: LDA 59.0%, SVM 79.2% Early detection: LDA 75.0%, SVM 40.6%.
Xie et al. (2017) [78]	Tomato—Gray mold disease	Leaf	KNN, C5.0, FR-KNN		√	Detection: 94.44%, 94.44%, 97.22% (KNN, C5.0, KNN) Early detection: 66.67%, 66.67%, 41.67%
Lu et al. (2017) [79]	Strawberry—Anthracnose crown rot	Leaf	SDA, FDA, KNN	√	√	SDA: 71.3%, FDA: 70.05%, KNN: 73.6%
Zhu et al. (2017) [20]	Tobacco—Mosaic virus	Leaf	SPA–GLCM–BPNN/ELM/SVM		√	95.00%
Huang et al. (2018) [76]	Wheat—Powdery mildew, stripe rust	Canopy	FLDA, SVM	√		FLDA: Powdery mildew 78.1%, stripe rust 95.6%
Weng et al. (2018) [80]	Citrus—Huanglongbing, Fe deficient	Leaf	LS–SVM			93.50%
Nagasubramanian et al. (2018) [81]	Sugar beet—Cercospora leaf spot	stem	GA–SVM		√	Detection: 97% Early detection: 90.91%
Abdulridha et al. (2019) [82]	Citrus—Canker	Leaf/canopy	RBF, KNN		√	Leaf: 94%, 95%, 96% (asymptomatic, early, and late symptoms) Canopy: 94%, 96%, 100%
Mahlein et al. (2019) [24]	Wheat—Fusarium head blight	Spikelet	SVM		√	Detection: 89.0% Early detection: 78.0%
Deng et al. (2019) [83]	Citrus—Huanglongbing	Leaf	SVM	√	√	Detection: 96% Early Detection: 90.8%
Lin et al. (2020) [84]	Rice—Sheath blight	Leaf	Relief decision tree			95.50%
Van de Vijver et al. (2020) [85]	Potato—Early blight	Leaf	PLS–DA, PCA–SVM, decision tree	√		92.78%
Karadag et al. (2020) [77]	Pepper—Fusarium disease, mycorrhizal fungus	Leaf	KNN, ANN, NB			KNN: 100%, ANN: 88.125%, NB: 82%
Zhang et al. (2020) [86]	Wheat—Fusarium head blight	Leaf	SPA-RF			96.44%
Gu et al. (2020) [87]	Tobacco—Tomato spotted wilt virus	Leaf	SPA–BRT		√	85.20%
Yuan et al. (2020) [74]	Tea—Anthracnose	Leaf	ISODATA			Pixel level: 98% Patch level: 94%
Calamita et al. (2021) [88]	Vitis vinifera—Fungi infection	Leaf	NB		√	Detection: 90% Early detection: 75%
Zhang et al. (2021) [73]	Rice—Sheath blight	Leaf	K-means–FLDA–HFPSSD			Pixel level: 98.42% Patch level: 95.92%
Zhao et al. (2022) [89]	Tea—Green leafhopper, anthracnose, sunburn	Leaf	CWA–K-means, SVM–RF			Green leafhopper: 93.99–94.20%, Anthracnose 94.12–94.28%, Sunburn 82.50–83.91%

However, the label samples of hyperspectral data are limited, and most machine learning methods based on spectral features are affected by the Hughes phenomenon. The classification results often have many discrete isolated points, which are seriously inconsistent with the actual land cover distribution [90].

Hyperspectral images are characterized by high dimension, high spectral redundancy, and mixed pixels. In the case of limited label samples, it is difficult to obtain accurate classification results by directly using spectral features as the classification basis. Therefore, in recent years, several groups have introduced the spatial information of hyperspectral images based on spectral information and made up for the deficiency of spectral information only through the dependence among the spatial pixels [91]. Ghasimi et al. [92] proposed a method that combined the hidden Markov random field and the SVM classifier. The method used the hidden Markov random field to generate segmentation results and the classification results generated by the SVM to determine the classification results of the final algorithm through majority voting.

Although the above classification methods based on the space spectrum combination achieved good results, they all needed to design the classification features manually through feature engineering to improve the classification accuracy, which required a lot of time and energy to repeatedly verify. Unlike ordinary machine learning, which requires a manual design to extract the classification features, deep learning methods can automatically extract the optimal classification features.

4.4. Deep Learning

Compared to ordinary machine learning methods, deep learning has the characteristics of automatic learning and a strong classification ability and has achieved great performance improvement in many fields, such as image classification, target detection, natural language processing, and so on. Due to the rapid development of deep learning in many fields, some remote sensing image processing algorithms based on deep learning have been slowly proposed in recent years. Common deep learning algorithms used in hyperspectral image classification mainly include the stacked auto encoder (SAE) [93], deep belief network (DBN) [94], and convolutional neural networks (CNNs) [95].

A stack autoencoder network (SAE) is a deep neural network model composed of several sparse autoencoders. Usually, the output of the former autoencoder serves as the input of the latter, and the final classification result is output by a logistic (for binary classification) or SoftMax (for multiple classification) classifier. The SAE algorithm is widely used in hyperspectral image classification. For example, Chen et al. [96] extracted null spectrum features in hyperspectral images by forming a deep network of multi-layer stacked autoencoders. Wei et al. [97] combined stacked denoising autoencoders and superpixels through space constraints to improve the classification accuracy with a novel structure.

Deng et al. [98] used a UAV hyperspectral remote sensing tool for the HLB rapid detection of citrus. The proposed HLB detection methods (based on the multi-feature fusion of the vegetation index) and canopy spectral feature parameters constructed (based on the feature band in the SAE) had a classification accuracy of 99.33% and a loss of 0.0783 for the training set, and a classification accuracy of 99.72% and a loss of 0.0585 for the validation set. The field-testing results showed that the model could effectively detect the HLB plants and output the distribution of the disease in the canopy, thus judging the plant disease level in a large area efficiently.

DBN, like the SAE algorithm, is composed of multiple layers of constrained Boltzmann machines, and its generation process is essentially a process of unsupervised pre-training for each constrained Boltzmann machine. Compared to SAEs, DBNs are used less often in hyperspectral classification. Zhao et al. [99] proposed a new feature extraction and image classification framework based on a DBN for hyperspectral image classification. Zhong et al. [100] regularized the pre-training and fine-tuning stages through the a priori of potential factors for diversity enhancement, to improve the original DBN classification effect.

Sun et al. [30] used the spectral and imaging information from hyperspectral reflectance (400 similar to 1000 nm) to evaluate and classify three kinds of common peach disease (*Botrytis cinerea, Rhizopus stolonifera, Colletotrichum acutatum*). Three decayed stages (slight, moderate, and severe decayed peaches) were considered for classification by DBN and PLSDA. The results showed that the DBN model had better classification results than the classification accuracy of the PLSDA model. The DBN model based on integrated information (494 features) showed the highest classification results for the three diseases, with accuracies of 82.5%, 92.5%, and 100% for slightly decayed, moderately decayed, and severely decayed samples, respectively.

SAEs and DBNs are deep neural networks based on full connection. With the deepening of the network depth, the number of parameters in the model becomes extremely high, which is not conducive to subsequent training and requires great computational power. In addition, although the above deep network model can effectively extract deep features in hyperspectral images to a certain extent to enhance the discrimination among different categories, the model cannot make good use of the spatial information of hyperspectral images by converting the input data into one-dimensional vectors. In this regard, the academic community adopted the classification method based on a CNN to avoid the above problems and improve the learning and expression ability of the network more efficiently.

Table 4 shows the comparison between an ordinary neural network and a convolutional neural network. Most traditional hyperspectral classification methods used to classify research are based on the characteristics of the project. This requires a priori knowledge, and the traits are mostly shallow characteristics, but in actual research, it is not known exactly what kind of characteristics exist in the image that are the most adaptable to the problem at hand, and it requires a lot of time and energy to determine them. However, the hyperspectral image algorithm based on a CNN can automatically learn and acquire hidden advanced features in the image by constructing a multi-layer neural network and can thus obtain the classification results more efficiently and accurately.

Table 4. Comparison of ANN and CNN.

Method	Advantages	Disadvantages
ANN	Able to learn advanced features of data automatically and has excellent classification performance.	(1) The model requires huge parameters and high computational force. (2) The input is a one-dimensional vector, which cannot make good use of map spatial information. (3) Difficult training, easy to produce gradient disappearance.
CNN	(1) Able to learn advanced features of data automatically and has excellent classification performance. (2) Compared to the ANN, the CNN model requires significantly fewer data parameters and trains faster.	The existence of a pooling layer may lead to the loss of valuable information while ignoring the relationship between the global and local scales.

Table 5 shows some relevant studies from recent years. For the detection of potato viruses, Polder et al. [101] designed a fully convolutional neural network that was adapted for hyperspectral images and trained on two experimental rows in the field. The trained network was validated on two other rows, with different potato cultivars. For three of the four row/date combinations, the precision and recall compared to conventional disease assessment exceeded 0.78 and 0.88, respectively. The classification of healthy and diseased wheat heads in a rapid and non-destructive manner for the early diagnosis of Fusarium head blight disease research is difficult. Zhang et al. [102] applied an improved two-dimensional CNN model to the pixels of a hyperspectral image to accurately discern the disease area. The results of the model show that the two-dimensional convolutional bidirectional gated recurrent unit neural network (2D-CNN-BidGRU) has an F1 score and accuracy of 0.75 and 0.743, respectively, which illustrates that the hybrid structure deep neural network is an excellent classification algorithm for the healthy and Fusarium head blight disease classifications in the field of hyperspectral imagery.

Table 5. Recent advances in plant disease detection based on deep learning.

Reference	Host–Pathogen System	Scale	Methods	Field Capable	Early Detection	Precision
Sun et al. (2018) [30]	Peach—Botrytis cinerea, Rhizopus stolonifera, Colletotrichum acutatum	Flesh	PCA–DBN		√	Detection: 92.5% Early Detection: 82.5%
Jin et al. (2018) [103]	Wheat—Fusarium head blight	Kernels	CNN		√	74.30%
Polder et al. (2019) [101]	Potato—Viruses	Leaf	CNN	√	√	78.00%
Zhang et al. (2019) [102]	Wheat—Yellow rust	canopy	DCNN	√		85%
Deng et al. (2020) [98]	Citrus—Huanglongbing	canopy	SAE	√		99.72%
Liang et al. (2020) [29]	Wheat—Fusarium head blight	Kernels/Flour	MSC-GA-SAE			Kernels: 100% Flour: 96%
Zhang et al. (2020) [15]	Rice—Bacterial blight	Leaf	CNN			82.60%
Feng et al. (2020) [104]	Rice—Leaf blight, rice blast, rice sheath blight	Leaf	SVM, LR, CNN			93.00%
Fazari et al. (2021) [100]	Olive—Anthracnose	Leaf	CNN		√	Detection: 100% Early Detection: 85%
Hernandez et al. (2021) [105]	Grape—Downy mildew	Leaf	KNN, CNN		√	CNN: 82% KNN: 66%
Nguyen et al. (2021) [106]	Grape—Grapevine vein-clearing virus	vine	2D CNN, 3D CNN		√	2D CNN:71% 3D CNN: 75%
Liu et al. (2022) [107]	Wheat—Fusarium head blight	kernels	ASSDN			98.31%

5. Summary

The application of hyperspectral imaging technology in agriculture makes full use of the advantages of hyperspectral mapping unity, which can precisely monitor crop growth and diseases. However, so far, hyperspectral cameras are still expensive, making them difficult to be widely applied in agriculture. The process of spectral acquisition is easily affected by environmental factors. It also takes a long time to acquire, analyze, and process the hyperspectral image data, which limits the application of hyperspectral imaging systems in agricultural real-time monitoring and online detection. When the pre-established model is applied to another index system, data analysis and model transformation are required. The characteristics of the sample may affect its classification results and prediction accuracy. In addition, hyperspectral imaging technologies have high redundancy in image processing. To reduce the time consumption of obtaining and processing hyperspectral data, it is often necessary to extract featured wavelengths for specific applications in different crops, which enables the multiple spectral imaging of the specific stresses or diseases and reduces the cost.

Other than pathogens, hyperspectral image information can also indicate stresses induced by abiotic environmental factors, crop growing process, and biological injury. With the combination of the latest machine learning models and hyperspectral data processing tools for plant disease detection, novel algorithms should be proposed to deal with complex natural conditions. Thus, other biotic or abiotic stress effects on plant spectrum features might be removed from plant disease identification models. As a result, this technology could be more useful for field practice. Meanwhile, plant disease identification models should effectively contribute to the decision support system to carry out real-time control strategies. Thus, the data processing speed should be improved with modern on-chip calculation systems such as FPGA. This means specific intellectual property kernels should be created for real-time plant disease identification and control strategy applications in later studies.

Author Contributions: Conceptualization, L.W., H.L., P.W., Y.Y. and C.L.; methodology, H.L., P.W. and Y.Y.; investigation, L.W., H.L. and P.W.; resources, A.W.; data curation, L.W. and H.L.; writing—original draft preparation, L.W. and H.L.; writing—review and editing, P.W., Y.Y. and C.L.; supervision, C.L. and P.W.; project administration, P.W.; funding acquisition, H.L., P.W. and A.W. All authors have read and agreed to the published version of the manuscript.

Funding: This research was funded by the National Natural Science Foundation of China, grant number 32001425; the Natural Science Foundation of Chongqing, China, grant numbers cstc2020jcyj-msxmX0414 and cstc2020jcyj-msxmX0459; the Key R&D Projects in the Artificial Intelligence Pilot Area of Chongqing, China, grant number cstc2021jscx-gksbX0067; the Open Funding of the Key Laboratory of Modern Agricultural Equipment and Technology (Jiangsu University), grant numbers MAET202105 and MAET202112; Local Financial Funds of the National Agricultural Science and Technology Center, Chengdu, grant number NASC2020KR05.

Acknowledgments: The authors would like to thank Lihong Wang and Qi Niu from Southwest University and Wei Ma from the Chinese Academy of Agricultural Sciences for their advice.

Conflicts of Interest: The authors declare no conflict of interest.

Abbreviations

ANN	Artificial Neural Network
ARI	Anthocyanin Reflectance Index
ASSDN	Architecture Self-Search Deep Network
BRT	Boosted Regression Tree
BPNN	Back Propagation Neural Network
CARI	Carotenoid Reflectance Index
CNN	Convolutional Neural Networks
CWA	Continuous Wavelet Analysis
DBN	Deep Belief Network
DCNN	Dynamic Convolution Neural Network)
ELM	Extreme Learning Machine
FDA	Fisher Discriminant Analysis
FLDA	Fisher's Linear Discrimination Analysis
FR	Features Ranking
GA	Genetic Algorithm
GLCM	Gray-level Co-occurrence Matrix
HFPSSD	Hyperspectral Feature Profile Scanning-based Scab Detection
HLB	HuangLongBing
ISODATA	Interactive Self-Organizing Data Analysis
KNN	K-Nearest Neighbor
LDA	Linear Discriminant Analysis
LR	Logistic Regression
LS	Least Squares
MLE	Maximum Likelihood Estimate
MSC	Multiplicative Scatter Correction
NB	Naive Bayes
NVDI	Normalized Difference Vegetation Index
PCA	Principal Component Analysis
PRI	Photochemical Reflectance Index
PSRI	Plant Senescence Reflectance Index
RBF	Radial Basis Function
RGB	Red, Green, Blue

RVI	Ratio Vegetation Index
SAE	Stacked Auto-Encoder
SDA	Stepwise Discriminant Analysis
SFDI	Spectral Fractal Dimension Index
SIPI	Structure Insensitive Pigment Index
SPA	Successive Projections Algorithm
SPAD	Soil and Plant Analyzer Development
SRC	Sparse Representation-based Classifier
SVM	Support Vector Machine
UAV	Unmanned Aerial Vehicle
VIS	Vegetation Indices
WI	Water Index

References

1. Savary, S.; Willocquet, L.; Pethybridge, S.J.; Esker, P.; McRoberts, N.; Nelson, A. The global burden of pathogens and pests on major food crops. *Nat. Ecol. Evol.* **2019**, *3*, 430–439. [CrossRef] [PubMed]
2. Fisher, M.C.; Henk, D.A.; Briggs, C.J.; Brownstein, J.S.; Madoff, L.C.; McCraw, S.L.; Gurr, S.J. Emerging fungal threats to animal, plant and ecosystem health. *Nature* **2012**, *484*, 186–194. [CrossRef] [PubMed]
3. Boyd, L.A.; Ridout, C.; O'Sullivan, D.M.; Leach, J.E.; Leung, H. Plant–pathogen interactions: Disease resistance in modern agriculture. *Trends. Genet.* **2013**, *29*, 233–240. [CrossRef]
4. Peralta, A.L.; Sun, Y.; McDaniel, M.D.; Lennon, J.T. Crop rotational diversity increases disease suppressive capacity of soil microbiomes. *Ecosphere* **2018**, *9*, e02235. [CrossRef]
5. Ristaino, J.B.; Anderson, P.K.; Bebber, D.P.; Brauman, K.A.; Wei, Q. The persistent threat of emerging plant disease pandemics to global food security. *Proc. Natl. Acad. Sci. USA* **2021**, *118*, e2022239118. [CrossRef]
6. Dan, E.; Ho, T.; Rwahnih, M.A.; Martin, R.R.; Tzanetakis, I. High throughput sequencing for plant virus detection and discovery. *Phytopathology* **2019**, *109*, 716–725.
7. Ma, Z.; Luo, Y.; Michailides, T.J. Nested pcr assays for detection of monilinia fructicola in stone fruit orchards and botryosphaeria dothidea from pistachios in california. *J. Phytopathol.* **2010**, *151*, 312–322. [CrossRef]
8. Singh, V.; Sharma, N.; Singh, S. A review of imaging techniques for plant disease detection. *Artif. Intell. Agric.* **2020**, *4*, 229–242. [CrossRef]
9. Zhu, W.; Chen, H.; Ciechanowska, I.; Spaner, D. Application of infrared thermal imaging for the rapid diagnosis of crop disease. *IFAC-PapersOnLine* **2018**, *51*, 424–430. [CrossRef]
10. Cen, H.; Weng, H.; Yao, J.; He, M.; Lv, J.; Hua, S.; Li, H.; He, Y. Chlorophyll fluorescence imaging uncovers photosynthetic fingerprint of citrus Huanglongbing. *Front. Plant Sci.* **2017**, *8*, 1509. [CrossRef]
11. Mahlein, A.K.; Kuska, M.T.; Behmann, J.; Polder, G.; Walter, A. Hyperspectral sensors and imaging technologies in phytopathology: State of the art. *Annu. Rev. Phytopathol.* **2018**, *56*, 535–558. [CrossRef] [PubMed]
12. Zhang, J.; Huang, Y.; Pu, R.; Gonzalez-Moreno, P.; Yuan, L.; Wu, K.; Huang, W. Monitoring plant diseases and pests through remote sensing technology: A review. *Comput. Electron. Agric.* **2019**, *165*, 104943. [CrossRef]
13. Zaneti, R.N.; Girardi, V.; Spilki, F.R.; Mena, K.; Etchepare, R. Quantitative microbial risk assessment of Sars-CoV-2 for workers in wastewater treatment plants. *Sci. Total Environ.* **2020**, *754*, 142163. [CrossRef]
14. Zhang, J.; Wang, B.; Zhang, X.; Liu, P.; Huang, W. Impact of spectral interval on wavelet features for detecting wheat yellow rust with hyperspectral data. *Int. J. Agr. Biolog. Eng.* **2018**, *11*, 138–144. [CrossRef]
15. Zhang, J.; Yang, Y.; Feng, X.; Xu, H.; He, Y. Identification of bacterial blight resistant rice seeds using terahertz imaging and hyperspectral imaging combined with convolutional neural network. *Front. Plant Sci.* **2020**, *11*, 821. [CrossRef] [PubMed]
16. Jia, L.; Wang, L.; Yang, F.; Yang, L. Spring corn leaf blight monitoring based on hyperspectral derivative index. *Chin. Agric. Sci. Bull.* **2019**, *35*, 143–150.
17. Lu, J.; Zhou, M.; Gao, Y.; Jiang, H. Using hyperspectral imaging to discriminate yellow leaf curl disease in tomato leaves. *Precis. Agric.* **2018**, *19*, 379–394. [CrossRef]
18. Nouri, M.; Gorretta, N.; Vaysse, P.; Giraud, M.; Germain, C.; Keresztes, B.; Roger, J.M. Near infrared hyperspectral dataset of healthy and infected apple tree leaves images for the early detection of apple scab disease. *Data Brief* **2018**, *16*, 967–971. [CrossRef]
19. Sandasi, M.; Vermaak, I.; Chen, W.; Viljoen, A. Skullcap and germander: Preventing potential toxicity through the application of hyperspectral imaging and multivariate image analysis as a novel quality control method. *Planta Med.* **2014**, *80*, 1329–1339. [CrossRef]
20. Zhu, H.; Chu, B.; Zhang, C.; Liu, F.; Jiang, L.; He, Y. Hyperspectral imaging for presymptomatic detection of tobacco disease with successive projections algorithm and machine-learning classifiers. *Sci. Rep.* **2017**, *7*, 4125. [CrossRef]
21. Hovmller, M.S.; Walter, S.; Bayles, R.A.; Hubbard, A.; Flath, K.; Sommerfeldt, N.; Leconte, M.; Czembor, P.; Rodriguez-Algaba, J.; Thach, T. Replacement of the European wheat yellow rust population by new races from the centre of diversity in the near-himalayan region. *Plant Pathol.* **2016**, *65*, 402–411. [CrossRef]

22. Huang, L.; Zhang, H.; Chao, R.; Huang, W.; Hu, T.; Zhao, J. Detection of scab in wheat ears using in situ hyperspectral data and support vector machine optimized by genetic algorithm. *Int. J. Agr. Biol. Eng.* **2020**, *13*, 7. [CrossRef]
23. Shi, Y.; Huang, W.; Zhou, X. Evaluation of wavelet spectral features in pathological detection and discrimination of yellow rust and powdery mildew in winter wheat with hyperspectral reflectance data. *J. Appl. Remote Sens.* **2017**, *11*, 026025. [CrossRef]
24. Mahlein, A.K.; Alisaac, E.; Masri, A.A.; Behmann, J.; Oerke, E.C. Comparison and combination of thermal, fluorescence, and hyperspectral imaging for monitoring fusarium head blight of wheat on spikelet scale. *Sensors* **2019**, *19*, 2281. [CrossRef]
25. Leucker, M.; Wahabzada, M.; Kersting, K.; Peter, M.; Beyer, W.; Steiner, U.; Mahlein, A.K.; Oerke, E.C. Hyperspectral imaging reveals the effect of sugar beet quantitative trait loci on cercospora leaf spot resistance. *Funct. Plant Bio.* **2016**, *44*, 1. [CrossRef] [PubMed]
26. Huang, S.; Qi, L.; Xue, K.; Wang, W.; Zhu, X. Hyperspectral image analysis based on bosw model for rice panicle blast grading. *Comput. Electron. Agric.* **2015**, *118*, 167–178. [CrossRef]
27. Shuaibu, M.; Lee, W.S.; Schueller, J.; Gader, P.; Hong, Y.; Kim, S. Unsupervised hyperspectral band selection for apple marssonina blotch detection. *Comput. Electron. Agric.* **2018**, *12*, 28. [CrossRef]
28. Ruszczak, B.; Smykaa, K. The detection of alternaria solani infection on tomatoes using ensemble learning. *J. Amb. Intel. Smart Environ.* **2020**, *12*, 407–418. [CrossRef]
29. Liang, K.; Huang, J.; He, R.; Wang, Q.; Chai, Y.; Shen, M. Comparison of Vis-NIR and SWIR hyperspectral imaging for the non-destructive detection of DON levels in Fusarium head blight wheat kernels and wheat flour. *Infrared Phys. Techn.* **2020**, *106*, 103281. [CrossRef]
30. Sun, Y.; Wang, K.; Liu, Q.; Pan, L.; Tu, K. Classification and discrimination of different fungal diseases of three infection levels on peaches using hyperspectral reflectance imaging analysis. *Sensors* **2018**, *18*, 1295. [CrossRef]
31. Wang, C.; Liu, B.; Liu, L.; Zhu, Y.; Li, X. A review of deep learning used in the hyperspectral image analysis for agriculture. *Artif. Intell. Rev.* **2021**, *54*, 5205–5253. [CrossRef]
32. Xing, H.; Feng, H.; Fu, J.; Xu, X.; Yang, G. *Development and Application of Hyperspectral Remote Sensing*; Springer: Berlin, Germany, 2017; Volume 546, pp. 271–282.
33. Telmo, A.O.; Joná, H.; Luís, P.; José, B.; Emanuel, P.; Raul, M.; Joaquim, S. Hyperspectral imaging: A review on UAV-based sensors, data processing and applications for agriculture and forestry. *Remote Sens.* **2017**, *9*, 1110.
34. Thomas, S.; Kuska, M.T.; Bohnenkamp, D.; Brugger, A.; Alisaac, E.; Wahabzada, M.; Behmann, J.; Mahlein, A.K. Benefits of hyperspectral imaging for plant disease detection and plant protection: A technical perspective. *J. Plant. Dis. Protect.* **2018**, *125*, 5–20. [CrossRef]
35. Gates, D.M.; Keegan, H.J.; Schleter, J.C. Sensorik für einen präzisierten Pflanzenschutz. *Gesunde Pflanz.* **2008**, *60*, 131–141.
36. Wang, L.; Jia, M.; Yin, D.; Tian, J. A review of remote sensing for mangrove forests: 1956–2018. *Remote Sens. Environ.* **2019**, *231*, 111223. [CrossRef]
37. Hennessy, A.; Clarke, K.; Lewis, M. Hyperspectral classification of plants: A review of waveband selection generalisability. *Remote Sens.* **2020**, *12*, 113. [CrossRef]
38. Olli, N.; Eija, H.; Sakari, T.; Niko, V.; Teemu, H.; Xiaowei, Y.; Juha, H.; Heikki, S.; Ilkka, P.L.N.; Nilton, I. Individual tree detection and classification with UAV-based photogrammetric point clouds and hyperspectral imaging. *Remote Sens.* **2017**, *9*, 185.
39. René, H.; Norbert, J.; André, G.; Jens, O. The effect of epidermal structures on leaf spectral signatures of ice plants (Aizoaceae). *Remote Sens.* **2015**, *7*, 16901.
40. Liu, Y.; Zhang, G.W.; Liu, D. Simultaneous measurement of chlorophyll and water content in navel orange leaves based on hyperspectral imaging. *Spectroscopy* **2014**, *29*, 40, 42–46.
41. Mutanga, O.; Van Aardt, J.; Kumar, L. Imaging spectroscopy (hyperspectral remote sensing) in Southern Africa: An overview. *S. Afr. J. Sci.* **2010**, *105*, 193–198. [CrossRef]
42. Wei, A.A.; Tian, L.; Chen, X.; Yu, Y. Retrieval and application of chlorophyll-a concentration in the Poyang Lake based on exhaustion method: A case study of Chinese Gaofen-5Satellitc AHSI data. *J. Huazhong Normal Univ.* **2020**, *54*, 447–453.
43. Qu, J.; Sun, D.; Cheng, J.; Pu, H. Mapping moisture contents in grass carp (ctenopharyngodon idella) slices under different freeze drying periods by vis-nir hyperspectral imaging. *LWT-Food Sci. Technol.* **2017**, *75*, 529–536. [CrossRef]
44. Oerke, E. Remote sensing of diseases. *Annu. Rev. Phytopathol.* **2020**, *58*, 225–252. [CrossRef] [PubMed]
45. Bonants, P.; Schoen, C.; Wolf, J.; Zijlstra, C. Developments in detection of plant pathogens and other plant-related organisms: Detection in the past towards detection in the future. *Mededelingen* **2001**, *66*, 25–37. [PubMed]
46. Pu, H.; Lin, L.; Sun, D. Principles of hyperspectral microscope imaging techniques and their applications in food quality and safety detection: A review. *Compr. Rev. Food Sci. F.* **2019**, *18*, 853–866. [CrossRef]
47. Gary, A.; Roth, S.; Tahiliani, N.M.; Neu-Baker, S.A. Hyperspectral microscopy as an analytical tool for nanomaterials. *Wires. Nanomed. Nanobi.* **2015**, *7*, 565–579.
48. Mahlein, A.K.; Rumpf, T.; Welke, P.; Dehne, H.W.; Plümer, L.; Steiner, U.; Oerke, E.C. Development of spectral indices for detecting and identifying plant diseases. *Remote Sens. Environ.* **2013**, *128*, 21–30. [CrossRef]
49. Mahlein, A.K.; Steiner, U.; Hillnhütter, C.; Dehne, H.W.; Oerke, E.C. Hyperspectral imaging for small-scale analysis of symptoms caused by different sugar beet diseases. *Plant Methods* **2012**, *8*, 3. [CrossRef]
50. Wahabzada, M.; Mahlein, A.K.; Bauckhage, C.; Steiner, U.; Oerke, E.C.; Kersting, K. Plant Phenotyping using Probabilistic Topic Models: Uncovering the Hyperspectral Language of Plants. *Sci. Rep.* **2016**, *6*, 22482. [CrossRef]

51. Behmann, J.; Mahlein, A.K.; Rumpf, T.; Romer, C.; Plumer, L. A review of advanced machine learning methods for the detection of biotic stress in precision crop protection. *Precis. Agric.* **2015**, *16*, 239–260. [CrossRef]
52. Lowe, A.; Harrison, N.; French, A.P. Hyperspectral image analysis techniques for the detection and classification of the early onset of plant disease and stress. *Plant Methods* **2017**, *13*, 80. [CrossRef] [PubMed]
53. Mahlein, A.K. Plant Disease Detection by Imaging Sensors-Parallels and Specific Demands for Precision Agriculture and Plant Phenotyping. *Plant Dis.* **2016**, *100*, 241–251. [CrossRef] [PubMed]
54. Fu, H.; Bian, L.; Cao, X.; Zhang, J. Hyperspectral imaging from a raw mosaic image with end-to-end learning. *Opt. Express* **2020**, *28*, 314–324. [CrossRef]
55. Liu, S.; Yang, G.; Zhou, H.; Jing, H.; Feng, H.; Xu, B.; Yang, H. Extraction of maize seedling number information based on UAV imagery. *Trans. CSAE* **2018**, *34*, 9.
56. Zhang, J.; Yang, C.; Zhao, B.; Song, H.; Hoffmann, W.C.; Shi, Y.; Zhang, D.; Zhang, G. Crop Classification and LAI Estimation Using Original and Resolution-Reduced Images from Two Consumer-Grade Cameras. *Remote Sens.* **2017**, *9*, 1054. [CrossRef]
57. Dai, J.; Zhang, G.; Guo, P.; Zeng, Y.; Cui, M.; Xue, J. Classification method of main crops in northern Xinjiang based on UAV visible waveband images. *Trans. CSAE* **2018**, *34*, 8.
58. He, Y.; Zhang, Y.; Li, J.; Wang, J. Estimation of stem biomass of individual Abies faxoniana through unmanned aerial vehicle remote sensing. *J. Beijing For. Univ.* **2016**, *38*, 8.
59. Nalepa, J.; Myller, M.; Kawulok, M. Transfer learning for segmenting dimensionally-reduced hyperspectral images. *IEEE Geosci. Remote Sens.* **2019**, *17*, 1228–1232. [CrossRef]
60. Cui, B.; Ma, X.; Xie, X.; Ren, G.; Ma, Y. Classification of visible and infrared hyperspectral images based on image segmentation and edge-preserving filtering. *Infrared Phys. Techn.* **2017**, *81*, 79–88. [CrossRef]
61. Ishiyama, T.; Tanaka, S.; Uchida, K.; Fujikawa, S.; Yamashita, Y.; Kato, M. Relationship among vegetation variables and vegetation features of arid lands derived from satellite data. *Adv. Space Res.* **2001**, *28*, 183–188. [CrossRef]
62. Vicente-Serrano, S.M. Evaluating the impact of drought using remote sensing in a mediterranean, semi-arid region. *Nat. Hazards* **2007**, *40*, 173–208. [CrossRef]
63. Zhao, X.; Zhou, D.; Fang, J. Satellite-based studies on large-scale vegetation changes in China. *J. Integr. Plant Bio.* **2012**, *54*, 713–728. [CrossRef] [PubMed]
64. Zhao, J. Research on Hyperspectral Remote Sensing Images Classification Based on K-means Clustering. *Geospat. Inform.* **2016**, *14*, 4.
65. Zhang, H.; Li, Y.; Jiang, H. Research status and Prospect of deep learning in hyperspectral image classification. *Acta Autom. Sin.* **2018**, *44*, 17.
66. Szulczewski, W.; Zyromski, A.; Jakubowski, W.; Biniak-Pierog, M. A new method for the estimation of biomass yield of giant miscanthus (*Miscanthus giganteus*) in the course of vegetation. *Renew. Sust. Energ. Rev.* **2018**, *82*, 1787–1795. [CrossRef]
67. Barriguinha, A.; Neto, M.D.; Gil, A. Vineyard yield estimation, prediction, and forecasting: A systematic literature review. *Agronomy* **2021**, *11*, 1789. [CrossRef]
68. Zou, L.; Cao, S.; Sanchez-Azofeifa, A. Evaluating the utility of various drought indices to monitor meteorological drought in tropical dry forests. *Int. J. Biometeorol.* **2020**, *64*, 701–711. [CrossRef]
69. Calzarano, F.; Pagnani, G.; Pisante, M.; Bellocci, M.; Cillo, G.; Metruccio, E.G.; Di Marco, S. Factors involved on tiger-stripe foliar symptom expression of esca of grapevine. *Plants* **2021**, *10*, 1041. [CrossRef]
70. Rumpf, T.; Mahlein, A.K.; Steiner, U.; Oerke, E.C.; De Hne, H.W.; Plümer, L. Early detection and classification of plant diseases with support vector machines based on hyperspectral reflectance. *Comput. Electron. Agr.* **2010**, *74*, 91–99. [CrossRef]
71. Cao, Y.F.; Xu, H.L.; Song, J.; Yang, Y.; Hu, X.H.; Wiyao, K.T.; Zhai, Z.Y. Applying spectral fractal dimension index to predict the spad value of rice leaves under bacterial blight disease stress. *Plant Methods* **2022**, *18*, 67. [CrossRef]
72. Sims, D.A.; Gamon, J.A. Relationships between leaf pigment content and spectral reflectance across a wide range of species, leaf structures and developmental stages. *Remote Sens. Environ.* **2002**, *81*, 337–354. [CrossRef]
73. Zhang, J.C.; Tian, Y.Y.; Yan, L.J.; Wang, B.; Wang, L.; Xu, J.F.; Wu, K.H. Diagnosing the symptoms of sheath blight disease on rice stalk with an in-situ hyperspectral imaging technique. *Biosyst. Eng.* **2021**, *209*, 94–105. [CrossRef]
74. Yuan, L.; Yan, P.; Han, W.; Huang, Y.; Bao, Z. Detection of anthracnose in tea plants based on hyperspectral imaging. *Comput. Electron. Agr.* **2019**, *167*, 105039. [CrossRef]
75. Calderon, R.; Navas-Cortes, J.A.; Zarco-Tejada, P.J. Early detection and quantification of verticillium wilt in olive using hyperspectral and thermal imagery over large areas. *Remote Sens.* **2015**, *7*, 5584–5610. [CrossRef]
76. Huang, W.; Lu, J.; Ye, H.; Kong, W.; Yue, S. Quantitative identification of crop disease and nitrogen-water stress in winter wheat using continuous wavelet analysis. *Int. J. Agr. Biol. Eng.* **2018**, *11*, 8. [CrossRef]
77. Karadag, K.; Tenekeci, M.E.; Tasaltin, R.; Bilgili, A. Detection of pepper fusarium disease using machine learning algorithms based on spectral reflectance. *Sustain. Comput Inform. Syst.* **2020**, *28*, 100299. [CrossRef]
78. Xie, C.; Yang, C.; He, Y. Hyperspectral imaging for classification of healthy and gray mold diseased tomato leaves with different infection severities. *Comput. Electron. Agr.* **2017**, *135*, 154–162. [CrossRef]
79. Lu, J.; Ehsani, R.; Shi, Y.; Abdulridha, J.; Castro, A.D.; Xu, Y. Field detection of anthracnose crown rot in strawberry using spectroscopy technology. *Comput. Electron. Agric.* **2017**, *135*, 289–299. [CrossRef]

80. Weng, H.Y.; Lv, J.W.; Cen, H.Y.; He, M.B.; Zeng, Y.B.; Hua, S.J.; Li, H.Y.; Meng, Y.Q.; Fang, H.; He, Y. Hyperspectral reflectance imaging combined with carbohydrate metabolism analysis for diagnosis of citrus Huanglongbing in different seasons and cultivars. *Sens. Actuators B-Chem.* **2018**, *275*, 50–60. [CrossRef]
81. Nagasubramanian, K.; Jones, S.; Sarkar, S.; Singh, A.K.; Singh, A.; Ganapathysubramanian, B. Hyperspectral band selection using genetic algorithm and support vector machines for early identification of charcoal rot disease in soybean stems. *Plant Methods* **2018**, *14*, 86. [CrossRef]
82. Abdulridha, J.; Batuman, O.; Ampatzidis, Y. Uav-based remote sensing technique to detect citrus canker disease utilizing hyperspectral imaging and machine learning. *Remote Sens.* **2019**, *11*, 1373. [CrossRef]
83. Deng, X.; Huang, Z.; Zheng, Z.; Lan, Y.; Dai, F. Field detection and classification of citrus huanglongbing based on hyperspectral reflectance. *Comput. Electron. Agr.* **2019**, *167*, 105006. [CrossRef]
84. Lin, F.; Guo, S.; Tan, C.; Zhou, X.; Zhang, D. Identification of rice sheath blight through spectral responses using hyperspectral images. *Sensors* **2020**, *20*, 6243. [CrossRef] [PubMed]
85. Vijver, R.; Mertens, K.; Heungens, K.; Somers, B.; Saeys, W. In-field detection of *Alternaria solani* in potato crops using hyperspectral imaging. *Comput. Electron. Agr.* **2019**, *168*, 105106. [CrossRef]
86. Zhang, D.; Chen, G.; Zhang, H.; Jin, N.; Chen, Y. Integration of spectroscopy and image for identifying fusarium damage in wheat kernels using hyperspectral imaging. *Spectrochim. Acta A* **2020**, *236*, 118344. [CrossRef] [PubMed]
87. Gu, Q.; Sheng, L.; Zhang, T.H.; Lu, Y.W.; Zhang, Z.J.; Zheng, K.F.; Hu, H.; Zhou, H.K. Early detection of tomato spotted wilt virus infection in tobacco using the hyperspectral imaging technique and machine learning algorithms. *Comput. Electron. Agr.* **2019**, *167*, 105066. [CrossRef]
88. Calamita, F.; Imran, H.A.; Vescovo, L.; Mekhalfi, M.L.; Porta, N.L. Early identification of root rot disease by using hyperspectral reflectance: The case of pathosystem grapevine/armillaria. *Remote Sens.* **2021**, *13*, 2436. [CrossRef]
89. Zhao, X.H.; Zhang, J.C.; Huang, Y.B.; Tian, Y.Y.; Yuan, L. Detection and discrimination of disease and insect stress of tea plants using hyperspectral imaging combined with wavelet analysis. *Comput. Electron. Agr.* **2022**, *193*, 106717. [CrossRef]
90. Du, P.; Xia, J.; Xue, Z.; Tan, K.; Su, H.; Bao, R. Research progress of hyperspectral remote sensing image classification. *Remote Sens. Bull.* **2016**, *20*, 21.
91. Hou, P.; Yao, M.; Jia, W.; Zhang, F.; Wang, D. Spatial spectrum discriminant analysis for hyperspectral image classification. *Opt. Precis. Eng.* **2018**, *26*, 450–460.
92. Ghasimi, P.; Benediktsson, J.A.; Ulfarsson, M.O. The spectral-spatial classification of hyperspectral images based on Hidden Markov Random Field. *IEEE T. Geosci. Remote Sens.* **2014**, *52*, 2565–2574.
93. Hinton, G.E.; Salakhutdinov, R.R. Reducing the Dimensionality of Data with Neural Networks. *Science* **2006**, *313*, 504–507. [CrossRef] [PubMed]
94. Hinton, G.E.; Osindero, S.; Teh, Y.W. A Fast Learning Algorithm for Deep Belief Nets. *Neural Comput.* **2014**, *18*, 1527–1554. [CrossRef]
95. Krizhevsky, A.; Sutskever, I.; Hinton, G.E. ImageNet Classification with Deep Convolutional Neural Networks. *Commun. ACM* **2012**, *25*, 1097–1105. [CrossRef]
96. Wei, F.; Li, S.; Fang, L. Spectral-spatial hyperspectral image classification via superpixel merging and sparse representation. *IEEE Geosci. Remote. Sens.* **2015**, *18*, 861–865.
97. Zhao, X.; Chen, Y.; Jia, X. Spectral-Spatial Classification of Hyperspectral Data Based on Deep Belief Network. *IEEE J. Sel. Top. Appl. Earth Obs. Remote Sens.* **2015**, *8*, 2381–2392.
98. Deng, X.L.; Zhu, Z.H.; Yang, J.C.; Zheng, Z.; Huang, Z.X.; Yin, X.B.; Wei, S.J.; Lan, Y.B. Detection of citrus huanglongbing based on multi-input neural network model of uav hyperspectral remote sensing. *Remote Sens.* **2020**, *12*, 2678. [CrossRef]
99. Zhong, P.; Gong, Z.; Li, S.; Schonlieb, C.B. Learning to Diversify Deep Belief Networks for Hyperspectral Image Classification. *IEEE T. Geosci. Remote* **2017**, *55*, 3516–3530. [CrossRef]
100. Fazari, A.; Pellicer-Valero, O.J.; Gomez-SancHs, J.; Bernardi, B.; Blasco, J. Application of deep convolutional neural networks for the detection of anthracnose in olives using vis/nir hyperspectral images. *Comput. Electron. Agr.* **2021**, *187*, 106252. [CrossRef]
101. Polder, G.; Blok, P.M.; Villiers, H.; Wolf, J.; Kamp, J. Potato virus detection in seed potatoes using deep learning on hyperspectral images. *Front. Plant Sci.* **2019**, *10*, 209. [CrossRef]
102. Zhang, X.; Han, L.; Dong, Y.; Shi, Y.; Sobeih, T. A deep learning-based approach for automated yellow rust disease detection from high-resolution hyperspectral uav images. *Remote Sens.* **2019**, *11*, 1554. [CrossRef]
103. Xiu, J.; Lu, J.; Shuai, W.; Hai, Q.; Shao, L. Classifying wheat hyperspectral pixels of healthy heads and fusarium head blight disease using a deep neural network in the wild field. *Remote Sens.* **2018**, *10*, 395.
104. Feng, L.; Wu, B.; Zhu, S.; Wang, J.; Zhang, C. Investigation on data fusion of multisource spectral data for rice leaf diseases identification using machine learning methods. *Front. Plant Sci.* **2020**, *11*, 577063. [CrossRef] [PubMed]
105. Hernandez, I.; Gutierrez, S.; Ceballos, S.; Iniguez, R.; Barrio, I.; Tardaguila, J. Artificial intelligence and novel sensing technologies for assessing downy mildew in grapevine. *Horticulturae* **2021**, *7*, 103. [CrossRef]

106. Nguyen, C.; Sagan, V.; Maimaitiyiming, M.; Maimaitijiang, M.; Bhadra, S.; Kwasniewski, M.T. Early detection of plant viral disease using hyperspectral imaging and deep learning. *Sensors* **2021**, *21*, 742. [CrossRef] [PubMed]
107. Lv, Y.P.; Lv, W.B.; Han, K.X.; Tao, W.T.; Zheng, L.; Weng, S.Z.; Huang, L.S. Determination of wheat kernels damaged by fusarium head blight using monochromatic images of effective wavelengths from hyperspectral imaging coupled with an architecture self-search deep network. *Food Control* **2022**, *135*, 108819.

agronomy

Article

Design and Spray Performance Evaluation of an Air–Ground Cooperation Stereoscopic Plant Protection System for Mango Orchards

Yangfan Li [1,2,3], Leng Han [1,2,3], Limin Liu [1,2,3], Zhan Huang [1,2,3], Changling Wang [1,2,3] and Xiongkui He [1,2,3,*]

1 College of Science, China Agricultural University, Beijing 100193, China; s20203101935@cau.edu.cn (Y.L.); wcl1991@cau.edu.cn (C.W.)
2 Centre for Chemicals Application Technology, China Agricultural University, Beijing 100193, China
3 College of Agricultural Unmanned System, China Agricultural University, Beijing 100193, China
* Correspondence: xiongkui@cau.edu.cn

Abstract: With the aim of solving the problems of high labor intensity, low operational efficiency, and poor deposition distribution uniformity in the mango canopy associated with traditional plant protection devices, an air-ground co-operation stereoscopic plant protection system consisting of an orchard caterpillar mist sprayer and a six-rotor plant protection UAV was designed to jointly undertake plant protection operations in mango orchards. We tested the spraying performance of the system on mango trees, compared with the single-machine operation, the air–ground co-operation system could significantly increase the droplet coverage on the upperside of mango leaves in each part of the canopy (a 14.7% increase for the mist sprayer and 12.9% for the UAV). This increased the active component deposition distribution uniformity in the mango canopy but could not significantly improve the deposition and coverage of droplets on the underside of leaves compared with the mist sprayer and plant protection UAV. Due to the characteristics of the mango canopy such as large leaf length and thickness and complex leaf inclination distribution, this led to poor deposition distribution uniformity of the two spray units, and the overall CV was over 150%. The pesticide active ingredients were almost uniformly distributed in the vertical direction when the application ratios (ground implements/plant protection drones) were 8/2 and 7/3, offering a promising protocol for reduced pesticide application in mango orchards. This study presents promising data that support the innovative integration of drones into crop protection programs for large canopy crops (e.g., mango) and provides guidance for the ACSPPS system in reduction and precision application research.

Keywords: smart mango orchard; co-operative plant protection; UAV; pest control

Citation: Li, Y.; Han, L.; Liu, L.; Huang, Z.; Wang, C.; He, X. Design and Spray Performance Evaluation of an Air–Ground Cooperation Stereoscopic Plant Protection System for Mango Orchards. *Agronomy* **2023**, *13*, 2007. https://doi.org/10.3390/agronomy13082007

Academic Editor: Xiangjun Zou

Received: 21 June 2023
Revised: 19 July 2023
Accepted: 25 July 2023
Published: 28 July 2023

1. Introduction

Asian mango production accounts for 72.9% of worldwide production, especially in some Southeast Asian countries, and their products occupy an important position in the export of agricultural products [1]. Mango plantations are widely distributed in tropical areas [2], with a high frequency of pest and disease outbreaks. The occurrence characteristics and laws of pest and disease differ between regions, producing difficulties in plant protection operations for mango trees. [3] In mango orchards, the common diseases included anthracnose, bacterial keratosis, and powdery mildew, and the common insect pests included thrips and weevils [4–6].

In Southeast Asian countries, manual knapsack sprayers and high-pressure line guns are still used as the main application equipment for chemical control [7], which means a low spray (i.e., pesticides) application efficiency for crops and poor deposition distribution uniformity [8]. The wind-assisted target orchard sprayer, with its higher spray volume and spray pressure, achieves an ideal pesticide deposition in the canopy of fruit trees to effectively control pests and diseases [9]. Therefore, this type of equipment is widely used in the plant protection operations in orchards. However, when spraying large-volume canopy fruit trees with orchard sprayers, pesticide deposition shows a gradual increase from the upper to the lower parts of the canopy [10]. In order to obtain sufficient pesticide deposition in the top canopy, it is necessary to increase the pesticide spray volume [11]. As a result, excessive pesticide deposition occurs in the middle and lower parts, greatly reducing the effective utilization rate of pesticides. To address these issues, smart orchard sprayer integrating sensory detection, intelligent decision making, and variable spraying were developed [12–14]. The orchard automatic variable-rate sprayer, based on LiDAR scanning detection technology, can adjust spray parameters in real-time based on canopy characteristics, effectively improving pesticide utilization and operational efficiency, and potentially saving up to 45.7% of the usage of pesticides [14]. In addition, using binocular cameras to detect real-time grape leaf canopy depth and combining it with the sprayer's forward speed, the orchard sprayer can be guided to perform variable-rate spraying on the grape canopy. However, these technologies involve high development costs and are difficult to implement in complex orchard environments for pesticide application [13].

Furthermore, the development of plant protection unmanned aerial vehicle (UAV) technology has provided another solution to the challenges of pesticide application in orchards. In East and Southeast Asia, plant protection UAVs are replacing the conventional plant protection equipment [7,15]. Plant protection UAV can be deployed quickly with high spraying efficiency and outstanding terrain adaptability [16,17]. By choosing the right operating parameters, UAV can effectively control the pest population to a level that does not cause serious crop yield reduction in orchards and forestry [18,19]. However, there are serious problems such as poor droplet deposition uniformity, a lack of droplet deposition on the backs of leaves [20], and spray drift for UAV [21,22]. Especially when dealing with large volume canopies such as apricot trees [23], citrus [24], etc., the UAV spraying operations show low droplet penetration in the lower and middle canopy, and poor distribution uniformity. The high concentration of pesticides often leads to damage in the targeted crops. Furthermore, there are certain limitations when facing different control objects [25].

Therefore, a co-operative spraying strategy applied using plant protection drones and orchard sprayers may achieve an all-round coverage of the fruit tree canopy. In particular, this strategy intends to combine the advantages of plant protection UAVs with strong terrain adaptability and more deposition of the liquid in the upper canopy by using an orchard sprayer with strong penetration ability and uniform liquid distribution in the lower canopy. In the agricultural field, current applications using aerial and ground devices to coordinate operations have been conducted to establish remote sensing information collection platforms and crop protection operations. Zhang et al. [26] have invented a method and control system for air–ground co-operation plant protection operation. This system can plan collaborative operation paths based on the characteristics of orchards and fruit trees, and control the coordinated operation of unmanned aerial vehicles and ground spraying vehicles to carry out spraying according to the planned paths. Jiang et al. [27] tested the spray performance of a stereoscopic plant protection system consisting of a small swing-arm sprayer and a T16 plant-protection UAV, and determined the optimum spray parameters using CFD simulation. The results show that the uniformity was 38.3% higher than conventional approaches. Current research mainly focuses on the control and improvement of equipment parameters of the co-operation plant protection system. When pesticides are sprayed using two different types of equipment, it is necessary to ensure the behavior of droplet deposition and the spraying ratio (such as the spraying volume of the two devices and the type of pesticide used) in order to achieve the uniform distribution of pesticides in the canopy of fruit trees. The rational distribution ratio of pesticide solution on two types of spraying equipment will help us reduce pesticide usage while achieving a uniform deposition of pesticides.

Our study established an air–ground co-operation stereoscopic plant protection system (ACSPPS) which consisted of an orchard mist sprayer and a six-rotor UASS. The synergistic mode of ACSPPS was selected for spraying mango trees. We conducted a study to evaluate the deposition performance of the system in the mango tree canopy using a tracer method. Additionally, we tested the droplet deposition distribution of the two spray units when applied separately to analyze the composition of droplet deposition for the ACSPPS. This analysis was performed to establish a deposition distribution index for the system. Finally, the optimal spraying ratio (the proportion of fixed doses of pesticide spray using two devices) of two spraying units was expected to be determined across the result of the droplet deposition index distribution. We expect to provide data support and practical guidance for green, precision, and intelligent plant protection in orchards.

2. Material and Methods

2.1. Design of ACSPPS

An ACSPPS consists of an agricultural aviation pesticide spraying device and a land-based plant protection vehicle. It is used on pesticide spraying of three-dimensional crops such as fruit trees, especially the canopy with large volume and high canopy density, for which it is difficult to obtain a good droplet deposition distribution. Generally, the aerial application is undertaken by a plant protection drone, and the land-based plant protection jobs are undertaken by a small and medium-sized orchard sprayer. The two devices spray the same plot, and a fixed dose of pesticide is distributed to the two spray units for spraying according to a certain ratio. Depending on the spray target and fruit tree growth information, ACSPPS supports the following operational modes:

(1) Single-machine operation: When the tree is in the early growth stage, the canopy with a small canopy volume and low leaf density and BBCH is 0–31 (pome fruit for example) [28], or the pests or diseases are in the early stages of outbreak, one spray unit can meet the current plant protection needs of fruit trees; the application operation can be carried out using a mist sprayer or UAV alone. This mode is suitable for standard orchards such as standard apples orchard or vineyards.

(2) Co-operative operation using two machines: The crop phenological stage is at full development at BBCH 32–91 [28]. It is difficult to achieve uniform deposition of pesticides in the canopy via one-machine operation, when the occurrence of pests and diseases is at its peak, or when pesticides need high coverage such as a protective fungicide (e.g., Bordeaux solution, which is shared by two machines, and both apply to trees) (Figure 1). The movement track cannot spray two trees at the same time. This mode is suitable for a fruit tree canopy with large volume and high canopy density and standardized planting mode, such as olive, mango, grapefruit, and apricot, etc. At the same time, this mode has the potential to reduce the usage of pesticides on the basis of achieving spatially uniform distribution of pesticide.

(3) Dual-machine hybrid operation: In the complex terrain orchard application operation, ground sprayers and UAVs operate in accordance with the preset route, and the ground implements cannot pass the specified place to wait for the end of the collaborative operation in the UAV to make up the application. At the same time, this mode can avoid the antagonistic effect of pesticide mixing, which can separate the configuration of pesticides to avoid chemical reactions between pesticides, such as insecticide spraying via drone and fungicide spraying using a mist sprayer. Different pesticide dosage forms can also be separated for spraying.

The ACSPPS of this study consisted of a six-rotor electric UASS AGRAS T30 (manufactured by SZ DJI Technology Co., Ltd., Shenzhen, China (Figure 2a)) and an orchard caterpillar mist sprayer (3WDZ-200D, Shanxi Nonggu Feinong Plant Protection Technology Co., Ltd., Datong, China (Figure 2b)). The UAV sprayer was equipped with a 30 L PPP container, with a total weight (excluding battery) of 36.5 kg and maximum take-off weight of 78 kg (at sea level). Sixteen extended-range flat-fan nozzles (SX11002VS, Spraying Systems Co., Wheaton, IL, USA), divided into eight couples, were equipped below the corresponding rotors of the UAV. The mist sprayer was equipped with a 200 L container. One XR95/02S (Yushengpenwu, China) nozzle was installed between two XR45/01S nozzles (spacing 40 cm, one side) and the spray angle of the mist sprayer in the orchard spraying mode was −22.5° to 90° from the horizontal plane. It is operated using a remote control. Both equipment parameters are shown in Table 1.

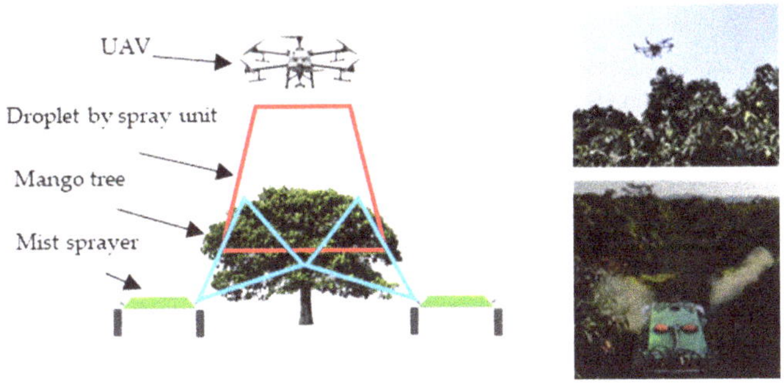

Figure 1. Air–ground co-operation stereoscopic plant protection system operation diagram.

Table 1. Two spray unit parameters.

Sprayer	Overall Size/mm × mm × mm	Maximum Payload/L	Nozzle Type × Number of Nozzles	Device Maximum Flow/(L/min)	Maximum Operating Speed/(m/s)	Spray Swath Width /m
UAV	2858 × 2685 × 790	30	TEEJET 11002VS × 16	9.0	7.0	9.0
Mist sprayer	1935 × 1140 × 690	200	XR45/01S × 4 XR95/02S × 2	16.0	1.25	8.0

UAV treatment is plant protection unmanned aerial vehicle spray treatment.

(**a**) Agras T30 UASS (**b**) 3WDZ-200D orchard caterpillar mist sprayer

Figure 2. Air–ground co-operation stereoscopic plant protection system component units.

2.2. Spray Test Site

This experiment was conducted from December 2021 to March 2022, with an average temperature of 30.4 °C, an average humidity of 41%, and an average wind speed of 0.6 m/s during the experiment. The test site was located in Dabuba Village (longitude: 109°18′9578″ E, latitude: 18°42′2322″ N), Sanya City, Hainan Province, China. The mango orchard was a traditional orchard, with high-pressure line guns and plant protection UAV that are used for daily pest control. The trees are 7–8-years-old, with an average plant height of 2.70–2.90 m, a crown diameter of 3.2 m, a plant spacing of 4.5 m, and a row spacing of 5 m. The mango trees were in good growth condition, with fully developed leaves, an umbrella-like distribution of canopy leaves, and almost no leaves inside the canopy, while the BBCH was at 74 [29]. Mango varieties included Mangifera indica Linn, and the mango trees were in the mango swelling stage during this experiment. The density of weeds in the test field was low and the height of weeds were below 0.3 m, which could be considered as having no effect on the spraying effect. The orchard was located on flat land with an open site free of obstacles (Figure 3).

Figure 3. Experimental orchard and test site.

2.3. *Experimental Design*

A plot measuring 30 m × 20 m was chosen as the experimental area within the orchard under investigation (Figure 4). Within this plot, three mango trees displaying uniform growth were selected for droplet sampling.

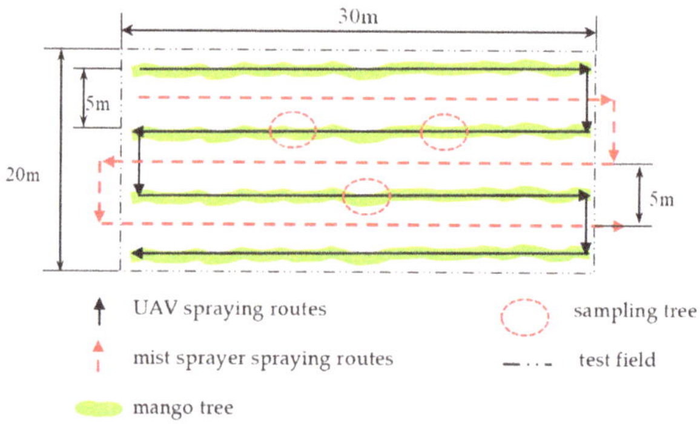

Figure 4. Test fields for spray test and work lines of UAV and caterpillar mist sprayer.

This study aims to test the droplet deposition distribution of the co-operative application mode of ACSPPS and determine a suitable spraying ratio, which refers to the pesticide distribution ratio between the two spray units. Initially, the droplet deposition distribution of the two spray units applied separately was tested to analyze the composition of the droplet deposition for the ACSPPS treatment. Subsequently, the deposition distributions of the two implements were combined, and the results were compared with the uniformity of the droplet deposition distribution achieved using the ACSPPS treatment. This comparison helped identify the optimal spray ratio for the two devices to work in synergy. To obtain the best spray ratio for the two machines to work together, three treatment groups were set up in this section: application using plant protection UAV, application using ground equipment, and ACSPPS. The parameter settings of the sprayer in each treatment are shown in Table 2. Three valid replicates were performed in each treatment group.

So as to avoid the mutual interference of flow fields during the operation of the two implements, the co-operation sprays using UAV firstly. The flight mode of the UAV is to mark the operational boundaries by using a remote control with an RTK module and then set the drone operation line distance and flight height for an autonomous operation. It can be ensured that the drone spraying path passed through the top of each mango tree. After the UAV operation was complete, the operator started the mist sprayer and controlled the advance of the implements using a remote control to start the operation until all fruit trees were treated. Tartrazine was selected as the tracer, and the spraying solution was the tartrazine solution.

Table 2. The mango tree spray test design and sprayers' parameters.

Treatment	Application Volume/(L/ha)	Tartrazine Concentration/(g/L)	Velocity/(m/s)	UAV Flight Height/(m)	Spray Route/Track
UAV	120	4.20	2.5	3	Top of the tree row
Mist sprayer	510	0.40	1.25	/	Spraying between tree rows
ACSPPS	UAV: 120 Mist sprayer: 510	UAV: 4.20 Mist sprayer: 0.40	UAV: 2.5 Mist sprayer: 1.25	3	/

UAV treatment is plant protection unmanned aerial vehicle spray treatment and ACSPPS is air–ground co-operation stereoscopic plant protection system treatment.

The meteorological parameters of each treatment during the experiment are shown in Table 3 (measured using a mobile weather station, YG-BX Weather Station, Yigu technology company, Shenzhen, China). All of the meteorological parameters provided a suitable meteorological condition for spraying.

Table 3. Weather conditions for each application.

Treatment	Wind Speed/(m/s)	Wind Direction/°	Temperature/(°C)	Humidity/(%)
UAV	0.6	156.8	31.6	46.5
Mist sprayer	0.8	203.7	28.4	50.8
ACSPPS	0.5	180.2	30.1	60.2

UAV treatment is plant protection unmanned aerial vehicle spray treatment and ACSPPS is air–ground co-operation stereoscopic plant protection system treatment.

In this study, droplet coverage and tartrazine deposition were used as the indexes for investigation, and tartrazine was regarded as the active ingredient of the pesticide solution. Droplets were collected using HD photographic paper (38 × 26 mm, Chenguang, China) and 7 cm filter paper (AL-1, Aokexingainian, Liaoyang, China). The diagram of the photo paper layout was shown in Figure 5, and the mango trees were divided into upper, middle, and lower layers: the outer layer of the lower was marked as O1, the inner was marked as I1, the outer layer of the middle was marked as O2, and the inner layer of the middle was marked as I2. Due to the characteristics of the mango tree canopy, the upper layer was not set up in the inner canopy sample, marked as O3. A length of 7 cm of filter paper arranged within 5 cm around the phase paper, both unobstructed from each other to detect the amount of deposition, while the front and back of the leaf were covered with phase paper and filter paper. Four filter papers were placed under O1 at the ground projection to collect ground loss. Three sets of replicates were set up for each treatment.

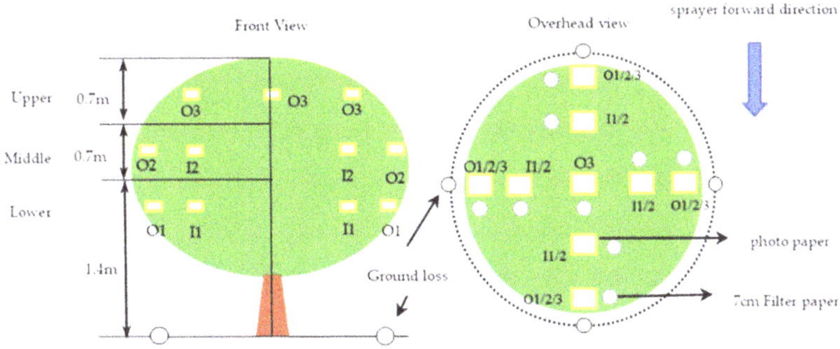

Figure 5. Sampler locations for assessing mango tree canopy spray deposition.

2.4. Droplet Site Measurement

Measurements of the droplet-size spectrum were carried out using a laser diffraction system (DP-02, Zhuhai Omeike, Zhuhai, China) in the Centre for Chemicals Application Technology, China Agricultural University according to ISO standard 25358 [30]. The tested nozzle of UAV, flat-fan nozzle SX11002VS, was mounted at a height of 50 cm above the laser analyzer vertically between the laser beam transmitter and the receiver lens. An electric diaphragm pump Shurflo 8000 series (Pentair, Golden Valley, MN, USA) and a 63 mm diameter pressure gauge (Lechler GmbH, Metzingen, Germany) were applied to maintain the pressure (3.0 bar) in accordance with the field trials. The tested nozzle of the mist sprayer, flat-fan nozzle XR45/01S, and XR95/02S were also mounted at a height of 50 cm above the laser analyzer; the nozzle is still connected to the spray system of the mist sprayer to maintain the pressure within the spray pressure in the field test state, by means of an extended pressure duct. In each test, the background measurement was performed for 5 s to eliminate the interference of other particles in the air, and the droplet-size spectrum measurements lasted for 10 s. The 10th percentile diameter (Dv10), volume median diameter (VMD, Dv50), 90th percentile diameter (Dv90), relative span (RS), and spray volume fractions generated using droplets finer than 75, 100, and 200 μm (V75, V100, and V200) were recorded for further analysis.

2.5. Data Analyses

Five minutes after spraying, all photo papers were collected in the corresponding position in the card holder and brought back to the laboratory within five days to be scanned using an (EPSON DS-1610, Tokyo, Japan) at 600 dpi and analyzed using DepositScan software (Version 1.38x) for droplet coverage and density. The filter paper was placed in a No. 5 self-sealing bag, eluted with 20 mL of deionized water (5 mL for the ground implement group) in the laboratory, and then analyzed using an enzyme standard (450 nm wavelength, iMark Analytical Instruments, Inc., Orange, LA, USA) to calculate the amount of deposition. A total of 20 mL of an initial solution for each sprayer was collected after each application and diluted 100 times. The uniformity in the deposit distribution of the spray coverage parameters on the tree canopy was expressed in the coefficient of variation (CV, %). A lower CV means a better homogeneity in the spray coverage distribution of the UAV spraying. The unit deposition was expressed as β, μg/cm^2, and calculated as follows:

$$\beta_{dep} = (C_t \times V_e)/(a \times A) \quad (1)$$

where β is the deposition volume, μg/cm^2; C_t is the tracer concentration, μg/mL; V_e is the eluent volume, mL; a is the filter paper area, cm^2; and A is the concentration of initial solution, g/L.

The homogeneity of the spray distribution over the upperside and the underside of the leaves was calculated for the spray coverage (HSC) as expressed in Equation (2):

$$HSC = \frac{SC_{und}}{SC_{upp}} \times 100\% \quad (2)$$

where SC_{upp} and SC_{und} are the mean values of the spray coverage obtained from the upper side and the underside of leaves, %, respectively.

The penetration coefficient in the vertical direction (K_V) of the mango tree, %, was calculated using the following equation:

$$K_V = \frac{\beta_a}{\beta_{b\&c}} \times 100\% \quad (3)$$

where β_a is the upper canopy deposition per area in the mist sprayer treatment and the lower canopy in the UAV treatment, μg/cm^2, and $\beta_{b\&c}$ is the middle and lower canopy deposition per area in the mist sprayer treatment and the upper and middle canopy in the UAV treatment, μg/cm^2.

The proportion of inner canopy deposition (R_I, %) could be calculated as:

$$R_I = \frac{\overline{\beta_I}}{\overline{\beta_I} + \overline{\beta_O}} \times 100\% \quad (4)$$

where $\overline{\beta_I}$ is the inner canopy droplet deposition per area, μg/cm^2, and $\overline{\beta_O}$ is the outer canopy droplet deposition per area, μg/cm^2.

The summation deposition per area (β_{add}, μg/cm^2) was used to calculate the deposition number of different concentrations of chemical solution in each canopy part using the synergistic operation and to construct the droplet deposition distribution index, which was calculated as

$$\beta_{add} = C_1 V_1 + C_2 V_2 \quad (5)$$

where C_1 and C_2 are the tartrazine concentrations of the mist sprayer and UAV treatment, g/L and V_1, V_2 is the deposition per area of the mist sprayer and UAV treatment, μL/cm^2.

The percentage distribution of the deposition unit area in the vertical direction (R_V, %) is the distribution ratio of the tartrazine in the vertical direction; when the value is closer to each part of the canopy, the more uniform the deposition of the solution in the vertical direction. This can be calculated as

$$R_V = \frac{\beta_{dep}}{\overline{\beta_U} + \overline{\beta_M} + \overline{\beta_L} + \overline{\beta_G}} \times 100\% \quad (6)$$

where $\overline{\beta_U}$, $\overline{\beta_M}$, $\overline{\beta_L}$, and $\overline{\beta_G}$ are the summation deposition of upper, middle, lower, and ground loss respectively, μg/cm^2.

All data were subjected to Anderson–Darling, Kolmogorov–Smirnov, and Lilliefors normality tests, and Levene and Bartlett tests for the equality of residual variances. Data from abnormal distributions were subjected to nonparametric Kruskal–Wallis tests at the 5% probability level. The result of the three application methods on droplet coverage and deposit density were compared using a three-way analysis of variance (ANOVA) at a significance level of 0.05 and Duncan's post hoc test at a significance level of 0.05 for the mean of each droplet deposit parameter. All statistical analyses were performed using SPSS Statistics Version 20 for Windows (IBM Corporation, Armonk, NY, USA).

3. Result

3.1. Droplet-Size Spectrum

The mean values of parameters Dv10, Dv50, Dv90, RS, V75, V100, and V200 and the spray quality classification for droplet-size spectrum characteristics are presented in Table 4. At the higher spraying pressure of the mist sprayer, both nozzles XR45/01S and XR95/02S produced droplets with VDM of no more than 65 μm, which was much smaller than UAV at 3-bar spray pressure (139 μm). Most of the droplet produced by the mister were below 100 μm, while the 75% droplet size of UAV is larger than 100 μm. The average RS value (0.7) of the mist sprayer for both nozzles of the implement was only 42.6% of that of the plant protection drone, indicating that this device has better fogging ability compared to UAV, which may increase the droplet deposition in the inner canopy.

Table 4. Droplet-size spectrum of the nozzles used in orchard tests.

Nozzle	Device	DV10 (μm)	DV50 (μm)	DV90 (μm)	RS	V75 (%)	V100 (%)	V150 (%)
SX11002VS	UAV	63.01	139.02	289.74	1.63	10.94	25.98	60.17
XR45/01S	Mist Sprayer	36.34	61.23	80.66	0.72	63.75	97.37	100.00
XR95/02S	Mist Sprayer	38.66	64.00	81.82	0.67	59.14	96.51	100.00

3.2. Characteristics of Deposition Distribution in The Mango Canopy

3.2.1. Spray Coverage and Deposit Density

Table 5 shows the distribution of droplet coverage and deposit density in the mango canopy using three application methods and the results of the significance analysis. The results showed that there was no significant difference ($p < 0.05$) in droplet coverage and deposit density in each layer of the canopy in ACSPPS treatment, and co-operation spray could improve the uniformity of droplet deposition distribution compared with the UAV and mist sprayer; however, there was poor deposition distribution uniformity on the underside of mango leaves in three application, and there was no statistical difference in the deposition on the underside of leaf ($p > 0.05$).

Table 5. Droplet deposition distribution characteristics of three applications.

Treatment	Sample Position	Coverage/%		Deposit Density/cm^2		HSC/%
		Upperside	Underside	Upperside	Underside	
Mist sprayer	O1	10.9 ± 2.0 a	2.8 ± 1 a	186.9 ± 23.2 a	36.6 ± 15.5 a	25.7
	O2	7.5 ± 2.3 a	5.2 ± 4.4 a	124.9 ± 30.1 b	26.9 ± 10.3 a	69.3
	O3	2.4 ± 0.9 b	0.5 ± 0.4 a	60.0 ± 16.3 c	15.5 ± 9.3 a	20.8
	I1	2.0 ± 1.0 b	0.6 ± 0.3 a	48.7 ± 17.6 c	19.8 ± 10.8 a	30.0
	I2	2.4 ± 1.7 b	0.1 ± 0.1 a	38.1 ± 17.7 c	4.8 ± 2.5 a	4.2
UAV	O1	14.6 ± 4.0 a	1.4 ± 1.1 a	22.1 ± 4.0 a	4.0 ± 2.2 ab	9.6
	O2	6.6 ± 2.7 b	5.2 ± 3.8 a	12.9 ± 4.7 ab	9.4 ± 4.2 ab	78.8
	O3	7.9 ± 2.0 ab	5.8 ± 2.7 a	16.1 ± 3.7 ab	11.9 ± 3.7 a	73.4
	I1	1.0 ± 0.4 b	0.2 ± 0.1 a	6.6 ± 2.3 bc	2.3 ± 0.6 b	20.0
	I2	4.1 ± 2.0 b	0.3 ± 0.3 a	1.8 ± 0.8 c	1.8 ± 0.8 b	7.3
ACSPPS	O1	27.8 ± 6.7 a	3.2 ± 1.6 a	120.5 ± 22.3 a	49.7 ± 19.3 a	11.5
	O2	25.8 ± 7.9 a	4.4 ± 2.6 a	116.7 ± 29.8 a	26.6 ± 10.9 b	17.0
	O3	21.1 ± 7.3 a	0.6 ± 0.3 a	89.1 ± 28.3 a	5.0 ± 1.3 b	2.8
	I1	24.0 ± 7.3 a	0.5 ± 0.2 a	124.1 ± 37.0 a	11.4 ± 3.4 b	2.0
	I2	8.9 ± 2.6 a	1.2 ± 1.0 a	79.9 ± 21.9 a	9.2 ± 6.2 b	13.5

Note: Data of coverage and deposit density were expressed as mean ± standard error. Different letters after the numbers indicate significant differences in different canopy positions in the same group of treatments at the $p < 0.05$ level using Duncan's test values.

Droplet coverage on the underside of the leaves is shown in Figure 6a. Co-operation spraying can significantly improve ($p < 0.05$) the underside droplet coverage of the mango canopy except for the innerside of the lower layer (I2) compared with mist sprayer and UAV treatment. There was no significant difference ($p > 0.05$) between the mist sprayer and UAV in terms of upperside coverage in all parts of the canopy. The deposition from bottom to top showed a more obvious decreasing trend in the mist sprayer treatment. Because of the umbrella shape of the mango canopy, the downwash airflow when the UAV passed the top of the mango tree would have a diffusion effect so that the droplet coverage did not show a significant decreasing trend in UAV treatment; the related penetration analysis was conducted in Section 3.2.2.

The underside coverage distribution is shown in Figure 6b, except for the upper layer (O3) where the UAV was significantly higher than the remaining two groups, and there was no significant difference between the treatment in others parts; according to Table 5, the mean value of the HSC in mist sprayer treatment was 9.36%, which were 30.0% and 37.82% in the mist sprayer and UAV treatment. Co-operative operation using two machines cannot significantly improve droplet coverage on the underside of mango leaves and only increases the droplet deposition rate on upperside of the leaves. According to Table 5, the mean value of underside coverage of the inner layer canopy (I1 and I2) did not exceed 1.2%, which meant that the two-sprays unit lacked the ability to deliver droplets to the underside of mango leaves.

Figure 6c,d shows the droplet density distribution on the upper and underside of mango leaves. The mist sprayer involved in spray treatment, for which the density on the upperside of all layers was significantly higher than that of the UAV treatment alone, which was due to the atomization performance of the mist sprayer under higher spray pressure. The droplet density on the upperside of the outer canopy of leaves could reach 92/cm^2 in the mist sprayer group, while the UAV only had 12/cm^2, for which most areas failed to meet the pest control requirement density of 20 deposit/cm^2. Through collaborative work, the value could be increased to 109/cm^2, and the distribution uniformity of droplet density in the canopy was improved. There was no significant difference in droplet density on upperside of leaves in each part of the canopy in this treatment. Except for O3, droplet density on the underside of other parts of the ACSPPS treatment group was improved compared with UAV, but without significant difference except for O1.

3.2.2. Spray Deposition and Penetration

The distribution of droplet deposition on the upper and the underside of mango leaves was shown in Figure 6e,f. In this test, using this application method, the mean deposition value for the mist sprayer was 0.26 μg/cm^2, while the UAV resulted in a deposition value of 0.68 μg/cm^2. Normalizing these values (deposition/spray volume), we obtain 0.12 (constant dimensionless) for the mist sprayer and 0.13 (constant dimensionless) for the UAV. There were no significant differences between the two methods in terms of the overall effective utilization of pesticides. The main difference was the distribution uniformity in the mango canopy. Except for O3, there was no significant difference in deposition per area on the upperside of leaves in the remaining parts. The tartrazine deposition on the underside of in ACSPPS treatment mainly came from UAV, and the deposition of UAV treatment at O1 was 2.6 times that of the mist sprayer, but there was no significant difference, because of the poor deposit distribution uniformity (CV value of 178.9%) of UAV on the underside of the blade, and deposition in some areas was slightly higher than that of the ACSPPS treatment.

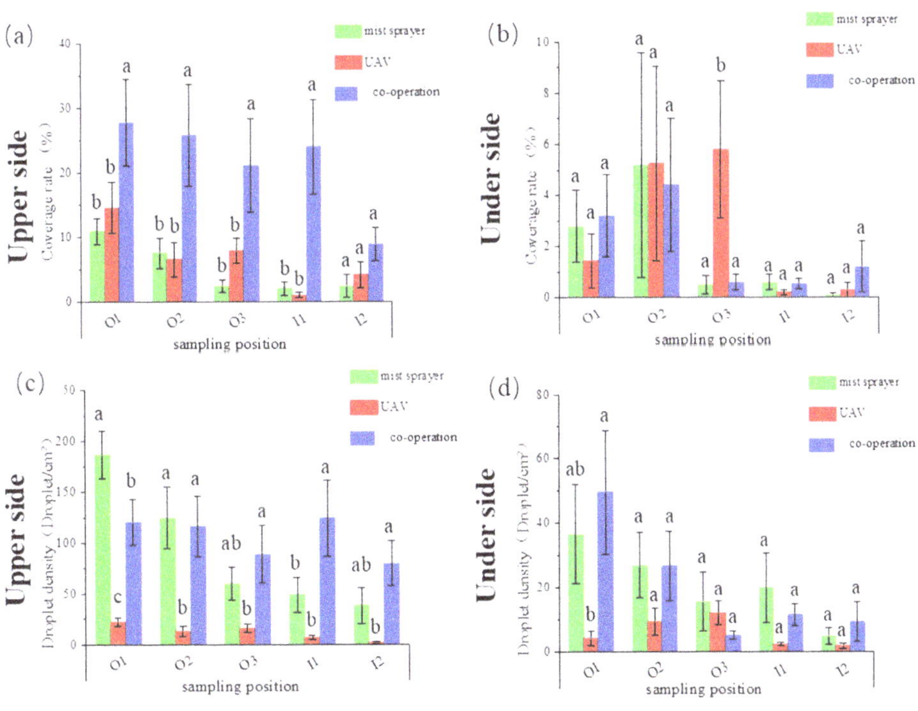

Figure 6. *Cont.*

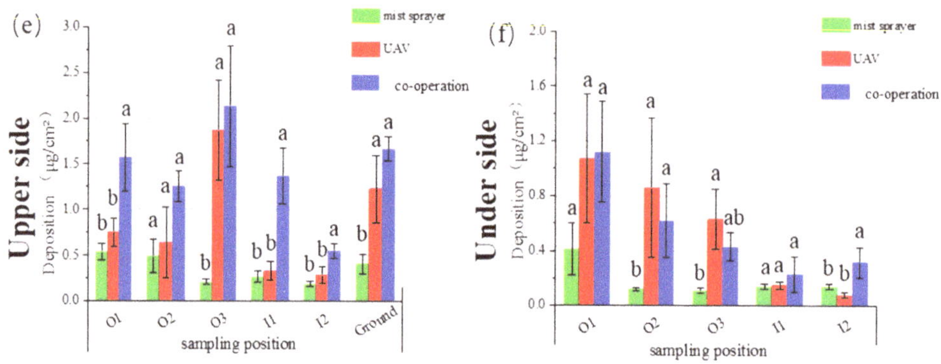

Figure 6. Droplet coverage (**a**,**b**), droplet density (**c**,**d**), and droplet deposition (**e**,**f**) on the upper- and underside of mango leaves in three application treatments. All data were expressed as mean ± standard error, and different letters on each bar indicate significantly different differences at the $p < 0.05$ level using Duncan's test values.

The mist sprayer treatment major deposition was concentrated in the outer part of the middle and lower canopy (a share of more than 50%). The deposition per area in the upper canopy was 0.16 μg/cm^2, which was only 47.0% of the average deposition, and the UAV treatment had the highest proportion of deposition in the upper layer (43.0%). According to Table 6, the proportion of the inner canopy deposition of the mist sprayer was 37.1%, while the UAV was only 18.0%. This was due to the large droplets (size after spreading, Dv50 = 631.4 μm) produced by the TEEJET 11002VS nozzle of UAV, which had limited penetration into the inner canopy and caused most of the droplets to deposit in the outside canopy, while the fine droplet produce by the mist sprayer was good at delivering the liquid into the interior, so that the proportion of the deposition to the inner canopy could be improved with the increase in the application ratio of the mist sprayer in ACSPPS plant protection work. The ground loss rate was about 21% for both of the spray units in the single-machine operation mode, which was basically the same as that of the co-operative operation (21.2%). Therefore, the ground loss ratio could not be reduced through co-operative operation.

3.3. Vertical Direction Droplet Deposition Distribution Index

The purpose of the air–ground co-operation stereoscopic plant protection system is to optimize the spatial distribution of the pesticide in the canopy and improve the effective utilization rate of the pesticide to achieve the purpose of reducing pesticide use. This section establishes a droplet deposition distribution index of pesticide deposition distribution in the mango canopy based on the depositional behavior of the mist sprayer and UAV and analyzes the droplet deposition distribution in mango trees with a different spray mass distribution ratio (which can be interpreted as the distribution of a certain amount of pesticide active ingredients using a two-spray unit sprayed evenly into a certain area of the orchard). The ratio was expressed as the ratio of application mass (g/ha) × tartrazine concentration (g/L) of two spray units; the approximate ratio of the mist sprayer to UAV application is 3.5:6.5 in this experiment. According to Equation (5), the added deposition of this experiment was calculated and compared with the actual deposition of the ACSPPS treatment (Figure 7).

The results of Figure 7 show that, except for the significant difference between the add deposition and the actual deposition at I1, the deviation between the theoretical add deposition and the actual value within 20%, with an average difference of 14.7%. The reason for the deviation may be that the travel routes of the mist sprayer cannot be fully overlapped, or the change in ambient wind speed has an impact on the spraying process of the UAV. Therefore, this error has little influence on the establishment of the add droplet deposition distribution index, and the add deposition of the spray mass distribution ratio (mist sprayer/UAV = 1/9, 2/8, 4/6, 5/5, 6/4, 7/3, 8/2, and 9/1) could be calculated according to this index, assuming that the deposition of the pesticide in all parts of the canopy in each co-operation was consistent with this experiment, and the add deposition of each application ratio could be calculated under the conditions of this application volume. The results of the spray mass distribution ratio add deposition are shown in Figure 8.

Table 6. Droplet deposition parameter.

Parameter		Treatment		
		Mist Sprayer	UAV	ACSPPS
Upperside droplet size	D_{v50upp} (μm)	189.0 ± 7.5 b	792.7 ± 69.9 a	968.6 ± 127.7 a
Underside droplet size	D_{v50und} (μm)	152.1 ± 9.6 b	470.0 ± 56.3 a	345.3 ± 45.3 a
Mean droplet size	D_{V50} (μm)	170.6 ± 8.6 b	631.4 ± 63.1 a	656.9 ± 76.0 a
Upperside coverage	SC_{upp} (%)	5.0 ± 0.8 b	6.8 ± 1.2 b	19.7 ± 4.5 a
Underside coverage	SC_{und} (%)	1.8 ± 0.8 a	2.5 ± 1.1 a	2.0 ± 0.6 a
Mean coverage	SC (%)	3.4 ± 0.6 b	4.6 ± 1.0 b	10.85 ± 2.3 a
Coverage distribution uniformity	CV_{Cov} (%)	165.9	158.0	124.8
Upperside deposit density	N_{dupp} (deposit/cm^2)	91.7 ± 11.1 a	11.5 ± 1.7 b	106.1 ± 12.1 a
Underside deposit density	N_{dund} (deposit/cm^2)	20.7 ± 5.4 a	5.9 ± 0.8 a	21.1 ± 4.9 a
Mean deposit density	N_d (deposit/cm^2)	56.2 ± 8.3 a	8.7 ± 1.2 b	67.6 ± 10.2 a
Upperside deposition	β_{upp} (μg/cm^2)	0.34 ± 0.1	0.77 ± 0.3	1.4 ± 0.4
Underside deposition	β_{und} (μg/cm^2)	0.18 ± 0.1	0.59 ± 0.3	0.54 ± 0.3
Mean deposition	B (μg/cm^2)	0.26 ± 0.1	0.68 ± 0.3	0.97 ± 0.3
Vertical penetration coefficient	K_V (%)	27.8	33.4	/
Proportion of inner canopy deposition	R_I (%)	37.1	18.0	34.2
Ground loss rate	$R_{Groundloss}$ (%)	21.8	21.2	21.2

Note: Data of coverage, deposit density, and deposition were expressed as mean ± standard error. Different letters after the numbers indicate significant different differences at the $p < 0.05$ level for Duncan's test values.

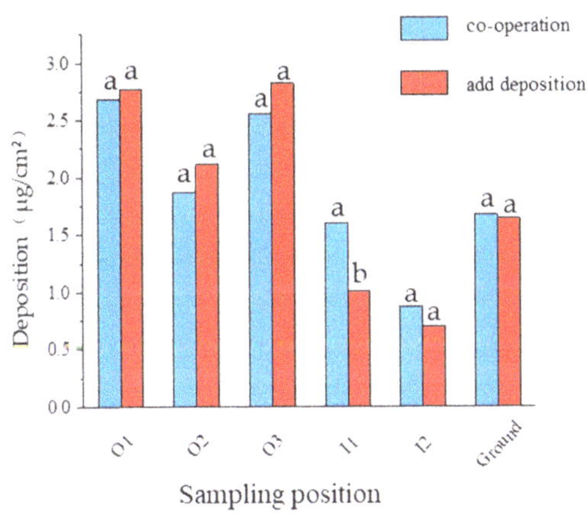

Figure 7. Add deposition versus actual deposition of ACSPPS, different letters on each bar indicate significantly different differences at the $p < 0.05$ level using Duncan's test values.

Based on Figure 8, it could be concluded that the percentage of ground loss basically remained between 21% and 22% when the co-operative operation was sprayed with different application ratios at the current spray volume. Changing the application ratios could not reduce the ground loss ratio. By increasing the application ratio of the mist sprayer, the proportion of deposition in the upper canopy layer decreased, while the deposition proportion in the middle and lower layers increased. Droplet deposition distribution uniformity (CV, %) in the vertical direction of the mango canopy via different application ratios are shown in Table 7. It can be conduct that, when the application ratios were 8/2 and

7/3 (0.92 g/L and 0.80 g/L for the tartrazine concentration of the mist sprayer and 1.29 g/L and 1.94 g/L for the UAV), the pesticide active ingredients were most evenly distributed in the vertical direction; the CV value of the percentage of vertical direction pesticide distribution in the canopy was the smallest among all application ratios at 13.64% and 14.34%. Therefore, the study of pesticide reduction could be carried out on the basis of these two applications ratios.

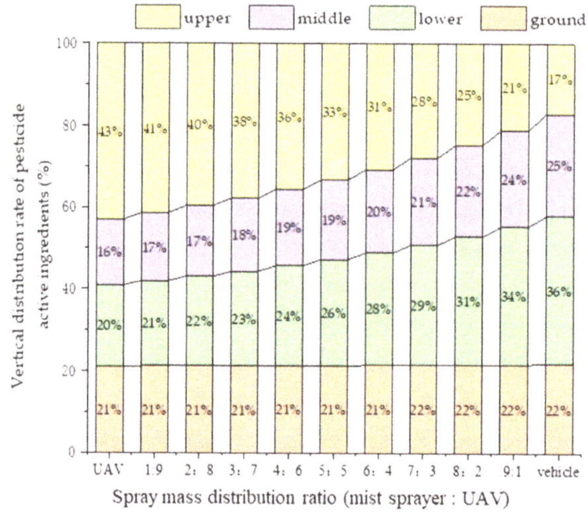

Figure 8. Vertical distribution rate of pesticide active ingredients of spray mass distribution ratio.

Table 7. Droplet deposition distribution uniformity in the vertical direction of the mango canopy via different application ratios.

Application Ratios	UAV	9:1	8:2	7:3	6:4	5:5	4:6	3:7	2:8	1:9	Vehicle
droplet deposition index/%	26.3 ± 11.9	26.3 ± 10.5	26.3 ± 9.9	26.2 ± 8.5	26.2 ± 7.1	26.2 ± 5.7	26.2 ± 4.6	26.1 ± 3.6	26.1 ± 3.7	26.1 ± 5.6	26.1 ± 7.8
CV/%	45.24	39.92	37.56	32.44	27.23	21.81	17.72	13.64	14.34	21.29	29.84

Note: Data of droplet deposition index was expressed as mean ± standard deviation.

4. Discussion

The deposition distribution uniformity test using two spray units shows that the application of the mist sprayer is faced with the problem of an insufficient deposition capacity in the upper layer canopy and underside of leaves. According to the result of the droplet-size spectrum, the mist sprayer has an outstanding atomization capacity which can deliver more pesticide to the inner canopy. While the droplets applied using the UAV were mainly deposited in the upper layer of the canopy, the droplets deposition density could not meet the pest control demand of 20 droplets/cm^2. The result aligns with what has been documented in recent literature. Specifically, the UAV encounters certain challenges, including inadequate droplet coverage density, a scarcity of droplets reaching the lower canopy, and inconsistent droplet distribution. These issues arise due to the lower spraying dosage of the plant protection UAV, exacerbated by the turbulent downwash airflow [20,23,24]. According to Table 6, the synergistic operation of two sprayers significantly improved the deposit coverage of the mango leaves and made up for the deficiency of an insufficient droplet deposition density of UAV application. Generally, only some large-scale air-assistant orchard sprayers can achieve more than 20% coverage on the surface of the canopy [11,31], but these sprayer are still difficult to adapt to spraying operations in complex orchard environments. Therefore, the small ground plant protection vehicle of ACSPPS in this study can be used to enhance the terrain adaptability of the system. However, due to the high canopy density of mango trees, and the high mutual cover-up rate of leaves, it is unfavorable for the droplets to penetrate outside the canopy to reach the inner canopy, and most of the droplets are deposited on the outside canopy. According to Figure 9, the characteristics of the mango canopy, such as large leaf length and thickness and complex leaf inclination distribution, led to poor deposition distribution uniformity of the two spray units; the overall CV was over 150%. The CV was improved by 24% via synergistic operation but could not solve the problem of less droplet deposition on the underside of mango leaves.

Figure 9. Test field mango trees' canopy characteristics.

The three-treatment droplet deposition scans results show that, the large droplets produced using the UAV (the average Dv50 value of 631 μm after spreading of the droplets) and the fine droplets produced using the mist sprayer (the average Dv50 value of 171 μm) on the leaves surface deposit action are shown in Figure 10. According to the study by Ebert et al. [32,33], the control effect can be significantly improved by increasing the droplet density under the same pesticide dose conditions. In addition to the biological optimum size theory (BODS theory) proposed by Himel and UK [34], the biological optimum size of foliar crawling pest larvae was 30~150 μm. Therefore, the mist sprayer had a better potential pest control ability than UAV. In actual orchard plant protection operations, the UAV was prone to damage the initial period fruit epidermis due to the higher application concentration. In addition, the mango tree's pests and diseases are prone to occur in the inner canopy and on tender shoots, but the ground-based orchard sprayer lacks the ability to deliver a deposit to the upper canopy. Through the co-operation of a two-spray unit, a high-concentration liquid can form a dilution and certain doses of the pesticide can be dispersed through the two machines, which can avoid causing damage to the crop while increasing the pesticide deposition proportion in inside and upper layer canopy and improving the deposition distribution uniformity in the vertical direction of the canopy. Through the droplet deposition distribution index, it can be found that increasing the application ratios of ground-based devices can increase the deposition uniformity of the canopy in the vertical direction, but the proportion is not as large as it should be. If the ground base device takes up too much application ratio it will result in insufficient deposition of the system in the upper canopy of the fruit tree. Therefore, when the application ratios were 8/2 and 7/3, the ACSPPS had the best deposition distribution index. It can be surmised that this ratio has the best potential for reduced application.

Research on the synergistic application of UAV using orchard sprayers is not yet mature. Jiang et al. [34] tested the spray performance of a stereoscopic plant protection system and determined the optimum spray parameters using CFD simulation. The results show that the uniformity was 38.3% higher than conventional approaches. But, it did not optimize the application ratio of the two spray units. This experiment verified the importance of air–ground co-operative plant protection operations to improve the pesticide distribution characteristics in the mango canopy and optimized application ratios for both spray units in the ACSPPS. This system was designed for tropical crops such as mango trees which have frequent pest outbreak cycles and high insect density, and its high droplet coverage and uniform distribution has obvious advantages when applying some pesticides such as contact insecticides and protective fungicides.

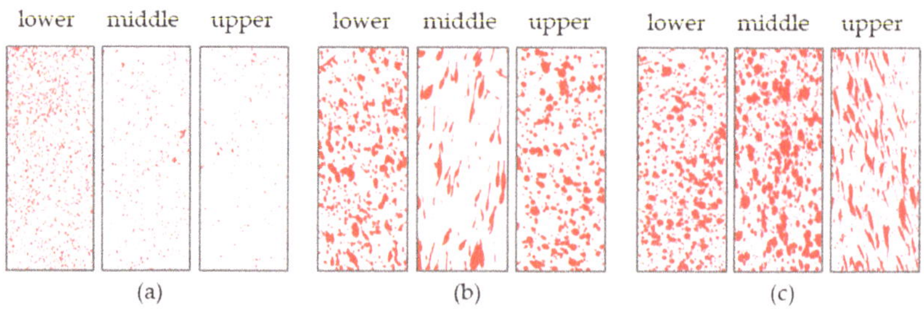

Figure 10. The scan results of mist sprayer (**a**), UAV (**b**), and ACSPPS (**c**) treatment droplet collecting cards.

However, there are still some shortcomings:

(1) In this study, only the droplet deposition analysis of the ACSPPS system was conducted under large spray volume conditions, which is not conducive to improving effective pesticide utilization.
(2) The field control efficiency of ACSPPS was not tested. The ACSPPS's strategies for reducing pesticide use have not been developed.

In a subsequent study, the focus will shift towards researching methods to reduce pesticide application while maintaining effective pest control. Building upon the findings of this experiment, further research can be conducted to explore strategies for reducing pesticide application. One aspect that will be investigated in later stages is the optimization of application ratios between the two spray units. This research will aim to reduce the water spray volume and pesticide dosage of the spray units while preserving the efficiency of the insecticide. For example, potential measures include halving the volume of spraying for both spray units and reducing pesticide usage by 30% by following a 8/2 or 7/3 (mist sprayer/UAV) application ratio for pest control. These adjustments will be based on the best application ratios derived from the results of this experiment, ensuring that the effectiveness of the insecticide is maintained while achieving a reduction in pesticide application.

With the maturity of artificial intelligence technology, intelligent agricultural equipment has become an inevitable trend for replacing traditional plant protection equipment. The two spray units in this study were retrofitted with autonomous navigation and variable spray technology. This system was applied to the management of smart orchards to develop a plant protection decision plan based on the detection information of the pest and disease sensing system, which was sent down to the variable spray system for accurate environmental pest management.

5. Conclusions

In this paper, an air–ground co-operation stereoscopic plant protection system combining a ground-based orchard mist sprayer with a plant protection UAV was established. The deposition distribution characteristics of three plant protection applications (spray using a mist sprayer, UAV, and ACSPPS application) in the mango canopy were evaluated, and a vertical droplet deposition index distribution of co-operative operation was constructed. The results show the following:

The air–ground co-operation stereoscopic plant protection system has shown significant improvements in droplet coverage on the upper side of mango leaves in all parts of the canopy (a 14.7% increase for the mist sprayer and a 12.9% increase for the UAV), as well as an increase in the uniform distribution of active components in the mango canopy. However, it cannot significantly improve the deposition and coverage of droplets on the underside of leaves compared with the mist sprayer and plant protection UAV.

Due to the characteristics of the mango canopy, such as large leaf length and thickness and complex leaf inclination distribution, this led to poor deposition distribution uniformity for the two spray units; the overall CV was over 150%.

Under the spraying volume in this experiment, it can be concluded from the vertical droplet deposition model that changing the spray mass distribution ratio of two devices cannot reduce the ground loss rate for the pesticide. Deposition in the upper canopy will be decreased and the ratio of deposition in the middle and lower layers will be increased with the increase in the application ratio of the mist sprayer. When the application ratios of a two-spray unit (mist sprayer/UAV) come to 8/2 and 7/3, the active ingredients in the pesticides are most evenly distributed in the vertical direction. A study of on pesticide reduction can be carried out on the basis of this these application ratios. This could be a new idea for mango orchard plant protection and provides a technical solution to improve spray uniformity in the canopy of fruit trees.

Author Contributions: Conceptualization, X.H.; methodology, Y.L., L.H. and L.L.; validation, X.H.; formal analysis, Y.L. and L.H.; investigation, Y.L., L.H. and Z.H.; resources, Z.H. and L.H.; data curation, Y.L., L.H. and L.L.; writing—original draft preparation, Y.L.; writing—review and editing, X.H. and C.W.; visualization, Y.L. and C.W.; supervision, X.H.; funding acquisition, X.H. All authors have read and agreed to the published version of the manuscript.

Funding: This research is funded by the Sanya Institute of China Agricultural University Guiding Fund Project, Grant (SYND-2021-06), the earmarked fund for China Agriculture Research System (CARS-28), and National Natural Science Foundation of China (No. 32202343).

Institutional Review Board Statement: Not applicable.

Informed Consent Statement: Not applicable.

Data Availability Statement: The data presented in this study are available on request from the corresponding author.

Acknowledgments: The authors would like to give special thanks to Penghui Agriculture Company for providing the UAV for this experiment.

Conflicts of Interest: The authors declare no conflict of interest.

References

1. Jahurul, M.; Zaidul, I.; Ghafoor, K.; Al-Juhaimi, F.Y.; Nyam, K.-L.; Norulaini, N.; Sahena, F.; Omar, A.M. Mango (*Mangifera indica* L.) by-products and their valuable components: A review. *Food Chem.* **2015**, *183*, 173–180. [CrossRef] [PubMed]
2. Galán Saúco, V. Trends in world mango production and marketing. In Proceedings of the XI International Mango Symposium 1183, Darwin, Australia, 28 September–2 October 2015; pp. 351–364.
3. Pena, J.; Mohyuddin, A.; Wysoki, M. A review of the pest management situation in mango agroecosystems. *Phytoparasitica* **1998**, *26*, 129–148. [CrossRef]
4. Ploetz, R. Diseases of mango. In *Diseases of Tropical Fruit Crops*; CABI Books: Wallingford, UK, 2003; pp. 327–363.
5. Rocha, F.H.; Infante, F.; Quilantán, J.; Goldarazena, A.; Funderburk, J.E. 'Ataulfo'Mango Flowers Contain a Diversity of Thrips (Thysanoptera). *Fla. Entomol.* **2012**, *95*, 171–178. [CrossRef]
6. Venkata Rami Reddy, P.; Gundappa, B.; Chakravarthy, A. Pests of mango. In *Pests and Their Management*; Springer: Berlin/Heidelberg, Germany, 2018; pp. 415–440.
7. Xiongkui, H.; Bonds, J.; Herbst, A.; Langenakens, J. Recent development of unmanned aerial vehicle for plant protection in East Asia. *Int. J. Agric. Biol. Eng.* **2017**, *10*, 18–30.
8. Gil, E.; Salcedo, R.; Soler, A.; Ortega, P.; Llop, J.; Campos, J.; Oliva, J. Relative efficiencies of experimental and conventional foliar sprayers and assessment of optimal LWA spray volumes in trellised wine grapes. *Pest Manag. Sci.* **2021**, *77*, 2462–2476. [CrossRef]
9. Xu, L.; Zhang, H.; Zhang, H.; Xu, Y.; Xu, M.; Jiang, X.; Zhang, H.; Jia, Z. Development and experiment of automatic target spray control system used in orchard sprayer. *Trans. Chin. Soc. Agric. Eng.* **2014**, *30*, 1–9.
10. Cross, J.; Walklate, P.; Murray, R.; Richardson, G. Spray deposits and losses in different sized apple trees from an axial fan orchard sprayer: 3. Effects of air volumetric flow rate. *Crop Prot.* **2003**, *22*, 381–394. [CrossRef]
11. Miranda-Fuentes, A.; Rodríguez-Lizana, A.; Gil, E.; Agüera-Vega, J.; Gil-Ribes, J.A. Influence of liquid-volume and airflow rates on spray application quality and homogeneity in super-intensive olive tree canopies. *Sci. Total Environ.* **2015**, *537*, 250–259. [CrossRef] [PubMed]
12. Shuran, S.; Jianze, C.; Tiansheng, H.; Cheng, Z.; Qiufang, D.; Xiuyun, X. Design and experiment of orchard flexible targeted spray device. *Trans. Chin. Soc. Agric. Eng.* **2015**, *31*, 57–63.
13. Yan, C.; Xu, L.; Yuan, Q.; Ma, S.; Niu, C.; Zhao, S. Design and experiments of vineyard variable spraying control system based on binocular vision. *Trans. CSAE* **2021**, *37*, 13–22.

14. Li, L.; He, X.; Song, J.; Wang, X.; Jia, X.; Liu, C. Design and experiment of automatic profiling orchard sprayer based on variable air volume and flow rate. *Trans. Chin. Soc. Agric. Eng.* **2017**, *33*, 70–76.
15. Richardson, B.; Rolando, C.A.; Somchit, C.; Dunker, C.; Strand, T.M.; Kimberley, M.O. Swath pattern analysis from a multi-rotor unmanned aerial vehicle configured for pesticide application. *Pest Manag. Sci.* **2020**, *76*, 1282–1290. [CrossRef]
16. Martinez-Guanter, J.; Agüera, P.; Agüera, J.; Pérez-Ruiz, M. Spray and economics assessment of a UAV-based ultra-low-volume application in olive and citrus orchards. *Precis. Agric.* **2020**, *21*, 226–243. [CrossRef]
17. Lan, Y.; Chen, S. Current status and trends of plant protection UAV and its spraying technology in China. *Int. J. Precis. Agric. Aviat.* **2018**, *1*, 1–9. [CrossRef]
18. Li, X.; Giles, D.K.; Andaloro, J.T.; Long, R.; Lang, E.B.; Watson, L.J.; Qandah, I. Comparison of UAV and fixed-wing aerial application for alfalfa insect pest control: Evaluating efficacy, residues, and spray quality. *Pest Manag. Sci.* **2021**, *77*, 4980–4992. [CrossRef] [PubMed]
19. Yao, W.; Guo, S.; Wang, J.; Chen, C.; Yu, F.; Li, X.; Xu, T.; Lan, Y. Droplet deposition and pest control efficacy on pine trees from aerial application. *Pest Manag. Sci.* **2022**, *78*, 3324–3336. [CrossRef] [PubMed]
20. Wang, C.; Liu, Y.; Zhang, Z.; Han, L.; Li, Y.; Zhang, H.; Wongsuk, S.; Li, Y.; Wu, X.; He, X. Spray performance evaluation of a six-rotor unmanned aerial vehicle sprayer for pesticide application using an orchard operation mode in apple orchards. *Pest Manag. Sci.* **2022**, *78*, 2449–2466. [CrossRef]
21. Wang, C.; Herbst, A.; Zeng, A.; Wongsuk, S.; Qiao, B.; Qi, P.; Bonds, J.; Overbeck, V.; Yang, Y.; Gao, W. Assessment of spray deposition, drift and mass balance from unmanned aerial vehicle sprayer using an artificial vineyard. *Sci. Total Environ.* **2021**, *777*, 146181. [CrossRef]
22. Huang, Z.; Wang, C.; Li, Y.; Zhang, H.; Zeng, A.; He, X. Field evaluation of spray drift and nontargeted soybean injury from unmanned aerial spraying system herbicide application under acceptable operation conditions. *Pest Manag. Sci.* **2023**, *79*, 1140–1153. [CrossRef]
23. Li, X.; Giles, D.K.; Niederholzer, F.J.; Andaloro, J.T.; Lang, E.B.; Watson, L.J. Evaluation of an unmanned aerial vehicle as a new method of pesticide application for almond crop protection. *Pest Manag. Sci.* **2021**, *77*, 527–537. [CrossRef]
24. Pan, Z.; Lie, D.; Qiang, L.; Shaolan, H.; Shilai, Y.; Yande, L.; Yongxu, Y.; Haiyang, P. Effects of citrus tree-shape and spraying height of small unmanned aerial vehicle on droplet distribution. *Int. J. Agric. Biol. Eng.* **2016**, *9*, 45–52.
25. Liu, Y.; Li, L.; Liu, Y.; He, X.; Song, J.; Zeng, A.; Wang, Z. Assessment of spray deposition and losses in an apple orchard with an unmanned agricultural aircraft system in China. *Trans. ASABE* **2020**, *63*, 619–627. [CrossRef]
26. Zhang, R.; Chen, L.; Li, L.; Zhang, L.; Tang, Q.; Li, X. An Air-Ground Synergistic Application Method and System. CN Patent 112965514A, 15 June 2021.
27. Jiang, S.; Chen, B.; Li, W.; Yang, S.; Zheng, Y.; Liu, X. Stereoscopic plant-protection system integrating UAVs and autonomous ground sprayers for orchards. *Front. Plant Sci.* **2022**, *13*, 1040808. [CrossRef] [PubMed]
28. Meier, U.; Graf, H.; Hack, H.; Hess, M.; Kennel, W.; Klose, R.; Mappes, D.; Seipp, D.; Stauss, R.; Streif, J.; et al. Phänologische Entwicklungsstadien des Kernobstes (*Malus domestica* Borkh. und *Pyrus communis* L.), des Steinobstes (*Prunus*-Arten), der Johannisbeere (*Ribes*-Arten) und der Erdbeere (*Fragaria* × *ananassa* Duch.). *Nachrichtenbl. Deut. Pflanzenschutzd.* **1994**, *46*, 141–153.
29. Agustí, M.; Zaragoza, S.; Bleiholder, H.; Buhr, L.; Hack, H.; Klose, R.; Staub, R. Escala BBCH para la descripción de los estadios fenológicos del desarrollo de los agrios (Gén. Citrus). *Levante Agrícola* **1995**, *332*, 189–199.
30. *ISO 25358:2018*; Crop Protection Equipment—Droplet-Size Spectra from Atomizers—Measurement and Classification. ISO: Geneva, Switzerland, 2018.
31. Miranda-Fuentes, A.; Rodríguez-Lizana, A.; Cuenca, A.; González-Sánchez, E.; Blanco-Roldán, G.; Gil-Ribes, J. Improving plant protection product applications in traditional and intensive olive orchards through the development of new prototype air-assisted sprayers. *Crop Prot.* **2017**, *94*, 44–58. [CrossRef]
32. Ebert, T.A.; Taylor, R.A.J.; Downer, R.A.; Hall, F.R. Deposit structure and efficacy of pesticide application. 1: Interactions between deposit size, toxicant concentration and deposit number. *Pestic. Sci.* **1999**, *55*, 783–792. [CrossRef]
33. Ebert, T.A.; Taylor, R.A.J.; Downer, R.A.; Hall, F.R. Deposit structure and efficacy of pesticide application. 2: Trichoplusia ni control on cabbage with fipronil. *Pestic. Sci.* **1999**, *55*, 793–798. [CrossRef]
34. Uk, S. Tracing insecticide spray droplets by sizes on natural surfaces. The state of the art and its value. *Pestic. Sci.* **1977**, *8*, 501–509. [CrossRef]

agronomy

Article

Open-Field Agrivoltaic System Impacts on Photothermal Environment and Light Environment Simulation Analysis in Eastern China

Long Zhang [1,2], Zhipeng Yang [3], Xue Wu [2], Wenju Wang [4], Chen Yang [5], Guijun Xu [5], Cuinan Wu [2,*] and Encai Bao [1,2,*]

1 School of Agricultural Engineering, Jiangsu University, Zhenjiang 212013, China
2 Institute of Agricultural Facilities and Equipment, Jiangsu Academy of Agricultural Sciences, Nanjing 210014, China
3 School of Engineering, Anhui Agricultural University, Hefei 230036, China
4 School of Energy and Power Engineering, Nanjing University of Science and Technology, Nanjing 210094, China; wangwenju1982@163.com
5 Shenzhen Energy Nanjing Holding Co., Ltd., Nanjing 210019, China
* Correspondence: wucuinan@126.com (C.W.); baoencai1990@163.com (E.B.); Tel.: +86-159-5188-7596 (C.W.); +86-177-6810-3527 (E.B.)

Abstract: In order to clarify the temporal and spatial changes in the internal photothermal environment in an open-field agrivoltaic system (OAVS), this paper took the OAVS in eastern China as the research object and divided the internal area into the southern area, middle area and northern area, according to the spatial structure. Further, a photothermal environment test was conducted in the above three areas in the summer and winter. The results showed that the summer average daylight rate ($R_{m\text{-}avg}$) in the middle area was 66.6%, while the $R_{m\text{-}avg}$ in the other two areas was about 20%, with no significant difference. In the winter, the light environment in the southern area was slightly better, and the $R_{m\text{-}avg}$ in the above three areas was 26.4%, 24.7% and 19.7%, respectively. On the whole, the relationship between the thermal environmental factors and the solar radiation intensity was consistent. Further, a 3D model of an OAVS was established using Autodesk Ecotect Analysis 2011, and the internal light environment was simulated. Compared with the measured values, the relative error was less than 10%, which verified the reliability of the OAVS model. Then, the model was used to reveal the temporal and spatial changes in the light environment of the OAVS. The simulation results showed that the daylighting rate in the summer from the ground to the height of the fig canopy inside the system was 20.7% to 61.5%. In the winter, the daylighting rate from the ground to the height of the fig canopy inside the system was 17.7% to 36.4%. The effectiveness of the OAVS in reducing the level of solar radiation intensity depended on the time of day and the angle of the sun. At the spatial scale, due to the strong consumption of light by photovoltaic panels, there was a strong horizontal and vertical light environment gradient inside the system. In conclusion, the photothermal environment research of an OAVS based on Autodesk Ecotect Analysis 2011 can not only provide a basis for agricultural production and structural design such as span, height and the laying density of PV panels, but also expand its application to regions with different latitudes and longitudes and specific climates.

Keywords: agrivoltaics; agricultural building; photothermal environment; ECOTECT; simulation

Citation: Zhang, L.; Yang, Z.; Wu, X.; Wang, W.; Yang, C.; Xu, G.; Wu, C.; Bao, E. Open-Field Agrivoltaic System Impacts on Photothermal Environment and Light Environment Simulation Analysis in Eastern China. *Agronomy* **2023**, *13*, 1820. https://doi.org/10.3390/agronomy13071820

Academic Editors: Xiongkui He, Fuzeng Yang and Baijing Qiu

Received: 12 June 2023
Revised: 3 July 2023
Accepted: 5 July 2023
Published: 8 July 2023

1. Introduction

The open-field agrivoltaic system (OAVS) is a modern integrated agricultural production system which combines PV power generation and agricultural production [1,2]. Under the condition that agricultural land and agricultural production are not affected, it can improve the efficiency of land use, which is of great significance to the development of the PV industry and the transformation of modern agriculture [3,4]. In the context of carbon neutrality, the development and utilization of solar energy has become an important "tuyere" [5]. The OAVS has been widely applied and promoted in China, which combines PV and field planting [6]. Relying on a good agricultural resource base, the OAVS project has developed rapidly in eastern China [7]. By the end of 2022, the cumulative grid-connected capacity of centralized PV power stations in this area reached 72.96 million kW, accounting for 19.8% of the national grid-connected capacity, among which the OAVS is the main application form of centralized PV power stations in this area [8]. At the same time, this area is the main vegetable-producing area in China, and has great potential for agrivoltaic implementation [9]. For example, the installed PV-saturated capacity of Jiangsu Province and Shandong Province in eastern China can reach 18.25 GW and 16.00 GW, respectively [7]. At present, the OAVS in East China covers a large area and is relatively concentrated in scale, making it easy to carry out large-scale, mechanized and intensive agricultural production. However, affected by the occlusion of photovoltaic panels, the OAVS is conducive to the growth and development of shad-tolerant and cold-loving crops [10]. Therefore, the OAVS is mostly planted with shad-tolerant, compact, high yield and high efficiency crop types and varieties. For example, it is suitable to plant corn, sorghum, potato, taro, yam and beans between the boards and under the eaves, and shady crops such as Chinese medicinal materials can be planted under the columns [11–14].

The development of the OAVS project can improve land utilization rate [15,16], promote industrial low carbon development [17,18], enrich field biomass [19], improve soil moisture [19,20] and increase comprehensive benefits [21,22]. However, there are some tricky problems in the OAVS in this area, such as the unclear internal photothermal environment and the poor coupling effect of agriculture and PV [23]. A large number of studies have shown that due to the shading of PV panels, there are phenomena such as uneven distribution and low intensity of light inside the system, which affect the normal growth of crops [24,25]. In order to clarify the variation rule in the internal photothermal environment of the OAVS, experts and scholars at home and abroad have conducted a lot of research on the photothermal environment of the system and similar PV buildings. Marrou et al. [26] monitored the microclimatic conditions of different PV panel laying densities in the OAVS in different seasons. The results showed that there was no significant difference in average daily air temperature and average daily relative humidity under different treatments, but there was a gradient difference in soil temperature. Gao et al. [27] studied the diurnal and annual variation characteristics of air temperature inside and outside of PV power stations. Through monitoring and the analysis of a single point inside and outside the power station, the results showed that the daytime air temperature inside and outside the power station was basically the same in the winter. In the spring, summer and autumn, the air temperature inside the power station was significantly higher than that outside, and the difference was the largest in the summer, 0.67 °C. Hassanpour Adeh et al. [19] studied the environmental effects of PV panels on unirrigated grasslands that frequently suffer from water stress, quantifying the effects of the presence of PV panels on microclimate and soil moisture. The results showed that the area under the PV panel maintained a higher soil moisture, and the biomass increased by 90% in the later period. Chang et al. [28] studied the thermal effect of PV panels on the temperature of adjacent air, and the results showed that the temperature of PV panels (height of the erection was 2 m) increased significantly, with an average annual increase of 9.7 °C and an average annual increase in adjacent air temperature of 3.8 °C. Sailor et al. [29] found through research that PV urban building systems heat the city during the day and cool it at night. Ezzaeri et al. [30] studied the effects of checkerboard form flexible PV panels on greenhouse microclimates and tomato yields in the summer and winter. The results showed that the PV panel reduced the air temperature in the greenhouse, and the occupancy rate of PV panel had no significant effect on the total yield of tomato.

In summary, the internal environment of the OAVS is relatively complex and has an important impact on agricultural production. However, there are no reports on the variation rule of the photothermal environment inside the system at a spatial scale. It is an important basis for the agricultural production of the OAVS to make clear the variation rule of internal environment and the photothermal performance. The accurate photothermal environment can not only provide a basis for the OAVS agricultural mode selection and production management, but also provide reference for the structural design of the system span, height and PV panel laying density.

As a civil building software, Autodesk Ecotect Analysis 2011 is able to conduct light environment analysis, energy consumption analysis and sunshine and shade analysis for buildings by setting building structural parameters, material parameters and environmental parameters [31–33]. Various studies show that the simulation accuracy of the software is good [34,35]. When applying the software for research and analysis, the boundary conditions are convenient to set, the operation process is relatively simple and the result visualization ability is strong [36,37]. Although the software is mainly used for the analysis of civil buildings, it has a broad prospect in the analysis and application of the light environment of agricultural buildings.

This paper will take the OAVS in eastern China as the research object and measure the solar radiation intensity, air temperature and soil temperature in the summer and winter. Further, the distribution of environmental parameters in different areas will be analyzed. Combined with Autodesk Ecotect Analysis 2011, a proportional agrivoltaic system model will be constructed to simulate and analyze the solar radiation, shading and sunshine hours in the system, so as to clarify its optical environment performance in space and time scales. The novelties of this study are as follows: (1) Combined with the sensor configuration strategy and spatial structure characteristics, the change rule of the photothermal environment in different areas inside the OAVS in the summer and winter is quantified. (2) The OAVS light environment model is established, revealing the spatial and temporal scales system interior light environment distribution characteristics.

2. Materials and Methods

2.1. Experimental Site Description

The experimental OAVS is located in Shenzhen Energy Nanjing Holding Co., Ltd. (31.62° latitude, 119.18° longitude, approx. 360 m above sea level), Lishui District, Nanjing City, Jiangsu Province, China. The power station was put into operation in 2016, covering an area of about 46.67 ha, with a total installed capacity of 20 MW and annual clean power generation of about 24 million kWh. It is also a base integrating photovoltaic power generation, ecological agriculture, sightseeing tourism, popular science exhibition and photovoltaic agricultural research and experiment (Figure 1a). Lishui district is in the transition zone from the north subtropical zone to the middle subtropical zone. The average annual temperature is 15.5 °C. The average annual sunshine duration is 2145.8 h and the average annual rainfall is 1036.9 mm.

The PV panels of the system face south and are supported by Φ300 concrete piles and an inclined supporting frame. The distance between concrete piles is 3 m and 6.8 m between the panel arrays. The PV module is made of 265 W polysilicon, with a photoelectric conversion efficiency of 16.3%. The size of a single PV module is 1640 × 992 × 35 mm, composed of 60 minimum cells. Four PV modules are arranged in a row and continuously laid along the east–west direction. The lowest edge of the PV modules is 3.2 m above the ground and has a tilt angle of 24°. One row of fig (*Ficus carica*) trees were planted in the middle of each span in the OAVS with a height of about 1.0 m and a root length of about 0.3 m. The figs were equally spaced with planting spacing of about 3.0 m, and their main growing season at this site is from April to October. The soil texture is classified as yellow loam. In addition, since the PV panes have the function of collecting rain, two drainage ditches in the east–west direction are formed inside the single span PV array (Figure 1b,c).

Figure 1. (**a**) Aerial view of OAVS; (**b**) spatial structure of experimental OAVS; (**c**) monitoring points in experimental OAVS.

2.2. Photothermal Environment Measurement and Daylighting Rate Calculation

According to the needs of the experimental design, the space occupied by the PV arrays is defined as the interior of the OAVS. The height of the space is the lowest height of the PV panel from the ground, which is about 3.2 m. The exterior of the OAVS is defined as a place far away from the OAVS and sufficiently open. According to the spatial structure of the OAVS, the single span PV array was divided into three areas along the north–south direction, namely, southern area, middle area and northern area. The width of each area is 1.8, 3.2 and 1.8 m, respectively (Figure 2). The environmental factors monitored inside and outside of this experimental OAVS include solar radiation intensity, air temperature and soil temperature. The arrangement of monitoring instruments in the three areas follows the following principles: (1) Reduce the sensor configuration redundancy under the condition of meeting the information collection requirements; (2) fully consider the shading of photovoltaic modules, plant height, plant root length and other factors; the expected test results can fully reflect the objective law to avoid a bad impact on the monitoring of the light temperature sensor. Based on this, the solar radiation intensity and air temperature measurement point was set at 1.0 m above the ground, the same as the height of the fig tree canopy, and the soil temperature measurement point was set at 0.15 m underground, which was in the middle position of fig root depth. The instruments were arranged 0.65 m away from the axis of the prefabricated pipe pile to fully reflect the photothermal performance of the OAVS.

The arrangement of measuring points is shown in Figure 2: (1) Solar radiation intensity and air temperature in the OAVS: Sensors for solar radiation intensity and air temperature were placed at south area, middle area and north area, respectively, which were 0.65, 3.40 and 6.15 m away from the concrete pile along the north–south direction and 1.0 m above the ground, one sensor for each area. (2) Soil temperature in the OAVS: Soil temperature measuring sensors were placed at depth of 0.15 m directly below the solar radiation intensity and air temperature measuring points, one sensor for each area. (3) All the measuring points of solar radiation intensity, air temperature and soil temperature inside the OAVS were arranged in the same plane along the north–south direction, and were not affected by side light. (4) The OAVS external environment measuring point: An environmental measuring point was set in open field far away from the PV arrays. The data of solar radiation intensity at a height of 1.0 m above the ground, air temperature and soil temperature at a depth of 0.15 m were collected. The monitoring scene of the experimental OAVS is shown in Figure 1c.

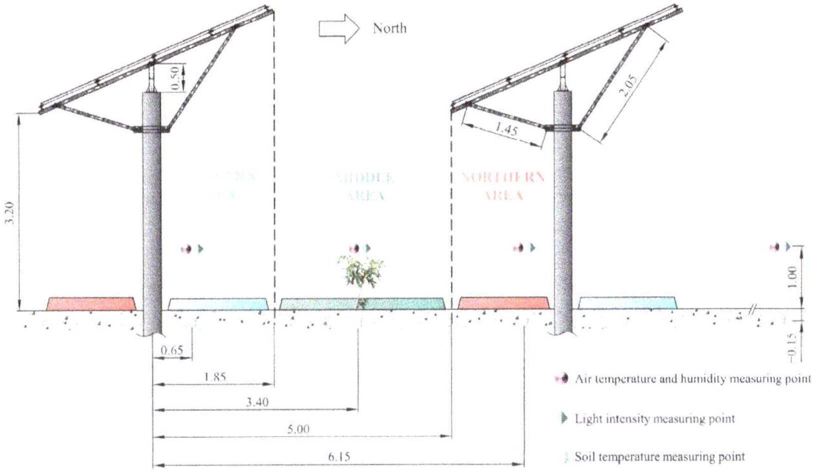

Figure 2. Layout of measuring points inside and outside the experimental OAVS.

Data were collected in the summer from 11 August 2022 to 25 August 2022 and in the winter from 15 December 2022 to 30 December 2022. The test instruments were HOBO series sensors, as shown in Table 1. The environmental data were recorded throughout the day at 30 min intervals. The HOBO UX100-011A air temperature sensor was equipped with a special meteorological radiation shield to eliminate the effects of solar radiation and complex outdoor environment on data reliability. The daytime period in the summer and winter was calculated from 06:00 to 18:00, and the night period was from 18:00 to $06{:}00^{+1}$.

Table 1. Instrument parameters.

Instrument Name	Manufacturer	Model	Range	Accuracy
HOBO temperature recorder	Onset Co., Ltd., Bourne, MA, USA	UX100-011A	−20~+70 °C	±0.2 °C
HOBO four-channel recorder		UX120-006M	−20~+70 °C	±0.2 °C
HOBO temperature sensor		TMC20-HD	−20~+100 °C	±0.15 °C
HOBO total solar radiation sensor		S-LIB-M003	0~+1280 W/m^2	±10 W/m^2

Daylighting rate, as a key evaluation index, was used in the analysis of the light environment of various buildings [38,39]. Sun et al. [40] took the ratio of the average solar radiation intensity inside the greenhouse and outdoor in a single moment as the average daylighting rate to evaluate the lighting performance of the greenhouse. Based on this, the ratio of the average daily internal solar radiation intensity ($E_{m\text{-}in}$) to the average daily outdoor solar radiation intensity ($E_{m\text{-}out}$) is prescribed in this paper as the daily daylighting rate ($R_{m\text{-}d}$). The average of daily daylighting rate of each measurement point during the whole recording period is described as the average daylighting rate ($R_{m\text{-}avg}$). The $R_{m\text{-}d}$ and $R_{m\text{-}avg}$ are used as the evaluation index of the internal light environment of the OAVS, and calculated according to Equations (1) and (2).

$$R_{m\text{-}d} = \frac{1}{k}\sum_{k=1}^{k} E_{m\text{-}in_k} \Big/ \frac{1}{k}\sum_{k=1}^{k} E_{m\text{-}out_k} \quad (1)$$

$$R_{m\text{-}avg} = \frac{1}{i}\sum_{i=1}^{i} R_{m\text{-}d_i} \quad (2)$$

where $E_{m\text{-}in}$ is the measured value of solar radiation intensity at a single time inside the OAVS (W/m^2); $E_{m\text{-}out}$ is the measured value of solar radiation intensity at a single time outside the OAVS (W/m^2); $R_{m\text{-}d}$ is the daylighting rate of the OAVS (%); $R_{m\text{-}avg}$ is the average daylighting rate (%); k is the number of daily records of solar radiation intensity data; and i is the number of experimental days (d).

2.3. Simulation of Light Environment

The internal photothermal environment of the OAVS is relatively complex according to previous research. Solar radiation intensity is the primary ecological factor in the OAVS and has strong interaction effects with other environmental factors. Therefore, Autodesk Ecotect Analysis 2011 was used to construct the light environment model of the OAVS to clarify the light environment performance in both spatial and temporal scales.

2.3.1. Model Construction

The 1:1 OAVS model (Figure 3) was established through a 3D framework (SOLIDWORKS 2020, France) and imported into Autodesk Ecotect Analysis 2011. The typical annual weather data of Nanjing was adopted, and relevant data were put into ECOTECT weather tools to automatically extract meteorological conditions such as direct/scattered solar radiation, annual temperature/humidity variation and annual wind speed variation. Then, the shading rate, sunlight hours and solar radiation inside the OAVS were elaborated based on the actual orientation and inclination. Building material attributes are not included in this paper, as solar radiation and other studied indicators are mainly related to solar altitude angle and azimuth angle. A flowchart for the OAVS simulation procedure was designed based on the model as shown in Figure 4.

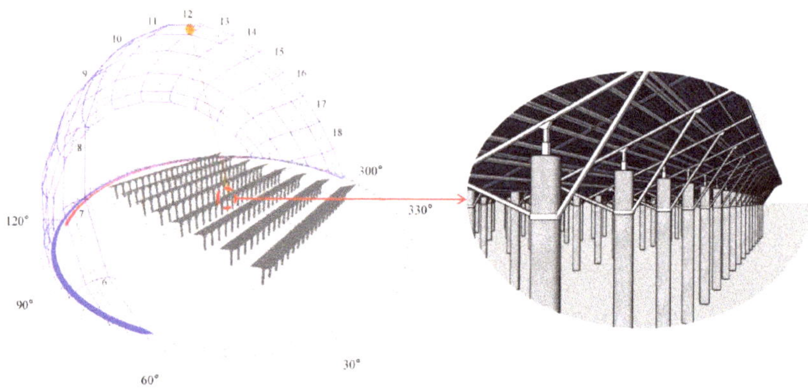

Figure 3. ECOTECT model of OAVS. "6-18"shows the hours of a day.

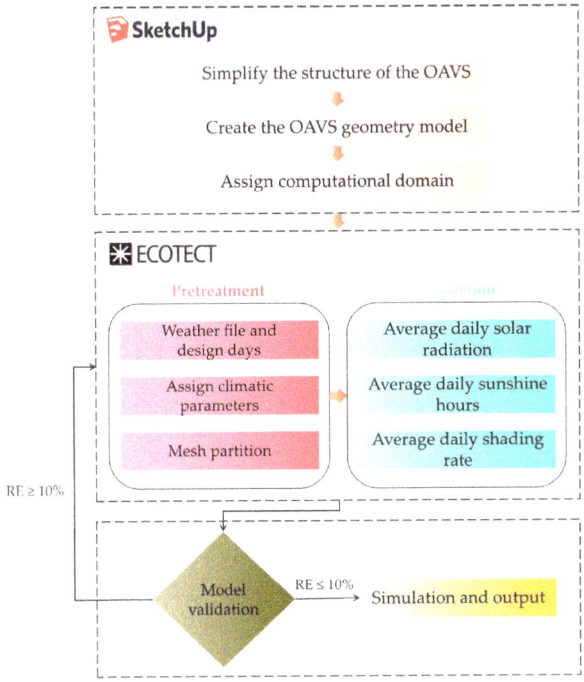

Figure 4. Flowchart for OAVS simulation procedure.

2.3.2. Setting of Numerical Simulation

The center area of the PV arrays was chosen as the calculation domain to eliminate the effect of side light. During grid division, the east–west profile was selected as the simulation plane, and the simulation plane was divided into grid units and calculated in turn. The "display analysis grid" was selected in the "analysis Grid" panel. The values of "2D slice position" were successively set to the required cross sections for calculation. On the basis of considering calculation accuracy and calculation speed, the mesh density should be increased as much as possible. The number of mesh set in this paper was 2048 ("64 × 32"). In Autodesk Ecotect Analysis 2011, we selected "Solar Access Analysis" in Calculate for relevant simulation settings (Figure 5). The simulated period was consistent with the actual test period.

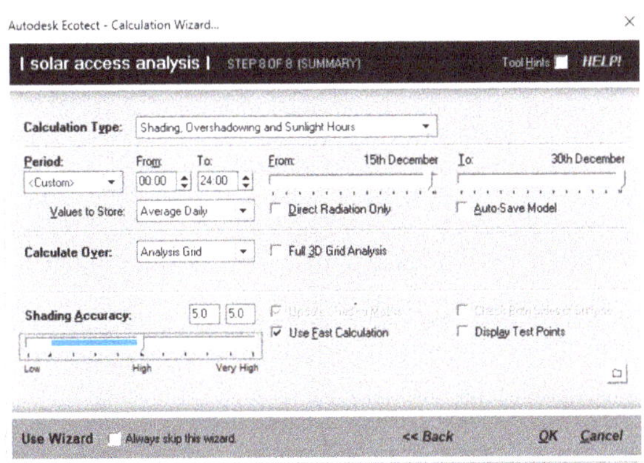

Figure 5. Simulation parameter setting diagram of ECOTET model.

2.4. Data Analysis

Excel 2016 was used for statistical processing and analysis of the experimental data, and Origin 2021 was used to make relevant charts.

3. Results

3.1. Analysis of Photothermal Environment in OAVS

3.1.1. Solar Radiation Intensity

As shown in Figure 6, the solar radiation intensity in the summer and winter differed greatly inside and outside of the OAVS. The solar radiation intensity outside of the OAVS was higher than that of the three measuring points inside the PV arrays. The maximum solar radiation intensity of all the points in the open field, southern area, middle area and northern area in the summer was 391.4, 69.4, 326.3 and 79.0 W/m^2 (Figure 6a), and the maximum solar radiation intensity of all the points in the open field, southern area, middle area and northern area in the winter was 250.3, 170.7, 56.3 and 63.4 W/m^2, respectively (Figure 6b). In the summer, the middle area had the highest solar radiation intensity inside the OAVS, followed by the northern area. The southern area was the lowest. The solar radiation intensity in the OAVS in the winter showed a different change pattern. In the winter, the solar radiation intensity on sunny days in the southern area was higher than that in the other two areas, while on cloudy days, the middle area had the highest solar radiation intensity.

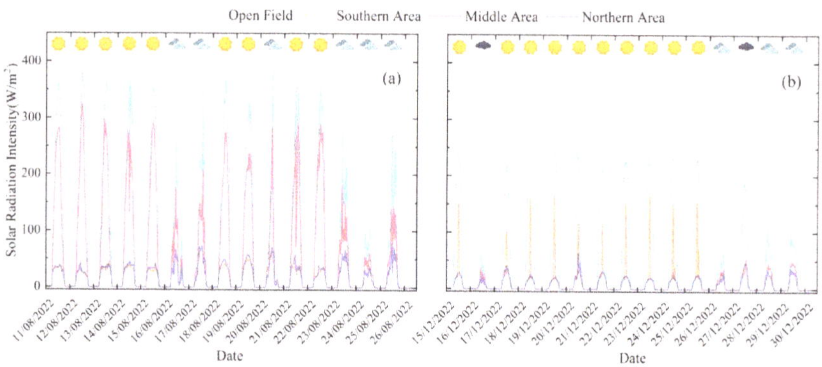

Figure 6. Variation in solar radiation intensity inside and outside OAVS; (**a**) summer; (**b**) winter.

In the summer, on both sunny and cloudy or rainy days, the solar radiation intensity curves of each measuring point had good consistency. On sunny days, the curves first rose from 6 am and reached the highest value at around 12 pm, and then they fell and went back to 0 after 6 pm. The curves of different measuring points reached the peak basically at the same time. On cloudy days, the curves presented more frequent fluctuation, which may be caused by the constant change in cloud thickness, leading to the constant change in direct solar radiation intensity. The change pattern of solar radiation intensity on cloudy days in the winter was similar to that in the summer. However, on sunny winter days, due to the change in solar altitude angle, solar radiation intensity in the southern area increased greatly around 12:00 am, while the other two areas stayed at a lower level.

As shown in Figure 7, the range of $E_{d\text{-}avg}$ in the southern area, middle area and northern area was 53.2~230.7, 18.6~37.3, 31.5~169.8 and 21.6~41.5 W/m^2, respectively. The range of $E_{d\text{-}avg}$ in the winter was 22.3~150.3, 8.7~34.2, 13.0~29.7 and 8.1~23.8 W/m^2, for the southern, middle and northern area, respectively. The $R_{m\text{-}avg}$ in the southern, middle and northern area was 20.5%, 66.6% and 22.7% (summer), and 26.4%, 24.7% and 19.7% (winter), respectively. It can be seen that the $E_{d\text{-}avg}$ and $R_{m\text{-}avg}$ of the OAVS in the summer were the highest in the middle area, and there is no significant difference between the other two areas. In the winter, the $E_{d\text{-}avg}$ and $R_{m\text{-}avg}$ of the OAVS were larger in the southern area, but the differences among the three areas were relatively small.

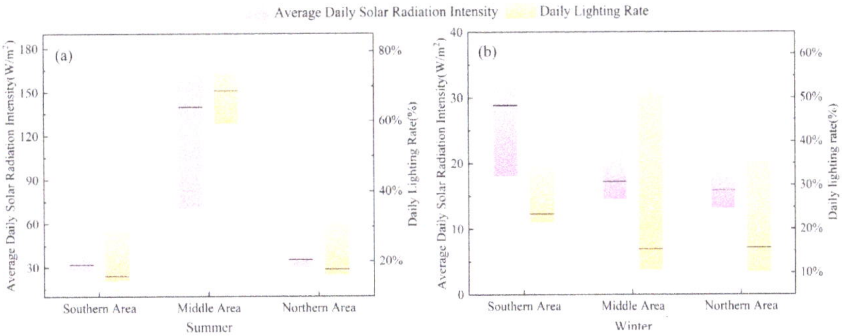

Figure 7. Box plots of average daily solar radiation intensity and daylighting rate in OAVS; (**a**) summer; (**b**) winter.

3.1.2. Air Temperature

Figure 8 shows the changes in air temperature inside and outside of the OAVS in the summer and winter. The statistical data show that the general trend in the air temperature curve inside and outside of the OAVS was basically the same, and the air temperature in the open field was significantly higher than the three measuring points inside the PV array. As shown in Figure 8a, the air temperature of all the measuring points in the open field, southern area, middle area and northern area in the summer ranged from 24.1 °C to 48.7 °C, 24.0 °C to 43.0 °C, 23.9 °C to 48.2 °C and 24.0~47.1 °C. As shown in Figure 8b, the air temperature varied from −8.7 to 16.7 °C, −8.5 to 16.9 °C, −9.0 to 13.7 °C and −8.7 to 14.7 °C at all measuring points in the open field, southern area, middle area and northern area in the winter. In addition, the air temperature distribution curve fluctuated greatly in the two seasons, and the difference between the daily peak value and the daily trough value was obvious, but the regularity was also obvious. In the summer, the air temperature in the three areas of the OAVS was as follows: middle area > northern area > southern area, and the air temperature in the middle area was much higher than that in the other two areas. The variation in air temperature in the three areas of the OAVS in the winter was somewhat different from that in the summer. The overall air temperature in the winter shows that the southern area is higher than the other two areas, and the other two areas are roughly equal.

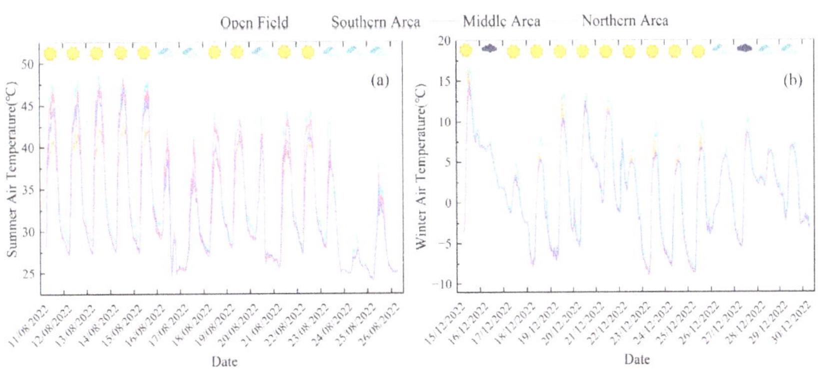

Figure 8. Air temperature variations inside and outside OAVS; (**a**) summer; (**b**) winter.

Combined with the weather types during the two seasons, whether it was sunny or cloudy or rainy, the curves of each measuring point had good consistency, showing a trend of first rising and then falling, and the corresponding time of peaks of each curve was basically the same. At all measuring points, the temperature basically maintained a gradual rising trend from 06:00 to 12:00, and reached the maximum temperature from 12:00 to 12:00, and gradually dropped from 14:30 to 6:00 the next day.

In order to better display the air temperature differences in different areas of the OAVS, this paper analyzed the distribution of four air temperature parameters. Figure 9 shows the changes in average daily air temperature, average daytime air temperature, average nighttime air temperature and daily extreme air temperature of the OAVS in the summer and winter. In the winter, the above four air temperature indices showed that there was no significant difference among the measuring points inside of the OAVS, and the daytime air temperature of the open field measuring points was slightly higher than that of other measuring points. Taking the average daytime air temperature as an example, the air temperatures in the open field, southern area, middle area and northern area were 5.2 °C, 4.5 °C, 4.2 °C and 4.3 °C, respectively. In the summer, the above three air temperature indices (excluding average nighttime air temperature) all showed the temperature relationship of four measuring points: open field > middle area > northern area > southern area. Taking the average daytime air temperature as an example, the temperature in the above four areas was 38.5 °C, 35.5 °C, 37.6 °C and 36.5 °C, respectively. In addition, there was no significant difference in the average nighttime air temperature at each measuring point in the summer. The air temperature in the above four areas was 29.1 °C, 28.5 °C, 28.4 °C and 28.5 °C, respectively.

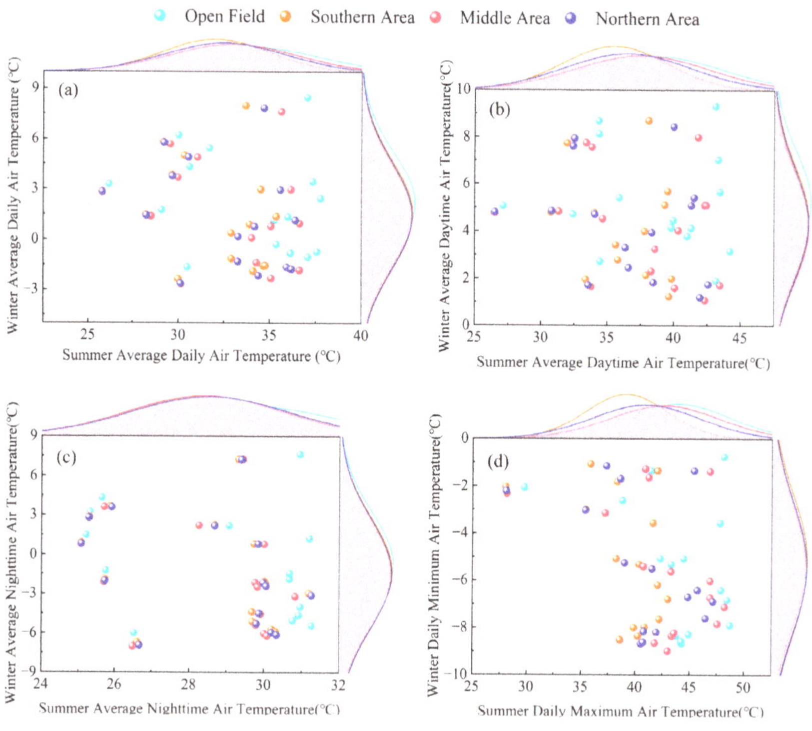

Figure 9. Marginal distribution curve data graph of air temperature in summer and winter; (**a**) average daily air temperature; (**b**) average daytime air temperature; (**c**) average nighttime air temperature; (**d**) daily extreme air temperature.

3.1.3. Soil Temperature

Figure 10 shows the changes in soil temperature inside and outside of the OAVS in the summer and winter. The statistical data show that the soil temperature curves inside and outside of the OAVS had the same general trend, and the soil temperature in the open field was significantly higher than that in the three measuring points inside the PV array. As shown in Figure 10a, the soil temperature of all measuring points in the open field, southern area, middle area and northern area in the summer ranged from 29.4 to 39.6 °C, 26.2 to 36.0 °C, 28.2 to 39.6 °C and 26.7 to 36.2 °C. As shown in Figure 10b, the soil temperature varied from 3.6 to 9.6 °C, 1.9 to 6.8 °C, 1.4 to 6.6 °C and 1.6 to 6.5 °C at all measuring points in the open field, southern area, middle area and northern area in the winter. In addition, the soil temperature distribution curve fluctuated greatly in the two seasons, and the difference between the daily peak value and the daily trough value in the four areas was obvious, but the regularity was also obvious. The relationship of soil temperature in the three areas of the OAVS in the summer was as follows: middle area > northern area > southern area, and the soil temperature in the middle area was much greater than that in the other two areas. In the winter, the soil temperature in the three areas of the OAVS was different from that in the summer. The soil temperature in middle area was greater than that in the southern area and the northern area, and the soil temperature in the southern and northern areas was roughly equal.

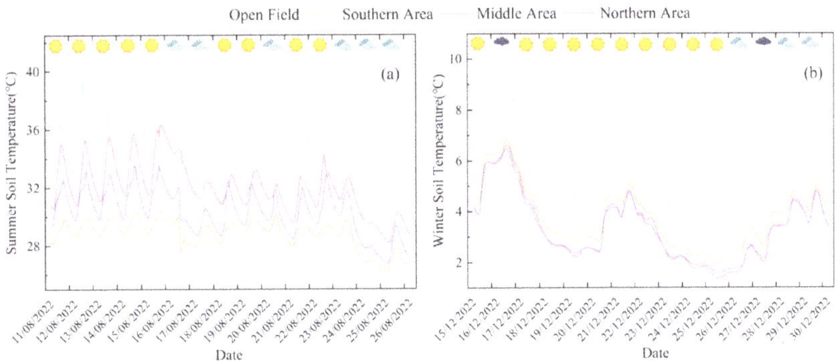

Figure 10. Soil temperature variations inside and outside OAVS; (**a**) summer; (**b**) winter.

Combined with the weather types during the two seasons, whether it was sunny or cloudy or rainy, the curves of each measuring point had good consistency, showing a trend of first rising and then falling, and the corresponding time of peaks of each curve was basically the same. All soil temperature measurement points basically maintained a gradually rising trend during the day, reached the maximum value at sunset and gradually decreased at night.

In order to better display the soil temperature differences in different areas of the OAVS, this paper analyzed the distribution of four soil temperature parameters. Figure 11 shows the variation in average daily soil temperature, average daytime soil temperature, average nighttime soil temperature and daily extreme soil temperature of the OAVS in the summer and winter, and the variation in each index was obvious. In the summer, the above four soil temperature indices all show the relationship between soil temperature in four areas: open field > middle area > northern area > southern area. Taking the average daytime soil temperature as an example, the average daytime soil temperature in the open field, southern area, middle area and northern area was 34.1 °C, 28.8 °C, 31.8 °C and 30.4 °C, respectively. In the winter, the above four soil temperature indices showed that there was no significant difference in soil temperature between the middle and northern areas of the OAVS, and the relationship of soil temperature between the four areas was as follows: open field > southern area > middle area ≈ northern area. Taking the average daytime soil temperature as an example, the soil temperature in the above four areas was 6.3 °C, 3.7 °C, 3.3 °C and 3.3 °C, respectively.

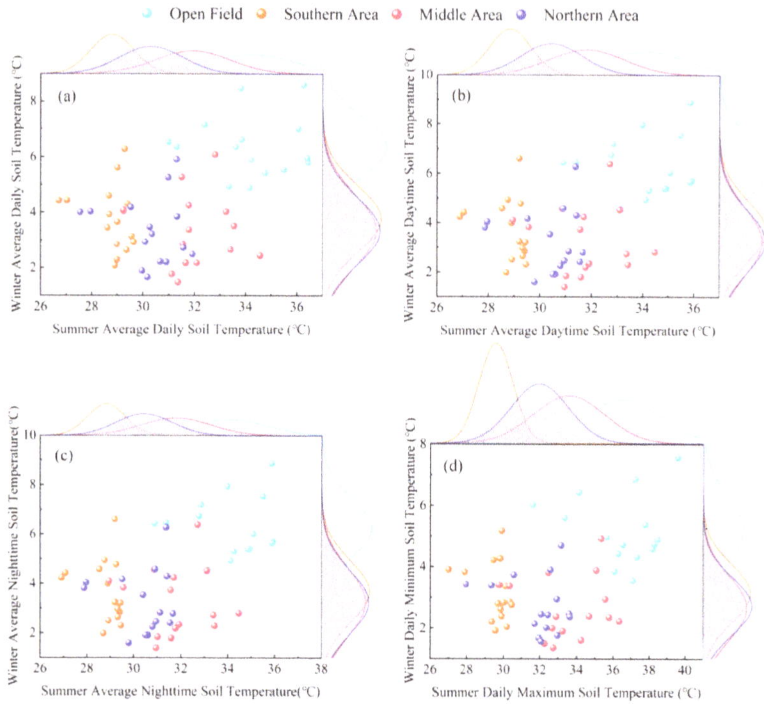

Figure 11. Marginal distribution curve data graph of soil temperature in summer and winter; (**a**) average daily soil temperature; (**b**) average daytime soil temperature; (**c**) average nighttime soil temperature; (**d**) daily extreme soil temperature.

3.2. Validation of ECOTECT Simulation

In order to verify the correctness and reliability of the calculated results of the established OAVS simulation model, combined with the previous test results, the measured values of each measurement point in the OAVS were compared with the simulated values. As Autodesk Ecotect Analysis 2011 is a simulation software based on the mean value of meteorological data for many years in the past, in order to avoid differences between the measured solar radiation intensity (E_m) and the average solar radiation intensity for many years, the average daylighting rate (R) was adopted for verification in this paper. The average daylighting rate ($R_{s\text{-}avg}$) and relative error (RE) obtained by simulation were calculated according to Equations (3) and (4).

$$R_{s\text{-}avg} = E_{s\text{-}in} / E_{s\text{-}out} \tag{3}$$

$$E = \left(R_{s\text{-}avg} - R_{m\text{-}avg}\right) / R_{m\text{-}avg} \tag{4}$$

where $E_{s\text{-}in}$ is the simulated value of the solar radiation intensity at the measuring point inside of the OAVS [MJ/(m^2·d)], $E_{s\text{-}out}$ is the simulated value of the solar radiation intensity at the measuring point outside of the OAVS [MJ/(m^2·d)] and $R_{s\text{-}avg}$ is the simulated value of the average daylighting rate (%).

Table 2 shows the comparison results between the measured values and simulated values. The measured values of each measurement point are in good agreement with the simulated values, and the RE between them are all less than 10%. The simulation results are relatively reliable, indicating that the established OAVS model meets the simulation requirements and can be used for subsequent simulation studies.

Table 2. Comparison between simulated and measured R at each measuring point.

Season(s)	Area(s)	Measured Value/%	Simulated Value/%	Relative Error/%
Summer	Southern area	20.5	22.1	7.8
	Middle area	66.6	60.8	−8.7
	Northern area	22.7	21.0	−7.5
Winter	Southern area	26.4	28.6	8.3
	Middle area	24.7	25.5	3.2
	Northern area	19.7	18.2	−2.5

3.3. Results of OAVS Light Environment Simulation Analysis

3.3.1. Average Daily Solar Radiation

Figure 12 shows the distribution of ™average daily solar radiation intensity in the north–south direction right below the OAVS PV panel and between the two adjacent PV panels. It can be seen from the figure that the average daily radiation intensity inside the OAVS has an obvious gradient change. In the summer, the intensity in the north and south areas was low and the intensity in middle area was high. In winter, the intensity in the southern area was high, while the intensity in the other two areas was low.

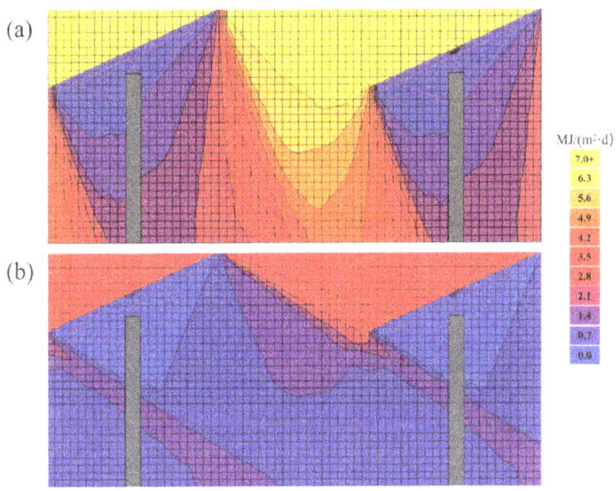

Figure 12. Cloud images of average daily solar intensity; (**a**) simulated cloud image in summer; (**b**) simulated cloud image in winter.

Figure 12a shows that the average daily radiation intensity in the canopy and below areas of the OAVS was between 1.88 and 5.58 MJ/(m^2·d) in the summer. The average daily radiation intensity in the southern, middle and northern areas was 1.94~2.72, 2.72~5.58 and 1.88~4.76 MJ/(m^2·d), respectively. On the whole, the average daily radiation intensity of the three internal areas was the highest in the middle area, and the average daily radiation intensity of the northern area of the middle area was greater, but there was no significant difference between the other two areas. Figure 12b shows that the average daily radiation intensity in the canopy and below areas of the OAVS that were not affected by side light was between 0.68 and 1.40 MJ/(m^2·d), and the overall average daily radiation intensity was small in the winter. The average daily radiation intensity in the southern, middle and northern areas was 0.73~1.40, 0.80~1.78 and 0.68~0.81 MJ/m^2·d), respectively. On the whole, the southern area had the highest daily average radiation intensity of the three internal areas, while the other two areas had no significant difference.

3.3.2. Average Daily Sunshine Hours

Figure 13 shows the distribution of the average daily sunshine hours in the north–south direction right below the PV panel of the OAVS and between the two adjacent PV panels. On the whole, the variation rule is similar to the average daily solar radiation. It can be seen from the figure that the average daily sunshine duration inside of the OAVS had an obvious gradient change. On the whole, the OAVS in the summer showed a rule of short duration in the northern and southern areas and long duration in the middle area. In the winter, the OAVS showed a pattern of slightly longer duration in the southern area and shorter duration in the other two areas.

Figure 13. Cloud images of average daily sunshine hours; (**a**) simulated cloud image in summer; (**b**) simulated cloud image in winter.

Figure 13a shows that the average daily sunshine duration in the OAVS was between 0.10 and 11.12 h in the areas below the crop canopy and unaffected by side light in the summer. The average daily sunshine duration in the southern, middle and northern areas was 0.78~2.57 h, 2.57~11.12 h and 0.10~9.27 h, respectively. In general, the average daily sunshine hours of the three internal areas were the largest in middle area, and the average daily sunshine hours in the northern part of the middle area were greater, while there was no significant difference between the other two areas. Figure 13b shows that the average daily sunshine hours of the crop canopy and the areas below the OAVS that were not affected by side light ranged from 0.00 to 2.72 h, and the overall average daily sunshine hours were small in the winter. Among them, the average daily sunshine hours in the southern, middle and northern areas are 0.00~2.72, 0.00~2.64 and 0.00~0.55 h, respectively. On the whole, the southern area had the largest average daily sunshine hours, while the other two areas had no significant difference.

3.3.3. Average Daily Shading Rate

The variation in solar radiation intensity in the OAVS caused by the shading of the PV panel obviously affects the microecological environment of the system, and the shading rate (S) can be used to characterize the shading rate of the PV panel on the solar radiation intensity of the crop canopy or ground. The internal shading rate of the OAVS was calculated according to Equation (5), where the sum of the shading and lighting rate is 1.

$$S = (E_{s\text{-}out} - E_{s\text{-}in}) / E_{s\text{-}out} \quad (5)$$

Figure 14 shows the distribution of the average daily shading rate in the summer and winter in the north–south direction directly below the OAVS PV panel and between the two adjacent PV panels. It can be seen that there was an obvious gradient change in the internal shading rate of the OAVS. In the summer, the shading rate in the northern and southern areas was large, while the shading rate in the middle was small. In the winter, the shading rate in the southern area was small, while the shading rate in the other two areas was large.

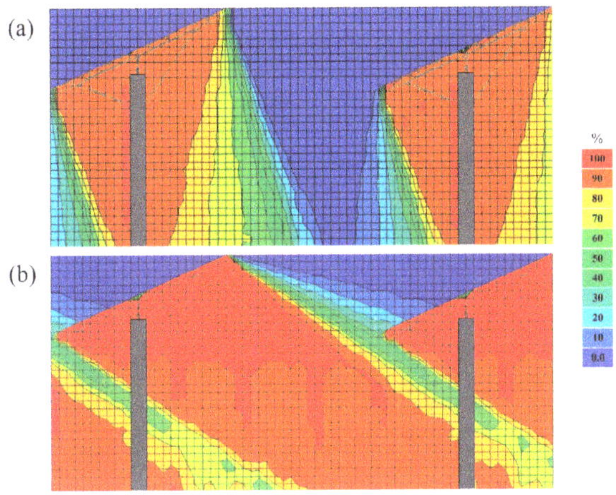

Figure 14. Cloud images of average daily shading rate; (**a**) simulated cloud image in summer; (**b**) simulated cloud image in winter.

Figure 14a shows that the shading rate of the canopy and the area below the crop in the OAVS which was not affected by side light ranged from 3.3% to 100% in the summer. Among them, the shading rate of the southern area was 77.6%~92.4%; the shading rate of the middle area was between 3.1% and 77.6% and the shading in the northern area was between 20.6% and 100%. On the whole, the southern area had the largest shading rate among the three interior areas, while the other two areas had no significant difference. Combined with agricultural production, the northern part of the middle area is more suitable for the growth of non-shade tolerant crops. Figure 14b shows that the shading rate of the OAVS which was not affected by side light ranged from 63.9% to 100%, and the overall shading rate was large in the winter. Among them, the shading rate of are area was 69.4%~100%; the shading rate in the middle area was 69.4%~100% and the shading in the northern area was 93.0%~100%. On the whole, the southern area was the least shaded among the three internal areas, while the other two areas had no significant difference.

4. Discussion

4.1. Temporal Variation in Light Environment in OAVS

For the OAVS, PV panels are an important factor affecting the light factor and have a significant impact on the agrivoltaic environment. The intensity of solar radiation is strongest at noon, weakest in the morning and evening and reaches its maximum around 12:00 noon. In addition, the solar radiation intensity is also affected by the weather and clouds. The variation in the solar radiation intensity fluctuates obviously on cloudy days, and the solar radiation intensity is significantly higher on sunny days than on cloudy days due to the influence of cloud thickness.

In the temperate regions of the northern hemisphere, the sunlight is strongest in the summer and weakest in the winter. The length of day and night is also different due to the change in the sun's height, and the solar radiation intensity changes periodically with the seasonal changes. Overall, OAVS PV panels have a significant impact on the interior light environment. In the summer, the lighting rate in the northern and southern areas is only about 20% of that in the open field, while in the middle area, it can reach about 65%. Winter is relatively small, only about 20% in three areas. The solar altitude angle in eastern China is low in the winter, and the solar radiation intensity inside the system is relatively weak, while the opposite is true in the summer. Our results support previous findings that the AVS has a cooling function [41] but has a large impact on daylighting capacity and thus agricultural production effectiveness (PV + kiwifruit [42], PV + coffee [43], PV + potato [14]). In conclusion, the effectiveness of the OAVS in reducing solar radiation intensity level depends on the time of day and solar angle under the condition of a certain spatial structure and shape.

4.2. Spatial Variation in Light Environment in OAVS

According to experiment and simulation results, the spatial light environment of the OAVS is similar to that of the forest light environment [44]. Because PV panels strongly consume light availability and produce a strong vertical light environment gradient, this vertical light environment gradient has an important impact on the light competition of crops under PV panels, and thus affects the growth of crops. Taking the middle area as an example, the higher the height, the higher the photosynthetic activity of crops, the better the ability to use bright light; the photosynthetic activity of crops in the middle and bottom of the canopy is low, but the ability to adapt to low light is strong. The vertical gradient light environment in the southern and northern areas is opposite to that in the middle area. The higher the height, the less suitable for crop growth.

On the horizontal gradient, due to the influence of the PV panel spatial structure, solar altitude angle and other factors, as well as the difference in surface roughness and terrain height caused by precipitation, the acceptance of solar radiation on the surface varies greatly. Our research results show that the OAVS can reduce air temperature and soil temperature to a certain extent, mainly because the PV panel shading reduces the solar radiation received by the ground and the solar radiation temperature, so that the air temperature and soil temperature are reduced to a certain extent. Our findings are consistent with many previous studies on the effect of shading structure on temperature. For example, Middel et al. found that different building shadows could reduce the daytime solar radiation temperature by more than 17 °C [45].

4.3. Limitations and Future Research Directions of This Study

The study has several limitations. Firstly, the authors only studied the influence of continuously laid PV panels on the internal photothermal environment of the OAVS under a certain span and height. Expanding the test scope to OAVS with different photovoltaic panel laying densities, photovoltaic array spans and heights will produce different photothermal environments, which will have a different guiding significance for agricultural production, and also contribute to the standardization of the photovoltaic agricultural industry. At present, there are few OAVS with different spatial structure forms, which will be realized by transforming existing OAVS in the later stage.

This paper only analyzed the photothermal environment and did not establish the response mechanism of agricultural production to the environment based on the environment. In the follow-up study, environmental analysis and agricultural production can be combined to explore further.

The meteorological data imported in Autodesk Ecotect Analysis 2011 are the average data of nearly 30 years, so there are some differences between the simulated data and the data of the year of the environmental test. Another limitation is that the software takes a long time to compute.

5. Conclusions

The $R_{m\text{-}avg}$ in the middle area was 66.6% in the summer, while the $R_{m\text{-}avg}$ in the other two areas was about 20%. In the winter, the light environment in the southern area was slightly better, and the $R_{m\text{-}avg}$ can reach 26.4%, while the $R_{m\text{-}avg}$ in the three areas was maintained at about 20%. The variation in the thermal environment in three areas was consistent with that in the light environment. The thermal environment of the OAVS middle area in the summer was better than that of the other two areas. The thermal environment of the southern area was better than that of the other two areas in the winter.

The OAVS optical environment model in eastern China was established, and the Simulated values were basically consistent with the experimental values at each test point, with the relative error within 10%. The constructed OAVS optical environment model can reflect the distribution of the optical environment factors in the OAVS. The simulation results show that the effectiveness of the OAVS in reducing the solar radiation intensity depends on the time of day and the solar angle. On the spatial scale, because the PV panels strongly consume the availability of light, the system produces a strong horizontal and vertical gradient of light environment.

The experimental method and simulation model can be applied to other regions of different latitudes and longitudes and specific climate conditions, so as to use the measured and simulated data to guide existing agricultural production. At the same time, using the simulation model, the optical performance of the photovoltaic agricultural system with a new structure can be obtained, so as to further guide the architectural design and agricultural production of the photovoltaic agricultural system.

Considering the effect of the OAVS agricultural production, the system's reasonable daylighting capacity is of great significance in engineering design. Because the power generation benefit is far greater than the agricultural production benefit, problems such as the optimal laying density of PV panels have not been solved. Therefore, this study can be used as a reference for further research on optimal PV panels laying density.

Author Contributions: Conceptualization, E.B. and C.W.; Methodology, L.Z. and C.W.; Software, Z.Y. and X.W.; Validation, Z.Y.; Formal analysis, Z.Y. and X.W.; Investigation, L.Z., W.W. and C.W.; Resources, C.Y.; Data curation, L.Z. and G.X.; Writing—original draft, L.Z. and G.X.; Writing—review and editing, W.W., X.W. and C.W.; Visualization, Z.Y. and C.Y.; Supervision, C.Y. and E.B.; Project administration, E.B.; Funding acquisition, E.B. All authors have read and agreed to the published version of the manuscript.

Funding: This research was funded by Key Laboratory of Farm Building in Structure and Construction, Ministry of Agriculture and Rural Affairs, China (202103).

Data Availability Statement: The data presented in this study are available upon request from the corresponding author.

Conflicts of Interest: The authors declare no conflict of interest.

References

1. Xue, J.L. Photovoltaic agriculture—New opportunity for photovoltaic applications in China. *Renew. Sustain. Energy Rev.* **2017**, *73*, 1–9. [CrossRef]
2. Muñoz-García, M.A.; Hernández-Callejo, L. Photovoltaics and electrification in agriculture. *Agronomy* **2021**, *12*, 44. [CrossRef]
3. Havrysh, V.; Kalinichenko, A.; Szafranek, E.; Hruban, V. Agricultural land: Crop production or photovoltaic power plants. *Sustainability* **2022**, *14*, 5099. [CrossRef]
4. Sarr, A.; Soro, Y.M.; Tossa, A.K.; Diop, L. Agrivoltaic, a synergistic co-location of agricultural and energy production in perpetual mutation: A comprehensive review. *Processes* **2023**, *11*, 948. [CrossRef]
5. Zhou, W.D.; Zhuang, G.Y.; Liu, L.B. Comprehensive assessment of energy supply-side and demand-side coordination on pathways to carbon neutrality of the yangtze river delta in China. *J. Clean. Prod.* **2023**, *404*, 136904. [CrossRef]
6. Miao, Q.Q.; Shi, C.Y.; Zhang, X.P. Photovoltaic technology under carbon neutrality. *Huagong Jinzhan* **2022**, *41*, 1125–1131.
7. Jing, R.; He, Y.; He, J.J.; Liu, Y.; Yang, S.B. Global sensitivity based prioritizing the parametric uncertainties in economic analysis when co-locating photovoltaic with agriculture and aquaculture in China. *Renew. Energy* **2022**, *194*, 1048–1059. [CrossRef]
8. China's Photovoltaic Power Generation Construction and Operation in 2022. Available online: http://www.nea.gov.cn/2023-02/17/c_1310698128.htm (accessed on 12 June 2023).
9. Wang, W.Y.; Zang, M.W.; Zhang, H.; Bai, Y.C.; Li, J.H.; Wang, D.Y.; Yuan, K.; Li, D. Current status of and development suggestions for food science and technology innovation power layout in China. *Sci. Food* **2022**, *43*, 336–341.
10. Dinesh, H.; Pearce, J.M. The potential of agrivoltaic systems. *Renew. Sustain. Energy Rev.* **2016**, *54*, 299–308. [CrossRef]

11. Geng, S.Q.; Wang, L. Planning of qianyan modern photovoltaic agricultural demonstration park basing on the photovoltaic model of medicinal materials. *Tianjin Agric. Sci.* **2021**, *27*, 75–79.
12. Campana, P.E.; Stridh, B.; Amaducci, S.; Colauzzi, M. Optimisation of vertically mounted agrivoltaic systems. *J. Clean. Prod.* **2021**, *325*, 129091. [CrossRef]
13. Kim, S.; Kim, S.; Yoon, C.Y. An efficient structure of an agrophotovoltaic system in a temperate climate region. *Agronomy* **2021**, *11*, 1584. [CrossRef]
14. Schulz, V.S.; Munz, S.; Stolzenburg, K.; Hartung, J.; Weisenburger, S.; Graeff-Hönninger, S. Impact of Different Shading Levels on Growth, Yield and Quality of Potato (*Solanum tuberosum* L.). *Agronomy* **2019**, *9*, 330. [CrossRef]
15. Edouard, S.; Combes, D.; Van Iseghem, M.; Tin, M.N.W.; Escobar-Gutierrez, A.J. Increasing land productivity with agriphotovoltaics: Application to an alfalfa field. *Appl. Energy* **2023**, *329*, 120207. [CrossRef]
16. Valle, B.; Simonneau, T.; Sourd, F.; Pechier, P.; Hamard, P.; Frisson, T.; Ryckewaert, M.; Christophe, A. Increasing the total productivity of a land by combining mobile photovoltaic panels and food crops. *Appl. Energy* **2017**, *206*, 1495–1507. [CrossRef]
17. Huang, K.; Shu, L.; Li, K.L.; Yang, F.; Han, G.J.; Wang, X.C.; Pearson, S. Photovoltaic agricultural internet of things towards realizing the next generation of smart farming. *IEEE Access* **2020**, *8*, 76300–76312. [CrossRef]
18. Leon, A.; Ishihara, K.N. Influence of allocation methods on the lc-co2 emission of an agrivoltaic system. *Resour. Conserv. Recy.* **2018**, *138*, 110–117. [CrossRef]
19. Hassanpour Adeh, E.; Selker, J.S.; Higgins, C.W. Remarkable agrivoltaic influence on soil moisture, micrometeorology and water-use efficiency. *PLoS ONE* **2018**, *13*, e0203256. [CrossRef]
20. Ali Abaker Omer, A.; Liu, W.; Li, M.; Zheng, J.; Zhang, F.; Zhang, X.; Osman Hamid Mohammed, S.; Fan, L.; Liu, Z.; Chen, F.; et al. Water evaporation reduction by the agrivoltaic systems development. *Sol. Energy* **2022**, *247*, 13–23. [CrossRef]
21. Feuerbacher, A.; Laub, M.; Högy, P.; Lippert, C.; Pataczek, L.; Schindele, S.; Wieck, C.; Zikeli, S. An analytical framework to estimate the economics and adoption potential of dual land-use systems: The case of agrivoltaics. *Agr. Syst.* **2021**, *192*, 103193. [CrossRef]
22. Giri, N.C.; Mohanty, R.C. Agrivoltaic system: Experimental analysis for enhancing land productivity and revenue of farmers. *Energy Sustain. Dev.* **2022**, *70*, 54–61. [CrossRef]
23. Chen, J.; Liu, Y.P.; Wang, L.J. Research on coupling coordination development for photovoltaic agriculture system in China. *Sustainability* **2019**, *11*, 1065. [CrossRef]
24. Li, Z.; Sun, X.; Zhou, J.W.; Wu, L.H.; Bi, D.; Zhao, J.; Zhu, R.F.; Christie, P. Sustainable phytoextraction of metal-polluted agricultural land used for commercial photovoltaic power generation. *J. Clean. Prod.* **2023**, *391*, 136093. [CrossRef]
25. Santra, P.; Meena, H.M.; Yadav, O.P. Spatial and temporal variation of photosynthetic photon flux density within agrivoltaic system in hot arid region of india. *Biosyst. Eng.* **2021**, *209*, 74–93. [CrossRef]
26. Marrou, H.; Guilioni, L.; Dufour, L.; Dupraz, C.; Wery, J. Microclimate under agrivoltaic systems: Is crop growth rate affected in the partial shade of solar panels? *Agric. For. Meteorol.* **2013**, *177*, 117–132. [CrossRef]
27. Gao, X.Q.; Yang, L.W.; Lyu, F.; Ma, L.Y.; Hui, X.Y.; Hou, X.Y.; Li, H.L. Observational study on the impact of the large solar farm on air temperature and humidity in desert areas of golmud. *Acta Energiae Solaris Sin.* **2016**, *37*, 2905–2919.
28. Chang, R.; Shen, Y.B.; Luo, Y.; Wang, B.; Yang, Z.B.; Guo, P. Observed surface radiation and temperature impacts from the large-scale deployment of photovoltaics in the barren area of gonghe, China. *Renew. Energy* **2018**, *118*, 131–137. [CrossRef]
29. Sailor, D.J.; Anand, J.; King, R.R. Photovoltaics in the built environment: A critical review. *Energy Build.* **2021**, *253*, 111479. [CrossRef]
30. Ezzaeri, K.; Fatnassi, H.; Bouharroud, R.; Gourdo, L.; Bazgaou, A.; Wifaya, A.; Demrati, H.; Bekkaoui, A.; Aharoune, A.; Poncet, C.; et al. The effect of photovoltaic panels on the microclimate and on the tomato production under photovoltaic canarian greenhouses. *Sol. Energy* **2018**, *173*, 1126–1134. [CrossRef]
31. Ayoub, M.; Elseragy, A. Parameterization of traditional domed-roofs insolation in hot-arid climates in aswan, Egypt. *Energy Environ.* **2017**, *29*, 109–130. [CrossRef]
32. Wu, Q.; Jo, H.K. A study on ecotect application of local climate at a residential area in chuncheon, korea. *J. Environ. Eng. Landsc.* **2015**, *23*, 94–101. [CrossRef]
33. Yang, L.; He, B.J.; Ye, M. Application research of ecotect in residential estate planning. *Energy Build.* **2014**, *72*, 195–202. [CrossRef]
34. Akbari, H.; Cherati, S.M.; Monazam, N.H.; Noguchi, M. Effect of courtyards' geometrical parameters on climate adaptability and shading performance in hot-arid climate of yazd (Iran). *Sustain. Energy Technol. Assess.* **2021**, *48*, 101594. [CrossRef]
35. Li, H.L.; Wu, D.; Zhou, J.Z. Effects of tubular daylight guidance systems on the daylighting performance and energy savings in office buildings under different climate zones. *J. Renew. Sustain. Energy* **2021**, *13*, 065102. [CrossRef]
36. Ahriz, A.; Mesloub, A.; Djeffal, L.; Alsolami, B.M.; Ghosh, A.; Abdelhafez, M.H.H. The use of double-skin façades to improve the energy consumption of high-rise office buildings in a mediterranean climate (csa). *Sustainability* **2022**, *14*, 6004. [CrossRef]
37. He, D.S.; Chang, J.G.; Han, X. Direct radiation model of louver shading in office building shade based on network optimization method. *Comput. Intell. Neurosci.* **2022**, *2022*, 5766448. [CrossRef]
38. Chi, F.A.; Borys, I.; Jin, L.; Zhu, Z.Z.; Bart, D. The strategies and effectiveness of climate adaptation for the thousand pillars dwelling based on passive elements and passive spaces. *Energy Build.* **2019**, *183*, 17–44. [CrossRef]
39. Iommi, M. Daylighting performances and visual comfort in le corbusier's architecture. The daylighting analysis of seven unrealized residential buildings. *Energy Build.* **2019**, *184*, 242–263. [CrossRef]

40. Sun, Z.P.; Huang, W.Y.; Li, T.L.; Tong, X.J.; Bai, Y.K.; Ma, J. Light and temperature performance of energy-saving solar greenhouse assembled with color plate. *Trans. Chin. Soc. Agric. Eng.* **2013**, *29*, 159–167.
41. Cho, J.; Park, S.M.; Park, A.R.; Lee, O.C.; Nam, G.; Ra, I.H. Application of photovoltaic systems for agriculture: A study on the relationship between power generation and farming for the improvement of photovoltaic applications in agriculture. *Energies* **2020**, *13*, 4815. [CrossRef]
42. Jiang, S.; Tang, D.; Zhao, L.; Liang, C.; Cui, N.; Gong, D.; Wang, Y.; Feng, Y.; Hu, X.; Peng, Y. Effects of different photovoltaic shading levels on kiwifruit growth, yield and water productivity under "agrivoltaic" system in southwest China. *Agric. Water Manag.* **2022**, *269*, 107675. [CrossRef]
43. Assis, B.D.P.; Gross, E.; Pereira, N.E.; Mielke, M.S.; Júnior, G.A.G. Growth response of four conilon coffee varieties (Coffea canephora Pierre ex A. Froehner) to different shading levels. *J. Agric. Sci.* **2019**, *11*, 29. [CrossRef]
44. Song, Y.; Ryu, Y. Seasonal changes in vertical canopy structure in a temperate broadleaved forest in korea. *Ecol. Res.* **2015**, *30*, 821–831. [CrossRef]
45. Middel, A.; Alkhaled, S.; Schneider, F.A.; Hagen, B.; Coseo, P. 50 grades of shade. *Bull. Am. Meteorol. Soc.* **2021**, *102*, E1805–E1820. [CrossRef]

agronomy

MDPI

Article

Comparison between Drift Test Bench and Other Techniques in Spray Drift Evaluation of an Eight-Rotor Unmanned Aerial Spraying System: The Influence of Meteorological Parameters and Nozzle Types

Changling Wang [1,2,3,*], Supakorn Wongsuk [1,2,3], Zhan Huang [1,2,3], Congwei Yu [1,2,3], Leng Han [1,2,3], Jun Zhang [1], Wenkang Sun [1], Aijun Zeng [1,2,3] and Xiongkui He [1,2,3,*]

[1] College of Science, China Agricultural University, Beijing 100193, China
[2] College of Agricultural Unmanned System, China Agricultural University, Beijing 100193, China
[3] Centre for Chemicals Application Technology, China Agricultural University, Beijing 100193, China
[*] Correspondence: wcl1991@cau.edu.cn (C.W.); xiongkui@cau.edu.cn (X.H.); Tel.: +86-62732830 (C.W.); +86-62731446 (X.H.)

Abstract: In the past decade, an unmanned aerial spraying system (UASS) was applied more and more widely for low-volume aerial pesticides spraying operations in China. However, UASS have a higher drift risk due to more fine droplets sprayed with a higher working height and a faster driving speed than ground sprayers. Study on UASS spray drift is a new hot spot within the field of pesticide application technology. The field test bench was originally designed and applied for the measurement of the spray drift potential of ground sprayers. No methodology using the test bench for UASS drift evaluation was reported. Based on our previous study, field drift measurements of an eight-rotor UASS were conducted using three techniques (test bench, ground petri dish, and airborne collection frame) in this study, and the effects of meteorological parameters and nozzle types were investigated, to explore the applicability and the feasibility of the test bench used in UASS field drift evaluation. The test bench is proven promising for direct drift determination of UASS and the described methodology enabled classification of different UASS configurations. Higher wind speeds and finer droplets produced higher drift values. The faster the wind speed and the lower the humidity, the more the spray drift. The test bench can reduce the site requirements and improve the efficiency of the field drift test.

Keywords: unmanned aerial spraying system (UASS); spray drift; test bench; petri dish; droplet size; wind speed; nozzle; cumulative drift percentage; correlation analysis

Citation: Wang, C.; Wongsuk, S.; Huang, Z.; Yu, C.; Han, L.; Zhang, J.; Sun, W.; Zeng, A.; He, X. Comparison between Drift Test Bench and Other Techniques in Spray Drift Evaluation of an Eight-Rotor Unmanned Aerial Spraying System: The Influence of Meteorological Parameters and Nozzle Types. *Agronomy* **2023**, *13*, 270. https://doi.org/10.3390/agronomy13010270

Academic Editor: Yanbo Huang

Received: 23 December 2022
Revised: 13 January 2023
Accepted: 14 January 2023
Published: 16 January 2023

1. Introduction

In the past 10–15 years, the unmanned aerial spraying system (UASS), or unmanned aerial vehicle (UAV) sprayer, was widely used worldwide as a new type of efficient plant protection machinery [1,2]. Especially in China, UASSs replaced manual sprayers with low efficiency, such as knapsack sprayers and sprayer guns. These aerial sprayers are flexible and easy to operate, and have a lot of unique advantages in scenarios where ground sprayers are difficult to access, i.e., paddy fields, hilly areas, and fruit trees planted in disorder. They can also reduce the chemical toxicity for operators effectively. According to the statistics from DJI and XAG, two leading Chinese manufacturers, the marketing holdings of agricultural UAV used for spraying and broadcasting exceeded 160 thousand and the annual working area reached 93 million ha all over China in 2021.

1.1. UASS Spray Drift

Under the combined influence of flying platform load capacity, spraying system performance, regulatory policies, development and maintenance costs, operation convenience, and flexibility, the max take-off weight of UASS does not exceed 150 kg and the liquid tank capacity does not exceed 100 L, currently. Therefore, a fine droplet (volume median diameter, VMD < 200 μm) and a low application volume (<150 L/ha) are needed to guarantee the pesticide active ingredients to be applied uniformly to targeted crops. UASS spraying operations are normally classified into ultra-low volume application (ULV) and very-low volume application (VLV) for field crops and tree crops. Compared to ground-based equipment, UASS aerial operation normally generate a finer droplet with a faster flight speed and a longer nozzle to target distance, contributing to an enormous risk of UASS's drift. Spray drift can be regarded as that part of a pesticide application that is deflected away from the target area by the action of the wind. Control of drift is important because of the potential exposure to pesticides of non-target organisms and structures outside a treatment zone and where there is the possibility of such organisms being sensitive to very small quantities of the pesticide materials [3]. Factors influencing the risk of spray drift vary, mainly including spray nozzle design and performance, nozzle to target distance, atmospheric variables, properties of the spray liquid, and the sprayer speed. Several studies on UASS spray drift evaluation were conducted to explore the effects of application parameters, meteorology conditions, and UASS configurations in different scenarios, such as field crops, orchards, and vineyards [4–12].

1.2. Drift Testing Method

The testing method of spray drift mainly consists of the direct field test and the indirect evaluation test. The field test is to measure the actual spray drift outdoors in typical field conditions or over a defined surface including grass turf following the ISO 22866 standard 'Methods for field measurement of spray drift' (all standards cited in this paper were introduced in Appendix A), [13] which is considered as the most realistic drift measurement method. On the other hand, the evaluation test includes wind tunnel [14], drift test bench [15,16], and droplets spectrum test [17,18]. The drift potential reduction percentages are calculated by measuring the spray deposition or the droplet size distribution for analysis [19–21].

1.3. Field Drift Test Bench

The field test bench was originally designed and applied for the measurement of the spray drift potential of ground sprayers, including boom sprayers for field crops and orchard airblast sprayers for tree crops. As an alternative methodology to simplify the assessment of spray drift risk for different equipment, the Department of Agricultural Forest and Environmental Economics and Engineering (DEIAFA) of the University of Turin developed the drift test bench in 2007 [22] X. Wang et al. tested the spray drift potential of six types of fan nozzles at two spraying pressures with a boom sprayer [23] Gil et al. evaluated the influence of wind velocity and wind direction on the drift potential value using the test bench [24]. In 2015, the ISO 22401 standard 'Method for measurement of potential spray drift from horizontal boom sprayers by the use of a test bench' was published [15], aiming at defining a test procedure to assess potential spray drift using the ad hoc test bench. Nuyttens et al. and Balsari et al. measured the effects of application speed, boom height, and spray characteristics on drift reduction potentials for boom sprayers using the drift test bench under controlled indoor conditions [25–27]. The applications of the test bench in quantifying the drift potential of airblast sprayers were also attempted by Grella et al. [28,29]. However, no methodology using the test bench for UASS spray drift evaluation was proposed and reported.

Under similar meteorological parameters, we compared the spray performance of three different typical commercial UASS types with two nozzles types with an artificial vineyard in 2019, and the characteristics of deposition, drift, and mass balance of unmanned aerial spraying were obtained in our previous work [30]. Based on this methodology, field drift measurements of an eight-rotor UASS were conducted using three techniques (test bench, ground petri dish and airborne collection frame) in this study, and the effects of meteorological parameters and nozzle types were investigated, to explore the applicability and the feasibility of the test bench used in UASS field drift evaluation.

2. Materials and Methods

2.1. Characteristics of Unmanned Aerial Spraying System

As shown in Figure 1, the unmanned aerial spraying system used in the field trials was an eight-rotor electric UAV sprayer AGRAS MG-1P (SZ DJI Technology Co., Ltd., Shenzhen, China), equipped with two types of nozzles: hollow cone nozzles (HC) TR 80-0067 and air induction flat fan nozzles (AI) IDK 120-015 (Lechler GmbH, Metzingen, Germany). The 8-rotor UASS has an unfolded size of 1460 mm long, 1460 mm width and 616 mm height, and a propeller diameter of 53 cm, equipped with a 10 L spray tank. The net weight (excluding battery) was 10 kg and the max take-off weight was 25 kg. Four nozzles were mounted below the four lateral rotors of the UASS with a horizontal spacing of 146 cm. Prior to all tests, the spraying system was calibrated to obtain a single nozzle flow rate of 0.30 L min^{-1} for the HC and 0.55–0.60 L min^{-1} for the AI, providing a nominal application volume of 45 L ha^{-1} and 90–100 L ha^{-1}, respectively. The UASS flew at an application speed of 2.0 m s^{-1} and an appropriate working width of 2.0 m during the tests.

Figure 1. DJI AGRAS MG-1P UASS applied in spray drift tests.

2.2. Field Drift Test Bench

The field drift test bench (AAMS-Salvarani BVBA, Maldegem, Belgium) is constructed according to the ISO standard 22401: 2015 about drift reduction. The test bench (TB) consists of several aluminium modules (2.0 m length and 0.5 m width) that can be connected to each other. Bench length depends on the configuration of the sprayer to be tested. With the help of the pneumatic valves and the slideable aluminium covers, the collection trays (0.5 × 0.2 m in size) positioned every 0.5 m along the bench can be closed and opened. These trays with artificial collectors (plastic petri dishes) collect the drifting droplets from the sprayer. The collection trays open when the boom of the ground sprayer touches the dedicated stick or the switch is turned on, and the compressor with the pressure tank then gets a signal to open them. The battery, compressor, and pressure tank are mounted on a convenient trolley with two wheels.

In this study, we used five modules to compose a 10.0 m long test bench (Figure 2). It was placed perpendicular to the flight direction of the UASS with the petri dishes (15 cm diameter, Nantong Fansibei Biological Technology Co., Ltd., Jiangsu, China) 0.3 m above the ground. The first two collectors were positioned 2.0 m away from the edge of the sprayed field (EOF). A total of 40 petri dishes from the test bench were obtained in a single test.

Figure 2. Field drift test bench and artificial collectors for UASS spray drift measurement.

2.3. Test Site and Sampling Layout

UASS spraying applications were performed at Beijing TT Aviation Technology Co. Ltd., Machikou town, Changping district, Beijing, China (40°11′30″ N; 116°10′10″ E). The test site was a flat open field with no obvious obstacles within 200 m, covering a total area of about 1000 m^2. The sampling layout is shown in Figure 3. An artificial vineyard in a size of 16 m length, 20 m width, and 2 m height, and three airborne drift collection frames (ACF) in a size of 5.5 × 2.0 m, built in our previous spray deposition, drift, and mass balance study [30], were utilized for these trials.

Except for the test bench, ground petri dish and airborne drift collection frame were applied in order to measure ground spray drift and airborne spray drift, respectively. They were arranged according to the procedure used by Wang et al. [30]. Ten 15 cm diameter plastic petri dishes were placed at 3, 5, 10, 15, and 20 m from the edge of the treated field in the downwind direction. Several metal plates were arranged on the ground to carry the petri dishes, ensuring they remained on the same plane. A total of 50 ground petri dishes (GPD) were collected for each application. Additionally, three frames were positioned parallel to the UASS path with an interval of 5.0 m at 2.0 m downwind from the EOF. Airborne drift collectors, polyethylene (PE) tubes (2 mm in diameter and 2.0 m long), were fixed horizontally on the frame from a height of 0.5 m with an interval of 0.5 m. Thus, thirty PE tubes were obtained for each application.

In order to make full use of the stable natural wind, the artificial vineyard row was set in a northwest–southeast direction according to the local historical wind data, and the metal plates carrying the ground petri dishes and the test bench were arranged symmetrically on both sides of the artificial vineyard. The sampling layout could be quickly switched to start the subsequent trials when the wind direction was reversed. Weather data, such as wind speed and direction, as well as air temperature and relative humidity (RH), were recorded at a sample rate of 1 Hz using a three-dimensional ultrasonic anemometer WindMaster (Gill Instruments, New Milton, UK) and a weather sensor 350-XL (Testo SE & Co. KGaA, Titisee-Neustadt, Germany) mounted 2 m above the ground.

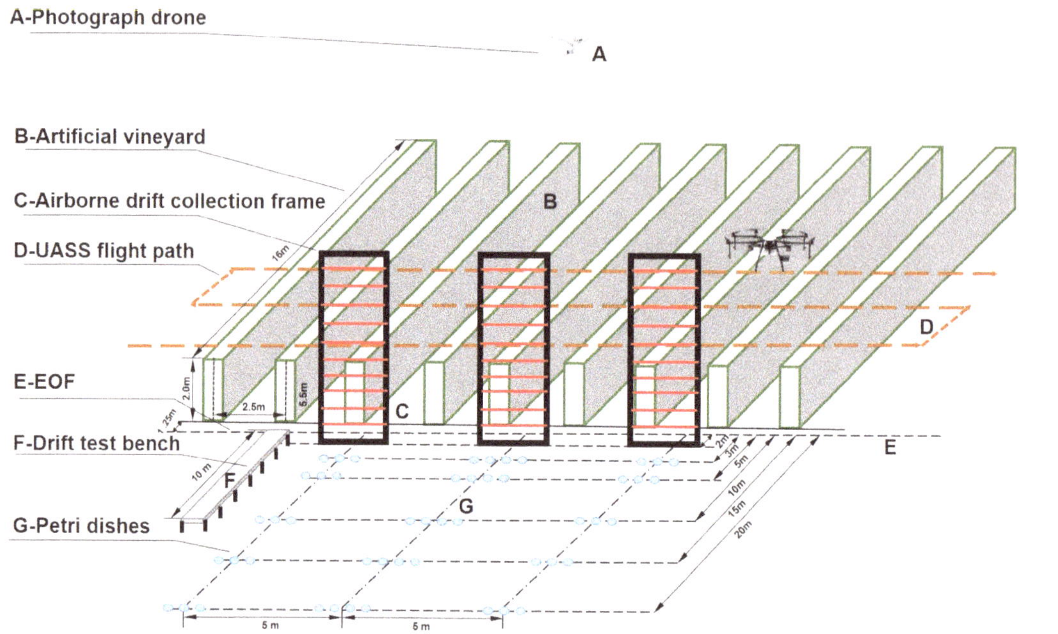

Figure 3. UASS spray drift measurement arrangement using three types of collectors: test bench, ground petri dish and airborne collection frame, and an artificial vineyard [30].

2.4. Experimental Methodology

In this work, we performed 13 tests for each nozzle type using the eight-rotor UAV sprayer MG-1P. Thus, twenty-six trials were totally conducted at wind speed from 1.61 m s^{-1} to 5.22 m s^{-1} and classified into three wind speed ranges: lower wind speed (LWS, <3.0 m s^{-1}), medium wind speed (MWS, 3.0 $\leq$ 4.0 m s^{-1}), and higher wind speed (HWS, $\geq$4.0 m s^{-1}). The flight height of UASS was 3.5 m above the ground and 1.5 m above the top of the artificial canopy. A fluorescent tracer dye pyranine (Simon & Werner, Germany) was dissolved in tap water at a concentration of 0.1% to prepare the spray liquid. The pyranine aqueous solution was sprayed at a pressure of 0.3 MPa during all trials. Spray drift measurements in this study were conducted in acceptable atmospheric conditions in the following ranges: (a) wind speeds of at least 1.0 m s^{-1} and no more than 10% of wind speed measurements should be less than this value; (b) mean wind direction at $90° \pm 30°$ to the UASS flight routes and no more than 30% of results shall be $90° \pm 45°$; and (c) temperatures of 5–35 °C.

Even if Grella et al. [28] suggested to use the test bench in nearly absence of wind, and more recently, authors found better results carrying trials indoors [31], due to the detrimental effect of environmental variables [24]; all experimental procedures were in according with both ISO 22866 standard [13] and our previous study in 2021 [30], and details would not be repeated in this paper. The test bench control system was activated to open the sliding cover while the UASS was in operation 20 m prior to the collector array. The test bench was then closed after the first route was performed, and only sampled the droplets from the first route, while the ground petri dishes and the airborne tubes collected the spray drift from all three passes during the application process. A four-rotor drone PHANTOM 4 PRO (SZ DJI Technology Co., Ltd., Shenzhen, China), equipped with a complementary metal-oxide-semiconductor (CMOS) sensor, with a size of 1 inch and a resolution of 20 megapixels, was hovering at 30 m height to record the whole process of flight. All samplers for spray drift were collected within less than 10 min after each test and stored in a box protected from light exposure in order to minimize degradation.

2.5. Sample Processing

In the laboratory, each petri dish collected from the test bench and the ground plates was filled with 60 mL of deionized water and was oscillated for 10 min using an orbital shaker TS-1000 (Haimen Kylin-Bell Lab Instruments Co., Ltd., Nantong, China) at 200 r min^{-1} (Figure 4). For an airborne drift collector, 100 mL of deionized water was added into the Ziplock bag containing PE tube, and then the bag was placed in an ultrasonic cleaner KM-36C (KJM Lab Instruments, China) for 5 min. The absorption of the tracer eluent was determined with a fluorescence spectrophotometer HITACHI F-2700 (Hitachi High-Technologies Corporation, Tokyo, Japan). The measuring configurations were set as follows: the voltage of 650 V; the excitation wavelength of 400 nm and emission wavelength of 505 nm; and the slit width of 15 nm.

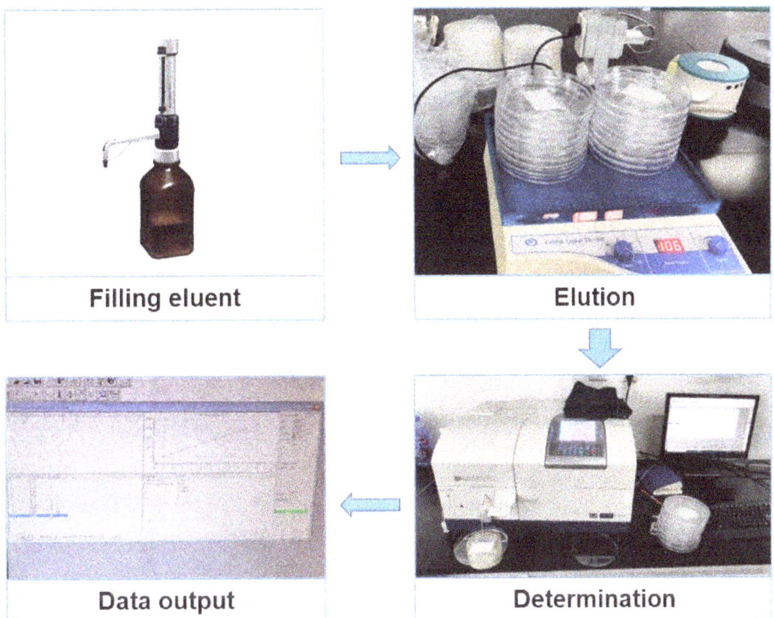

Figure 4. Sample processing procedures of UASS spray drift collectors.

2.6. Droplet Size Measurement

Based on the ISO standard 25358 [17], the droplet size spectrum was measured using a laser diffraction system SprayTec (Malvern Panalytical Ltd., Malvern, UK) in the College of Agricultural Unmanned System, China Agricultural University after field trials. The two types of tested nozzle were fixed 0.5 m above the analyzer between the laser transmitter and the receiver. The spray liquid same with field test was sprayed at 2.5, 3.0, and 4.0 bar with at least 5 valid replicates. The 10th percentile diameter ($D_{v0.1}$), VMD ($D_{v0.5}$), 90th percentile diameter ($D_{v0.9}$), relative span (RS), and spray volume fractions generated with droplets finer than 75, 100, and 200 μm (V75, V100, and V200) were obtained via Spraytec software for Windows V3.30 (Malvern Panalytical Ltd., Malvern, UK).

2.7. Calculation of Spray Drift Values

2.7.1. Spray Drift Percentage

According to ISO standards [13,15], the spray drift percentage (DP) accounted for the actual applied volume from the reading of the fluorimeter of each collector can be calculated in accordance with Equations (1)–(3).

$$D_{ij} = \frac{\left(\rho_{smpl} - \rho_{blk}\right) \times V_{dil}}{\rho_{spray} \times A_{col}} \tag{1}$$

$$DP_{ij} = \frac{D_{ij}}{(D_V/100)} \times 100\% \tag{2}$$

$$D_V = \frac{q \times 10000}{B \times v \times 60} \times 100\% \tag{3}$$

where D_i is the spray drift deposit on a single collector i of collector type j ($\mu L \cdot cm^{-2}$); DP_{ij} is the DP on a single collector i of collector type j (%); D_V is the applied volume ($L \cdot ha^{-1}$); ρ_{smpl} is the fluorimeter reading of the sample; ρ_{blk} is the is the fluorimeter reading of the blank control; V_{dil} is the dilution liquid volume used to solute tracer from collector (L); ρ_{spray} is the fluorimeter reading of the tank sample; A_{col} is the collector area (cm^2); q is the total flow rate ($L \cdot min^{-1}$); B is the swath width (m); and v is the flight speed ($m \cdot s^{-1}$).

2.7.2. Cumulative Drift Percentage

Cumulative drift percentages (CDP) of the three types of collector can be calculated according to Equation (4) [13]. For this study, CDP_B and CDP_G are the cumulative drift percentages obtained from the field drift test bench and the ground petri dish, respectively, while CDP_A is the cumulative airborne drift percentage from the PE tube of the collection frame.

$$CDP_j = \int_a^b DP(x)_j dx \tag{4}$$

where CDP_j is the cumulative drift percentage for the collector type j; x is the downwind distance from the EOF or the height from the ground (m); $DP(x)_j$ is the drift percentage as a function of downwind distance or height for the collector j (%); and a and b are the start and the end point of the sampling interval, respectively (m).

2.7.3. Drift Reduction Percentage

The drift reduction percentage (DRP, %) derived from CDP value was calculated via the following expression, according to the ISO 22369-1 [32].

$$DRP = \left(1 - \frac{CDP_{tst}}{CDP_{ref}}\right) \times 100\% \tag{5}$$

where CDP_{tst} was the CDP value for each tested configuration; CDP_{ref} was the CDP value for the reference configuration, which is hollow cone nozzle trial for nozzle type comparison. The reduction class was also defined by the ISO 22369-1 as follows: A $\geq$ 99%, B 95 $\leq$ 99%, C 90 $\leq$ 95%, D 75 $\leq$ 90%, E 50 $\leq$ 75%, and F 25 $\leq$ 50%.

2.8. Statistical Analysis

All the statistical analyses were performed using IBM SPSS Statistics for Windows V22 (IBM Corp., Armonk, NY, USA). Both two-way and three-way analysis of variance (ANOVA) were applied to investigate the effects of downwind distance, nozzle type, and wind speed via the two different collectors. In all trials, the mean values were compared using the Duncan's post hoc test. Statistical significance in all cases was when $p < 0.05$.

3. Results

3.1. Meteorology Conditions and UASS Operation Parameters

Table 1 presents mean values of meteorological conditions and actual operation parameters for each treatment and replicate. During field tests, the temperature ranged from 26.4 °C to 34.3 °C and the RH was between 11.8% and 35.5%. The average wind speed recorded was 1.61–5.22 m s^{-1}, with an average wind direction (degrees from the ideal direction) between −35.4° and 42.6°. Field trials were classified into lower wind speed (HC1-HC4 and AI1-AI7), medium wind speed (HC5-HC10 and AI8-AI10), and higher wind speed (HC11-HC13 and AI11-AI13) according to the mean value of wind speed for further analysis. Consequently, atmospheric conditions monitored met the requirements of ISO standard [13].

Table 1. Meteorological conditions and experimental parameters of all spray drift tests.

Treatment and Replicates	Temperature (°C)	Relative Humidity (%)	Wind Speed (m·s^{-1})	Wind Direction (Degrees from the Ideal Direction)	Total Output (L min^{-1})	Flight Speed (m·s^{-1})	Applied Volume (L·ha^{-1})
HC1	33.9	33.0	2.19	−1.5	1.77	2.15	45.7
HC2	34.1	31.5	2.19	28.8	1.21	2.13	47.3
HC3	32.4	33.8	2.63	4.9	1.21	2.10	48.1
HC4	33.2	25.7	2.65	2.8	1.21	2.16	46.6
HC5	31.7	18.7	3.15	3.7	1.24	3.08	33.5
HC6	29.9	13.4	3.28	28.7	1.24	2.68	38.5
HC7	34.1	28.9	3.35	38.4	1.77	2.16	45.5
HC8	32.2	30.6	3.81	9.4	1.77	2.09	47.0
HC9	33.1	29.1	3.93	31.2	1.77	2.16	45.5
HC10	32.2	31.4	3.97	22.4	1.77	2.12	46.3
HC11	33.7	20.9	4.77	31.0	1.21	2.11	47.9
HC12	31.2	23.7	5.11	23.9	1.21	2.07	48.8
HC13	32.3	16.8	5.22	42.6	1.24	2.61	39.6
AI1	30.3	35.1	1.61	−25.1	3.29	2.16	84.6
AI2	30.2	35.5	1.75	−16.9	3.29	2.12	86.2
AI3	32.0	32.9	1.96	−25.4	3.29	2.11	86.4
AI4	34.2	17.1	2.32	−32.5	2.37	2.05	96.2
AI5	33.6	34.7	2.59	−35.4	3.29	2.15	85.0
AI6	34.3	19.4	2.65	15.8	2.37	2.04	97.0
AI7	33.7	16.9	2.82	−12.2	2.37	2.00	98.6
AI8	31.6	11.8	3.79	−35.3	2.40	2.00	100.0
AI9	30.7	14.4	3.89	−23.3	2.40	2.96	67.5
AI10	34.2	17.1	3.93	6.3	2.37	2.12	93.0
AI11	32.3	18.1	4.41	6.3	2.37	2.11	93.8
AI12	33.3	13.0	4.43	−16.4	2.40	2.00	100.0
AI13	26.4	22.6	4.48	35.9	3.29	2.00	91.4

From the operation video captured by the photograph drone, flight speed for most of trials ranged from 2.0 m s^{-1} to 2.2 m s^{-1}, except for HC5, HC6, HC13, and AI9, with a bit higher speed. UASS's total output was found between 1.21 L min^{-1} and 1.77 L min^{-1} for HC trials, and between 2.37 L min^{-1} and 3.29 L min^{-1} for AI trials. Therefore, the applied volume was achieved at 38.5–48.8 L ha^{-1} for HC trials and at 84.6–100.0 L ha^{-1} for AI trials, while higher flight speed contributed to lower applied volume for two trials (HC5: 35.5 L ha^{-1} and AI9: 67.5 L ha^{-1}).

3.2. Droplet Size Spectrum

Table 2 shows the mean values of droplet size spectrum parameters $D_{v0\cdot1}$, $D_{v0\cdot5}$, $D_{v0\cdot9}$, RS, V75, V100, V200, and the spray quality classification. Droplet size ($D_{v0\cdot1}$, $D_{v0\cdot5}$, and $D_{v0\cdot9}$) varied with the nozzle type and the pressure. The highest value of VMD (408.4 μm) was obtained at 2.5 bar for HC, while the lowest value (94.6 μm) was obtained at 4.0 bar for AI, indicating a huge difference between these two nozzles. The RS values were found higher for AI (1.64) than HC (1.22). Huge differences were also found between the two nozzles for droplet size distribution parameters V75, V100, and V200, which were proven to be instructive indicators for evaluating spray drift potential in several publications [4,7,18,33]. For all the HC replicates, most of droplets were smaller than 200 μm in diameter (97.6%), and more than half (54.7%) of droplets have a size lower than 100 μm, and 31.8% of droplets were smaller than 75 μm in diameter. For the AI nozzle, V75 and V100 did not reach a proportion of 10%, and only about 20% of droplets had a diameter lower than 200 μm. The spray quality of the HC nozzle at 4.0 bar, and the AI nozzle at 2.5 and 3.0 bar was classified into three categories, 'Very fine' (VF), 'Very Course' (VC), and 'Course' (C), respectively.

Table 2. Droplet size spectrum and spray quality classification of the hollow cone nozzle and the air induction nozzle used in tests.

Nozzle Model	Replicate No.	Pressure (Bar)	DV0.1 (μm)	DV0.5 (μm)	DV0.9 (μm)	RS	V75 (%)	V100 (%)	V200 (%)	Spray Quality [1]
TR 80-0067, HC	1–13	4.0	45.5	94.6	161.4	1.22	31.8	54.7	97.6	VF
IDK 120-015, AI	1/2/3/5/13	2.5	130.3	408.4	798.1	1.64	2.9	5.9	19.1	VC
	4/6/7/8/9/10/11/12	3.0	124.8	352.6	702.9	1.64	3.1	6.1	23.0	C

[1] Spray quality was classified according to the ISO standard 25358: extremely fine (EF), VF, very fine (VF); fine (F); medium (M); coarse (C); very coarse (VC); extremely coarse (EC); and ultra coarse (UC) [17].

3.3. *Spray Drift Percentage*

The spray drift percentage profiles at different downwind distances obtained from the test bench using UASS with HC and AI nozzles at different wind speed ranges are shown in Figure 5. Regardless of which nozzle was assembled, drift percentage from TB mostly peaked at 2.0–5.0 m, and then declined sharply at 5.0–12.0 m downwind with a decreasing decay rate for medium and higher wind speed; under lower wind speed, DP peaked within 3.0 m, followed by a progressive decrease. In the case of the HC nozzle, the highest DP value at LWS ranged from 5% to 6% and DP dropped to less than 1% at 6.0–8.0 m. Buffer zone width can be simply estimated with the criterion that a maximum drift fallout of 1% was allowed, based on deltamethrin data by the Swedish EPA [34]. At MWS and HWS, the peak of DP both exceeded 10% (MWS: 11.9%; HWS: 13.8%), but large differences and poor repeatability between tests were found within a same wind speed range, highlighting the uncertainty of the field spray drift measurements; DP could not fall to less than 1% until the downwind distance reached 6.0 m or farther for most tests. Regarding the AI nozzle, DP values did not surpass 2% for LWS, 5% for MWS, and 8% for HWS, and decreased to less than 1% within 6.0 m. For both two-nozzle types, DP at LWS was obviously reduced at different downwind locations compared to MWS and HWS. In addition, it can be found from the flight parameters in Table 1 that UASS flied faster for HC5, HC13, and AI9, which may be the main factor leading to a higher DP. Moreover, lower spray drift percentage under HWS (AI13) could be obtained with a larger droplet size based on the droplet size spectrum results presented in Table 2.

Figure 6 shows the spray drift percentage profiles at different downwind distances obtained from the test bench using UASS with HC and AI nozzles at lower, medium, and higher wind speed. In all trials, the spray drift percentage decreased continuously from the EOF to 20 m downwind, with a sharp decline at the first 5 m and a steady decline after 10 m. Unlike the test bench, DP from GPD did not show a process of rising to a peak value and then falling. In terms of HC nozzle, at LWS, the maximum value of DP ranged from 4% to 8%, and DP fell to less than 1% at the distance range of 5–10 m; under MWS and HWS, the highest DP value rose to 7–16%, the value dropped below 1% at 15 m downwind or a farther distance. After the replacement of the AI nozzle, drift percentage at LWS did not exceed 2% and declined to below 1% within 5 m; DP at MWS and HWS increased but could not reach 7%, and decreased to less than 1% at 5–10 m. Similar to the TB results, the trial AI13 also had a better anti-drift performance under the highest lateral wind speed, and there was no obvious correlation between the drift value from GPD and the wind speed at the same WSR, as well.

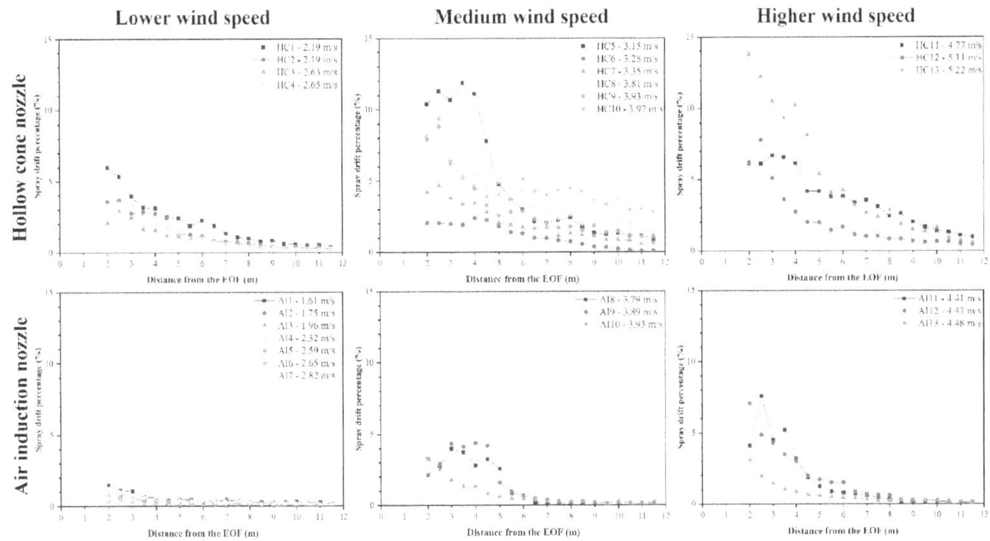

Figure 5. Spray drift percentage profiles obtained from the test bench using UASS with HC and AI nozzles at lower wind speed (LWS, <3.0 m s^{-1}), medium wind speed (MWS, 3.0 $\leq$ 4.0 m s^{-1}), and higher wind speed (HWS, $\geq$4.0 m s^{-1}).

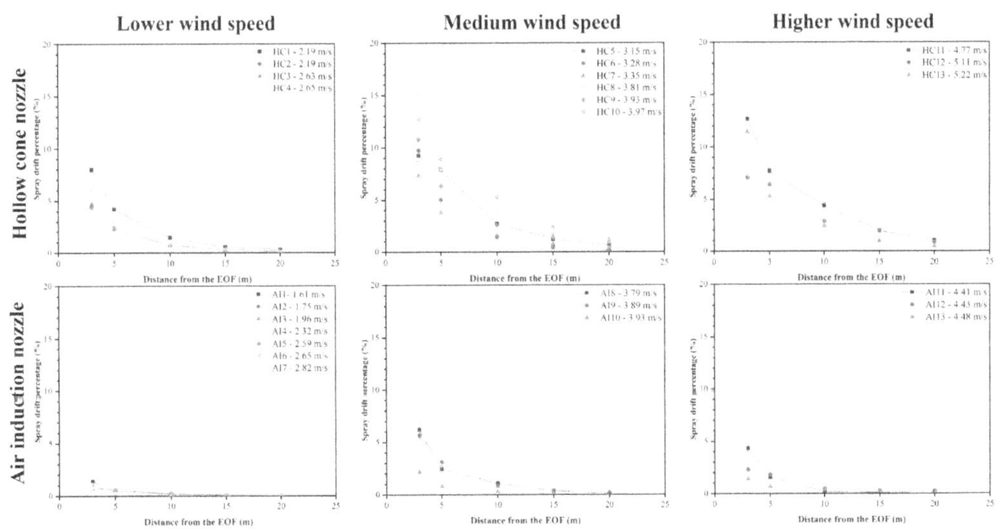

Figure 6. Spray drift profiles obtained from the ground petri dish using UASS with HC and AI nozzles at LWS, MWS, and HWS.

Airborne spray drift percentage profiles at different heights obtained from the airborne collection frame for all the field trials are presented in Figure 7. In all tests, the highest DP was found at the bottom of the frame, and the DP value decreased sharply from 0.5 to 2.0 m above the ground. Even though within a same wind speed range, DP values varied a lot for different tests. With regard to HC nozzle, airborne DP for lower wind speed was found more than 100% near the ground and fell to less than 1% at the height of 2.5–4.0 m; DP for MWS and HWS increased markedly relative to LWS, with a peak value for HC9 close to 300% and values at all collected height more than 1% for HC5/6/7/9/13, implying the severe drift risk for fine droplets. In the case of the AI nozzle, the airborne drift was reduced significantly, with DP value below 7% at LWS and below 20% at MWS and HWS. Meanwhile, the measured airborne drift was also found to be higher for AI9 and lower for AI13, in accordance with the results from the TB and GPD.

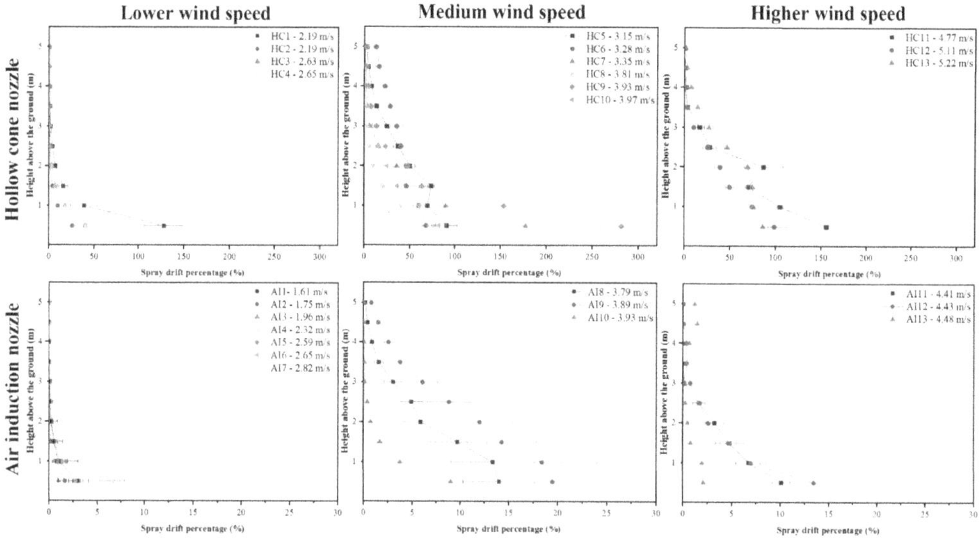

Figure 7. Airborne spray drift profiles obtained from the airborne drift collection frame using UASS with HC and AI nozzles at LWS, MWS, and HWS.

The results obtained from the three-way ANOVA analysis are demonstrated in Table 3. The effects of crosswind speed range, nozzle type, and sampling position on DPs were extremely significant for all three techniques ($p < 0.001$), suggesting the abilities of TB, GPD, and ACF to describe spray drift percentage were generally consistent.

Table 3. Significance obtained from three-way ANOVAs for DPs as affected by wind speed, nozzle type, and sampling position (downwind distance or height above the ground) using UASS based on test bench, ground petri dish, and airborne collection frame.

Sources of Variation	Test Bench		Ground Petri Dish		Airborne Collection Frame	
	p (>F)	Sign.	p (>F)	Sign.	p (>F)	Sign.
Wind speed range (WSR)	0.000	*** [1,2]	0.000	***	0.000	***
Nozzle type (NT)	0.000	***	0.000	***	0.000	***
Sampling position (SP)	0.000	***	0.000	***	0.000	***
WSR × NT	0.000	***	0.001	**	0.000	***
WSR × SP	0.000	***	0.000	***	0.013	*
NT × SP	0.000	***	0.000	***	0.000	***
WSR × NT × SP	1.000	NS	0.880	NS	0.174	NS

[1] Statistical significance level: NS $p > 0.05$, * $p < 0.05$, ** $p < 0.01$, and *** $p < 0.001$; [2] 0.000 represents p value lower than 10^{-4}.

3.4. Cumulative Drift Percentage

Figure 8 provides cumulative drift percentages and drift values calculated from drift results obtained from test bench and ground petri dish with HC and AI nozzles at LWS, MWS, and HWS. The orders of the mean values under different combinations of wind speed and nozzle type measured by the three methods were basically the same. Higher CDP values were observed for HC-HWS and HC-MWS trials, followed by HC-LWS, AI-MWS, and AI-HWS with similar values, and the lowest CDP was always found for AI-LWS. The mean CDP of both HC-HWS and HC-MWS were significantly higher than the other configurations, and no significant difference was found between any other two configurations. With respect to ACF, the CDP values with the HC nozzle were much higher than the other two collection methods at the corresponding wind speeds (at least three times the values obtained from TB and PD). However, this huge difference was not presented when AI nozzles were used, although higher mean CDP was even found at MWS instead of HWS, in agreement with the GPD results.

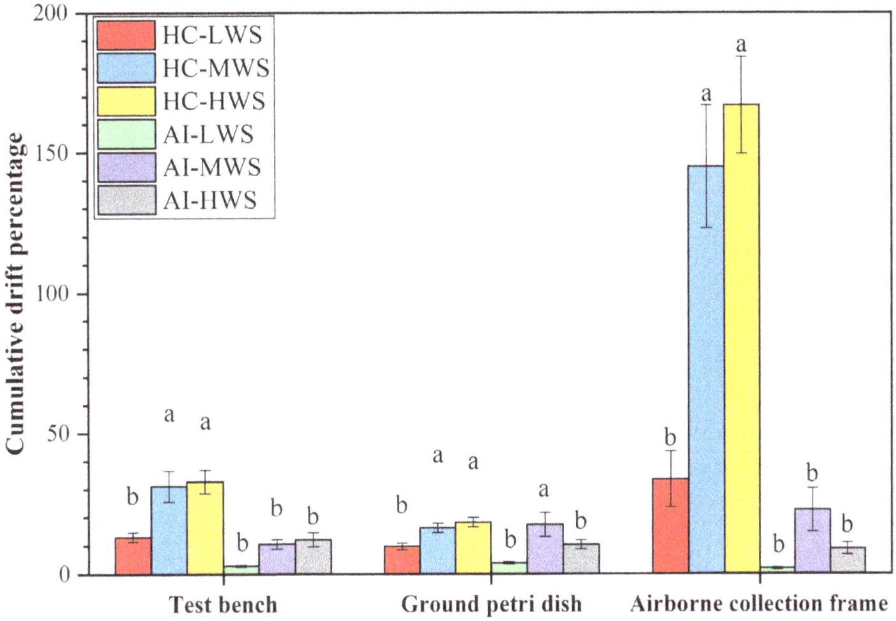

Figure 8. Cumulative drift percentage in mean ± standard error obtained from test bench, ground petri dish, and airborne collection frame with HC and AI nozzles at LWS, MWS, and HWS. Different letters above the bars represent significant differences among different configurations for each measuring method (Duncan test, α = 0.05).

The two-way ANOVA results for CDPs as affected by wind speed and nozzle type via the three measuring techniques in Table 4 show that WSR had a significant influence on CDP value obtained from both TB and GPD ($p < 0.05$), but its effect on airborne CDP was not significant ($p > 0.05$), while a good significant relationship was found between nozzle type and CDP values measured via all three collectors ($p < 0.01$).

Table 4. Significance obtained from two-way ANOVAs for CDP_B, CDP_G, and CDP_A, as affected by wind speed, nozzle type based on test bench, ground petri dish, and airborne drift collection frame.

Sources of Variation	CDP_B		CDP_G		CDP_A	
	p (>F)	Sign.	p (>F)	Sign.	p (>F)	Sign.
Wind speed range (WSR)	0.016	* [1]	0.029	*	0.243	NS
Nozzle type (NT)	0.000	*** [2]	0.001	**	0.000	***
WSR × NT	0.166	NS	0.746	NS	0.387	NS

[1] Statistical significance level: NS $p > 0.05$, * $p < 0.05$, ** $p < 0.01$, and *** $p < 0.001$; [2] 0.000 represents p value lower than 10^{-4}.

3.5. Drift Reduction Percentage

Comparisons of drift reduction results calculated using CDP between the HC and AI nozzle obtained from the three types of collectors can be seen in Table 5. Regardless of tested wind speed, the highest drift reduction percentage was always found for the airborne collection frame, followed by the test bench and the ground petri dish. For sedimenting drift, DRP results show lower values (less than 70%) at medium and higher wind speeds (>3.0 m/s), indicating a better anti-drift performance was usually achieved at lower wind speed when using the AI nozzle. However, there is not any similar trend for airborne spray drift, and the AI nozzle can save 84.4–94.6% of spray drift compared to the HC nozzle.

Table 5. Drift reduction percentage and classification when using the AI nozzle based on test bench, ground petri dish and airborne collection frame for the tested wind speed ranges.

DRP Source	Wind Speed Range	DRP (%) [a]	Reduction Class [b]
Test bench	LWS	78.1	D
	MWS	66.2	E
	HWS	63.1	E
Ground petri dish	LWS	60.7	E
	MWS	9.2	None
	HWS	43.4	F
Airborne collection frame	LWS	94.2	C
	MWS	84.4	D
	HWS	94.6	C

a Drift reduction percentage values were calculated based on the HC treatment as the reference configuration.
b Classification determined by ISO 22369-1: A $\geq$ 99%, B 95 $\leq$ 99%, C 90 $\leq$ 95%, D 75 $\leq$ 90%, E 50 $\leq$ 75%, and F 25 $\leq$ 50%.

3.6. *Correlation Analysis Results*

The results of the correlational analysis between CDP values and meteorology parameters for the two different nozzle type are presented in Figure 9 and Table A1. For HC nozzles, CDP measured by all three techniques showed a significant positive correlation with wind speed ($p < 0.05$, $\rho > 0$), i.e., the higher the wind speed, the higher the cumulative drift percentage. In terms of relative humidity (RH) and wind direction, significant correlations were only observed with the airborne CDP. The airborne CDP, meanwhile, increased as the RH decreased ($p < 0.05$, $\rho < 0$) and as the degree from the ideal wind direction increased ($p < 0.05$, $\rho > 0$), as well. For AI nozzles, both ground and bench values were significantly positively correlated with wind speed ($p < 0.05$, $\rho > 0$), except for airborne percentage. There was a significant negative correlation between CDP value and RH for all three types of collectors ($p < 0.05$, $\rho < 0$), consistent with the airborne drift results of HC nozzles. In general, under the test conditions of this study, the correlations between CDPs and crosswind speed and RH were more pronounced, and significant positive and negative correlations were presented, respectively. No relationships ($p > 0.05$) were detected between CDP values and temperature for all settings in this study, and there was no significant correlation ($p > 0.05$) between CDP and wind direction under most of circumstances. Correlations between CDPs achieved from two ground collectors and meteorological parameters were comparable. By contrast, the correlation results of airborne drift differed due to the differences in the behaviors and distribution characteristics of airborne droplets.

In addition, Table 6 provides the intercorrelations among the three types of collectors for spray drift. For HC nozzle trials, there was only a significant positive correlation between CDP values obtained from TB and those from GPD ($p < 0.01$, $\rho > 0$), but no significant correlation was found between TB results and ACF results and between GPD results and ACF results ($p > 0.05$). For AI nozzle trials, all three combinations of results were found to be significantly positive correlated with each other ($p < 0.01$, $\rho > 0$), showing a good consistency of CDP value for the three types of collectors. Overall, these results suggest that the correlation between each two-drift assessment technique differed when using different types of nozzles.

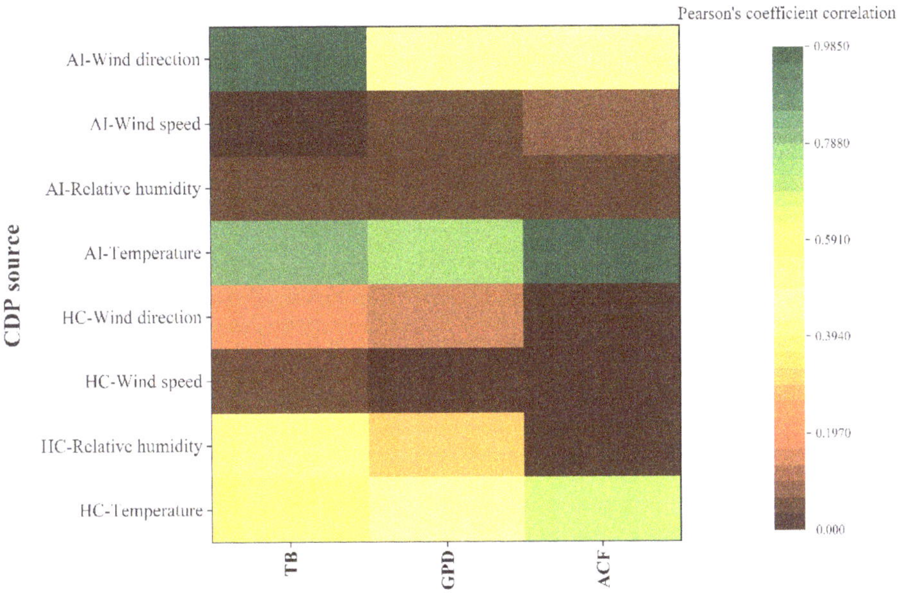

Figure 9. Heat map of Pearson's coefficient correlation among cumulative drift percentages (respectively, test bench, ground petri dish, and airborne collection frame) and meteorology characteristics (temperature, relative humidity, wind speed, and wind direction).

Table 6. Correlation among CDP values obtained from different collectors (respectively, test bench, ground petri dish, and airborne collection frame) for the HC and AI nozzle.

Nozzle Type	TB and GPD			TB and ACF			GPD and ACF		
	p (>F)	Sign.[a]	ρ [b]	p (>F)	Sign.	ρ	p (>F)	Sign.	ρ
HC	0.001	**	0.812	0.134	NS	0.438	0.069	NS	0.519
AI	0.001	**	0.816	0.005	**	0.729	0.000	***	0.946

a Statistical significance level: NS $p > 0.05$, ** $p < 0.01$, and *** $p < 0.001$. b Pearson's coefficient of correlation.

4. Discussion and Conclusions

Filed drift measurements using the three techniques prove that the test bench is also promising for direct drift determination of the unmanned aerial sprayer in windy conditions (higher than 1 m/s), in addition to being used for spray drift potential evaluation of boom sprayers and airblast sprayers when used in nearly absence of wind. The described methodology enabled classification of different UAV sprayer configurations.

Regarding the distribution characteristics of spray deposition percentage, the results measured by the three collectors were generally consistent, i.e., higher wind speeds and finer droplets produced higher DP values and longer safe downwind distances or higher safe heights. These results are in agreement with those obtained in previous studies related to UASS spray drift [8,30,35]. However, unlike ground petri dish, DP from test bench showed a process of rising to a peak value and then falling under high wind speeds, and meanwhile, in the case of collecting only one route's spray drift, did not decrease substantially. This result may be explained by the fact that the sampling interval of TB is small and more samples can be obtained for each test. Even though only a single route was collected, a larger amount of direct drifted droplets could be captured to make up for the shortfall in total applied volume, reflecting a higher collection efficiency of the field test bench.

Similar to the DP results, wind speed range and nozzle type had a significant effect on the CDP value measured by the three techniques. Particularly, the WSR did not affect the airborne CDP value significantly, and the aerial CDP of the HC nozzle was very high, at least three times the values obtained from the other two techniques and six times the AI nozzle results. A possible explanation for these results may be fine droplet behaviors. Fine droplets tend to float in the air, and at a downwind distance of more than 2 m, most of airborne spray would be in droplets < 100 μm [36]. Moreover, the collection efficiency of the PE tube for small droplets produced by the HC nozzle is considerably higher than that for large droplets from the AI nozzle. This leads to a too high airborne drift result for the HC nozzle, so the significant correlation between WSR and airborne CDP cannot be distinguished in the two-way ANOVA.

For sedimenting DRP results, a better anti-drift performance was usually achieved at lower wind speed when changing nozzles to provide a coarser spray. However, the findings of the current study do not support the previous research [29,37,38]. This study defined three wind speed ranges. Lower spray drift was observed at low wind speeds, but when the crosswind speed increased to 3.0 m/s, the effect of increased wind speed on the increased drift value was not obvious. Tiny difference could be obtained between MWS and HWS, and AI nozzle's CDP measured by the GPD at MWS was even higher than that at HWS. This resulted in a decrease in the DRP value when the wind speed increased. However, more replicates are needed to verify whether this result is generalizable.

In general, both wind speed and humidity had significant effects on UASS spray drift at different droplet sizes (Figure 9 and Table A1). The faster the wind speed and the lower the humidity, the higher the drift value. These results are in line with those of previous studies [4,39–42]. Furthermore, the correlations between the ground drift values and various meteorological factors were consistent for TB and GPD, demonstrating the feasibility of using TB to describe the ground drift characteristics. It is noteworthy that the two ground CDPs were not correlated with RH for the HC nozzle's fine spray, while a significant negative correlation was found for the AI nozzle's coarse spray. This could be attributed to the combined effects of droplet size and humidity. Under the tested low humidity conditions (11.8–35.5%), fine droplets evaporate faster, but they have longer settling time, so regardless of how the RH changes, ground collector cannot collect those evaporated droplets. Conversely, for coarse droplets with a longer lifetime, evaporation would bring a reduction in droplet size during movement, increasing the amount of drift [43]. In addition, the ACF is closer to where the droplet is released, and the fine droplets are not completely evaporated when reaching the collector, so the amount of evaporation is still directly related to the humidity. Moreover, as another important factor impacting droplet evaporation, temperature has no significant correlation with CDP values in this study (Table A1). This may be due to the narrow range of the temperature change during the field tests (<10 °C).

With respect to the discordance of the correlation among CDPs obtained from different collectors, there was no significant correlation between sedimenting drift value (TB or GPD) and airborne drift value (ACF) when using HC nozzles (Table 6). These results may be explained by the fact that fine droplets are prone to suspend in the air, drift with the wind, and evaporate, and then the number of droplets can be collected under the influence of changing meteorological conditions is also varying. The trends of the spray drift distribution between the ground and the airborne collector are unlikely to be coincident. Nevertheless, almost all droplets produced by AI nozzle sizes are greater than 100 μm (Table 2). Coarse droplets tend to sediment quickly, so the majority of drifting droplets can be sampled by ground and airborne collectors.

In summary, the test bench can be closed in time after the end of one route's operation to reduce dust contamination and avoid downwash airflow interference during tests, and the airborne frame has similar advantages. In addition, the TB has higher collection efficiency and reflects drift characteristics at a close downwind distance more comprehensively, making the influence of different factors on the droplet distribution more obvious. Therefore, the use of TB and ACF is able to diminish the sampling distance when comparing the drift risk of different sprayers and configurations, which can not only reduce the requirements for the test site, but also improve the efficiency of the field test.

The external meteorological conditions normally have a serious impact on the completion of the field trials, which eventually leads to a reduction in the number of valid replicates. In this study, more than 20 trials were conducted for each nozzle, but those trials that did not meet the requirements of meteorological conditions, flight parameters, and applied volume were discarded after careful evaluation. Therefore, the accuracy of the field drift tests need to be improved in the further study by increasing the number of tests, finding sites with more stable meteorological conditions, and strengthening pre-planning and inspection during the tests. However, field drift test conditions are ultimately uncontrollable, and direct measurements consume too much time, labor, and material costs. Further work is required to explore the applicability and feasibility of TB in determining spray drift potential of UASS under static wind conditions. A part of experiments was conducted and the results are being processed and analyzed. It can be expected that TB used in this method will further reduce the test conditions required for the UASS drift evaluation.

Author Contributions: Conceptualization, C.W. and X.H.; methodology, C.W., S.W., Z.H. and A.Z.; software, C.W. and L.H.; validation, C.W., C.Y. and J.Z.; formal analysis, C.W.; investigation, W.S.; resources, C.W. and X.H.; data curation, C.W. and A.Z.; writing—original draft preparation, C.W.; writing—review and editing, X.H.; visualization, C.W.; supervision, X.H.; project administration, X.H.; funding acquisition, C.W. and X.H. All authors have read and agreed to the published version of the manuscript.

Funding: This study was supported by the earmarked fund for Project 32202343 supported by National Natural Science Foundation of China, China Agriculture Research System (CARS-28), Chinese Universities Scientific Fund (Grant No. 2022TC128), and the 2115 Talent Development Program of China Agricultural University.

Data Availability Statement: The data presented in this study are available upon request from the corresponding author.

Acknowledgments: The authors would like to thank Yi Yang of Beijing TT Aviation Technology Co. Ltd., Andreas Herbst of Julius Kühn Institute, Tian Li and all other staff of CCAT and CAUS, China Agricultural University for their great contributions to this work.

Conflicts of Interest: The authors declare no conflict of interest.

Appendix A

ISO 22866: This International Standard establishes principles for the measurement of droplet drift from all types of equipment designed for applying plant protection products. Detailed specifications relate to tractor-mounted, trailed, and self-propelled agricultural sprayers operating in arable field crops (boom sprayers) and in bush and tree (including vines, hops, and fruit) crops (including broadcast air-assisted sprayers).

ISO 22856: This International Standard establishes general principles for the measurement of spray drift potential in wind tunnels under controlled laboratory conditions, and it is applicable where comparative assessment or classification of the relative spray drift potential from spray generators or spray liquids is needed.

ISO 22401: This International Standard provides a test method to measure spray sedimentation from horizontal boom sprayers using a test bench. The sedimentation measure gives a value for potential spray drift. These measurements can be used to compare different sprayer setups on the same sprayer.

ISO 25358: This standard specifies procedures for classifying droplet size spectra from atomizers used in spraying for crop protection. It provides a reference system for defining classes of droplet size spectra. Depending on their function principle and individual setup, measuring systems for droplet sizing can give different results. This document provides a means of comparing measured droplet size spectra to reference spectra and enables relative comparisons of droplet size spectra obtained from different measuring systems.

ISO 22369: All parts specify the drift classification of spraying equipment. ISO 22369-1 defines the spray drift reduction classes. The other parts specify the test procedures. The object of these standards is to provide uniform procedures for the determination of the drift reducing performance of spraying equipment.

Appendix B

Table A1. Relationships among cumulative drift percentages (respectively, test bench, ground petri dish, airborne collection frame) and meteorology characteristics (temperature, relative humidity, wind speed, and wind direction) for each nozzle's 13 field trials ($n = 13$).

Nozzle Type	CDP Source	Temperature			Relative Humidity			Wind Speed			Wind Direction		
		p (>F)	Sign. [a]	P [b]	p (>F)	Sign.	ρ	p(>F)	Sign.	ρ	p (>F)	Sign.	ρ
HC	TB	0.576	NS	0.171	0.514	NS	−0.199	0.045	*	0.564	0.188	NS	0.539
	GPD	0.431	NS	0.239	0.319	NS	−0.300	0.031	*	0.598	0.134	NS	0.663
	ACF	0.685	NS	0.125	0.029	*	−0.604	0.018	*	0.640	0.032	*	0.596
AI	TB	0.808	NS	−0.075	0.038	*	−0.579	0.002	**	0.771	0.962	NS	0.015
	GPD	0.746	NS	−0.100	0.044	*	−0.565	0.038	*	0.579	0.447	NS	−0.231
	ACF	0.984	NS	−0.006	0.038	*	−0.580	0.070	NS	0.517	0.471	NS	−0.220

a Statistical significance level: NS $p > 0.05$, * $p < 0.05$, ** $p < 0.01$. b Pearson's coefficient of correlation.

References

1. He, X. Rapid development of unmanned aerial vehicles (UAV) for plant protection and application technology in China. *Outlooks Pest Manag.* **2018**, *29*, 162–167. [CrossRef]
2. He, X.; Bonds, J.; Herbst, A.; Langenakens, J. Recent development of unmanned aerial vehicle for plant protection in East Asia. *Int. J. Agric. Biol. Eng.* **2017**, *10*, 18–30. [CrossRef]
3. Wang, C.; Liu, Y.; Zhang, Z.; Han, L.; Li, Y.; Zhang, H.; Wongsuk, S.; Li, Y.; Wu, X.; He, X. Spray performance evaluation of a six-rotor unmanned aerial vehicle sprayer for pesticide application using an orchard operation mode in apple orchards. *Pest Manag. Sci.* **2022**, *78*, 2449–2466. [CrossRef]
4. Wang, X.; He, X.; Wang, C.; Wang, Z.; Li, L.; Wang, S.; Bonds, J.; Herbst, A.; Wang, Z. Spray drift characteristics of fuel powered single-rotor UAV for plant protection. *Trans. Chin. Soc. Agric. Eng. Trans. CSAE* **2017**, *33*, 117–123. [CrossRef]
5. Wang, X.; He, X.; Song, J.; Wang, Z.; Wang, C.; Wang, S.; Wu, R.; Meng, Y. Drift potential of UAV with adjuvants in aerial applications. *Int. J. Agric. Biol. Eng.* **2018**, *11*, 54–58. [CrossRef]
6. Wang, C.; He, X.; Wang, X.; Wang, Z.; Wang, S.; Li, L.; Bonds, J.; Herbst, A.; Wang, Z. Testing method and distribution characteristics of spatial pesticide spraying deposition quality balance for unmanned aerial vehicle. *Int. J. Agric. Biol. Eng.* **2018**, *11*, 18–26. [CrossRef]
7. Wang, C.; Zeng, A.; He, X.; Song, J.; Herbst, A.; Gao, W. Spray drift characteristics test of unmanned aerial vehicle spray unit under wind tunnel conditions. *Int. J. Agric. Biol. Eng.* **2020**, *13*, 13–21. [CrossRef]
8. Herbst, A.; Bonds, J.; Wang, Z.; Zeng, A.; He, X.; Goff, P. The influence of unmanned agricultural aircraft s ystem design on spray drift. *J. Kult.* **2020**, *72*, 1–11. [CrossRef]
9. Wang, G.; Han, Y.; Li, X.; Andaloro, J.; Chen, P.; Hoffmann, W.C.; Han, X.; Chen, S.; Lan, Y. Field evaluation of spray drift and environmental impact using an agricultural unmanned aerial vehicle (UAV) sprayer. *Sci. Total Environ.* **2020**, *737*, 139793. [CrossRef]
10. Brown, C.R.; Giles, D.K. Measurement of pesticide drift from unmanned aerial vehicle application to a vineyard. *Trans. ASABE* **2018**, *61*, 1539–1546. [CrossRef]
11. Li, L.; Hu, Z.; Liu, Q.; Yi, T.; Han, P.; Zhang, R.; Pan, L. Effect of flight velocity on droplet deposition and drift of combined pesticides sprayed using an unmanned aerial vehicle sprayer in a peach orchard. *Front. Plant Sci.* **2022**, *13*, 981494. [CrossRef]
12. Biglia, A.; Grella, M.; Bloise, N.; Comba, L.; Mozzanini, E.; Sopegno, A.; Pittarello, M.; Dicembrini, E.; Alcatrão, L.E.; Guglieri, G.; et al. UAV-spray application in vineyards: Flight modes and spray system adjustment effects on canopy deposit, coverage, and off-target losses. *Sci. Total Environ.* **2022**, *845*, 157292. [CrossRef] [PubMed]
13. *ISO 22866*; Equipment for Crop Protection—Methods for Field Measurement of Spray Drift. ISO: Geneva, Switzerland, 2005.
14. *ISO 22856*; Equipment for Crop Protection—Methods for the Laboratory Measurement of Spray Drift—Wind Tunnels. ISO: Geneva, Switzerland, 2008.
15. *ISO 22401*; Equipment for Crop Protection—Method for Measurement of Potential Spray Drift from Horizontal Boom Sprayers by the Use of a Test Bench. ISO: Geneva, Switzerland, 2015.
16. Douzals, J.-P.; Tinet, C.; Goddard, R. Use of a flexible drop counter for a better comparability of potential spray drift measurement protocols in wind tunnels. *Asp. Appl. Biol. Int. Adv. Pestic. Appl.* **2018**, *137*, 277–284.
17. *ISO 25358*; Crop Protection Equipment-Droplet-Size Spectra from Atomizers—Measurement and Classification. ISO: Geneva, Switzerland, 2018.

18. Van de Zande, J.C.; Holterman, H.J.; Wenneker, M. Nozzle Classification for Drift Reduction in Orchard Spraying; Identification of Drift Reduction Class Threshold Nozzles. *Agric. Eng. Int. CIGR E J.* **2008**, *X*, 253–260. [CrossRef]
19. Sousa Alves, G.; Kruger, G.R.; da Cunha, J.P.A.R.; Vieira, B.C.; Henry, R.S.; Obradovic, A.; Grujic, M. Spray Drift from Dicamba and Glyphosate Applications in a Wind Tunnel. *Weed Technol.* **2017**, *31*, 387–395. [CrossRef]
20. Brusselman, E.; Van Driessen, K.; Steurbaut, W.; Gabriels, D.; Cornelis, W.; Nuyttens, D.; Sonck, B.; Baetens, K.; Nicolai, B.; Verboven, P.; et al. Wind tunnel evaluation of several tracer and collection techniques for the measurement of spray drift. *Commun. Agric. Appl. Biol. Sci.* **2004**, *69*, 829–836. [CrossRef]
21. Torrent, X.; Garcerá, C.; Moltó, E.; Chueca, P.; Abad, R.; Grafulla, C.; Román, C.; Planas, S. Comparison between standard and drift reducing nozzles for pesticide application in citrus: Part I. Effects on wind tunnel and field spray drift. *Crop Prot.* **2017**, *96*, 130–143. [CrossRef]
22. Balsari, P.; Marucco, P.; Tamagnone, M. A test bench for the classification of boom sprayers according to drift risk. *Crop Prot.* **2007**, *26*, 1482–1489. [CrossRef]
23. Wang, X.; He, X.; Herbst, A.; Langenakens, J.; Zheng, J.; Li, Y. Development and performance test of spray drift test system for sprayer with bart. *Trans. Chin. Soc. Agric. Eng. Trans. CSAE* **2014**, *30*, 55–62. [CrossRef]
24. Gil, E.; Gallart, M.; Balsari, P.; Marucco, P.; Almajano, M.P.; Llop, J. Influence of wind velocity and wind direction on measurements of spray drift potential of boom sprayers using drift test bench. *Agric. For. Meteorol.* **2015**, *202*, 94–101. [CrossRef]
25. Nuyttens, D.; Zwertvaegher, I.; Dekeyser, D. Comparison between drift test bench results and other drift assessment techniques. *Int. Adv. Pestic. Appl. Appl. Biol.* **2014**, *122*, 293–302.
26. Nuyttens, D.; Zwertvaegher, I.K.A.; Dekeyser, D. Spray drift assessment of different application techniques using a drift test bench and comparison with other assessment methods. *Biosyst. Eng.* **2017**, *154*, 14–24. [CrossRef]
27. Balsari, P.; Gil, E.; Marucco, P.; van de Zande, J.C.; Nuyttens, D.; Herbst, A.; Gallart, M. Field-crop-sprayer potential drift measured using test bench: Effects of boom height and nozzle type. *Biosyst. Eng.* **2017**, *154*, 3–13. [CrossRef]
28. Grella, M.; Gil, E.; Balsari, P.; Marucco, P.; Gallart, M. Advances in developing a new test method to assess spray drift potential from air blast sprayers. *Span. J. Agric. Res.* **2017**, *15*, e0207. [CrossRef]
29. Grella, M.; Marucco, P.; Balsari, P. Toward a new method to classify the airblast sprayers according to their potential drift reduction: Comparison of direct and new indirect measurement methods. *Pest Manag. Sci.* **2019**, *75*, 2219–2235. [CrossRef]
30. Wang, C.; Herbst, A.; Zeng, A.; Wongsuk, S.; Qiao, B.; Qi, P.; Bonds, J.; Overbeck, V.; Yang, Y.; Gao, W.; et al. Assessment of spray deposition, drift and mass balance from unmanned aerial vehicle sprayer using an artificial vineyard. *Sci. Total Environ.* **2021**, *777*, 146181. [CrossRef]
31. Grella, M.; Marucco, P.; Manzone, M.; Gallo, R.; Mazzetto, F.; Balsari, P. Indoor test bench measurements of potential spray drift generated by multi-row sprayers. In Proceedings of the 2021 IEEE International Workshop on Metrology for Agriculture and Forestry, Trento, Italy, 3–5 November 2021; pp. 356–361. [CrossRef]
32. *ISO 22369-1*; Crop Protection Equipment—Drift Classification of Spraying Equipment—Part 1: Classes. ISO: Geneva, Switzerland, 2006.
33. Tang, Q.; Zhang, R.; Xu, M.; Xu, G.; Yi, T.; Zhang, B.; Chen, L. Droplet spectra and high-speed wind tunnel evaluation of air induction nozzles. *Front. Agric. Sci. Eng.* **2018**, *5*, 442–454. [CrossRef]
34. Nilsson, E.; Svensson, S.A. Buffer zones when using plant protection products—A Swedish approach. *Annu. Rev. Agric. Eng.* **2005**, *4*, 143–150.
35. Hunter, J.E.; Gannon, T.W.; Richardson, R.J.; Yelverton, F.H.; Leon, R.G. Coverage and Drift Potential Associated with Nozzle and Speed Selection for Herbicide Applications using an Unmanned Aerial Sprayer. *Weed Technol.* **2019**, *34*, 235–240. [CrossRef]
36. Miller, P. Spray drift and its measurement. In *Application Technology for Crop Protection*; Matthews, G.A., Hislop, E.C., Eds.; Centre for Agricultural Bioscience International: Wallingford, UK, 1993; pp. 101–122.
37. Bird, S.L.; Esterly, D.M.; Perry, S.G. Off-Target Deposition of Pesticides from Agricultural Aerial Spray Applications. *J. Environ. Qual.* **1996**, *25*, 1095–1104. [CrossRef]
38. Grella, M.; Gallart, M.; Marucco, P.; Balsari, P.; Gil, E. Ground Deposition and Airborne Spray Drift Assessment in Vineyard and Orchard: The Influence of Environmental Variables and Sprayer Settings. *Sustainability* **2017**, *9*, 728. [CrossRef]
39. Miller, D.R.; Yendol, W.E.; Ducharme, K.M.; Maczuga, S.; Reardon, R.C.; McManus, M.A. Drift of aerially applied diflubenzuron over an oak forest. *Agric. For. Meteorol.* **1996**, *80*, 165–176. [CrossRef]
40. Miller, D.R.; Stoughton, T.E. Response of spray drift from aerial applications at a forest edge to atmospheric stability. *Agric. For. Meteorol.* **2000**, *100*, 49–58. [CrossRef]
41. Parkin, C.S.; Walklate, P.J.; Nicholls, J.W. Effect of drop evaporation on spray drift and buffer zones. In *Proceedings of the BCPC International Conference—Crop Science and Technology 3C-3, Glasgow, UK, 10–12 November 2003*; BCPC: Farnham, UK; pp. 261–266.
42. Nuyttens, D.; de Schampheleire, M.; Steurbaut, W. Experimental study of factors influencing the risk of drift from field sprayers. Part 1: Meteorological conditions. *Int. Adv. Pestic. Appl. Asp. Appl. Biol.* **2006**, *77*, 321–329.
43. Spillman, J.J. Evaporation from freely falling droplets. *Aeronaut. J.* **1984**, *88*, 181–185. [CrossRef]

agronomy

Article

Comparison of Weather Acquisition Periods Influencing a Statistical Model of Aerial Pesticide Drift

Steven J. Thomson [1] and Yanbo Huang [2,*]

[1] USDA National Institute of Food and Agriculture, Kansas City, MO 64105, USA
[2] USDA Agricultural Research Service, Genetics and Sustainable Agriculture Research Unit, Mississippi State, MS 39762, USA
* Correspondence: yanbo.huang@usda.gov

Abstract: Off-target drift of crop protection materials from aerial spraying can be detrimental to sensitive crops, beneficial insects, and the environment. So, it is very important to accurately characterize weather effects for accurate recommendations on drift mitigation. Wind is the single-most important weather factor influencing localized off-target drift of crop protection materials. In drift sampling experiments, it is difficult to accurately characterize wind speed and direction at a drift sampling location, owing to the natural variability of spray movement on the way to the sampling target. Although it is difficult or impossible to exactly track wind movement to a target, much information can be gained by altering the way wind speed and tracking is characterized from field experiments and analyzed using statistical models of spray drift. In this study two methods of characterizing weather were compared to see how they affect results from a statistical model of downwind spray drift using field data. Use of a method that implemented weather averages over the length of a spray run resulted in model-based estimates for spray tracer concentration that compared well with field data averages. Model results using this method showed only a slight sensitivity to changes in wind speed, and this difference was more pronounced further downwind. The degree of this effect was consistent with field results. Another method that used single weather values obtained at the beginning of each run resulted in an unexpected inverse relationship of residue concentration with respect to increases in wind speed by sensitivity analysis and would thus not be recommended for use in a statistical model of downwind spray drift. This study could provide a guideline for general agricultural aviation analysis and unmanned aerial vehicle spray application studies.

Keywords: aerial application; drift control; agricultural aircraft; pesticide application

Citation: Thomson, S.J.; Huang, Y. Comparison of Weather Acquisition Periods Influencing a Statistical Model of Aerial Pesticide Drift. *Agronomy* **2023**, *13*, 213. https://doi.org/10.3390/agronomy13010213

Academic Editor: Francisco Manzano Agugliaro

Received: 31 October 2022
Revised: 3 January 2023
Accepted: 7 January 2023
Published: 10 January 2023

1. Introduction

Modern agriculture has seen significant innovation led by precision and site-specific crop management since the late 1980s, primarily based on improvements in agricultural mechanization that integrates information from global positioning systems (GPS), remote sensing, and geographic information systems (GIS). Spatial detection methods can now better characterize within-field variability of soils, nutrients, weed and insect pests, disease, and yield. With the advancement of information and electronics technologies integrated with artificial intelligence, automation and big data, precision agriculture is moving forward to the next stage: smart agriculture [1]. In smart agriculture, agricultural production and management advances to a new level supported by supercomputing, inter-networking, machine or artificial intelligence, automation, and massive data resources in integrated fashion to optimize crop production systems and resources.

Advanced technologies have been developed for spraying systems that target pests precisely, thereby saving on pesticide costs with commensurate reduction in environmental pressure. Even so, aerial spraying systems have not caught up with ground-based spraying systems in their effectiveness for targeted spraying. By their nature, the best one can hope for is to limit spray application only to patches or small field areas. Advanced methods are still required to mitigate off target spray drift, no matter which advanced technologies are employed for site specific application.

Agricultural aviation began in 1906 by John Chaytor, who spread seed over a swamped valley floor in Wairoa, New Zealand, using a hot air balloon with mobile tethers [2]. In 1924, the first commercial operation of crop dusting was conducted in Macon, Georgia, USA, by a company named Huff-Daland Crop Dusting, co-founded by a pilot Lt. Harold R. Harris, who tested the system in McCook Field, Dayton, Ohio, USA [3]. Presently, aerial application is used in the US for plant protection and is especially useful for covering large and remote areas or where wet weather conditions preclude timely use of ground sprayers, primarily across the Midwest and Midsouth US. This makes aerial application an indispensable tool for ensuring high yields [4]. The National Agricultural Aviation Association (NAAA) (Alexandria, Virginia, USA) defines this as an industry that "consists of small businesses and pilots that use aircraft to aid farmers in producing a safe, affordable and abundant supply of food, fiber and biofuel" while "aerial application is a critical component of high-yield agriculture" (https://www.agaviation.org/aboutagaviation (accessed on 5 January 2023)).

There are some limits or challenges in aerial application of plant protection materials. Off-target drift is one of them, which may cause economic loss and environmental damage. So, in the past two decades scientists and engineers have investigated the factors and impact of off-target spray drift in plant protection. For example, in the Mississippi Delta USDA (United States Department of Agriculture), scientists have evaluated various nozzles for drift control and tested a hydraulic pump with automatic flow controller for aerial variable-rate spray studies [5]. The authors also collected meteorological data and studied the daily change and probability of atmospheric stability for the likelihood of surface temperature inversions [6–8]. Using these data, an algorithm was developed for a website, published to guide aerial applicators on the timing to conduct spray application for avoidance of off-target movement of spray parcels caused by temperature inversions [9,10].

Minimization of off-target drift from aerial- or ground- based spray systems [11] is critical to reduce damage to off targeted crops and to mitigate environmental effects [12]. Drift minimization practices can also work to increase application efficiency. Methods to mitigate off target movement of aerially applied agricultural materials have been well documented [13–22]. Some research efforts focus on the prediction of drift [23–25] from aerial spraying with several standards having been established [26,27]. Others are based on optimized drift modeling from design of experiments [28,29].

Experimental assessment of off-target drift of crop protection materials continues to be a challenge due in large part to highly variable environmental conditions. In their review of spray drift assessment methods, Felsot et al. [30] outlined two spray drift situations: the first is characterized as localized movement of agrochemical residues from target to non-target sites. In this aspect, the in-swath spray deposition of a low-drift nozzle was determined on the booms of Air Tractor 402B agricultural aircraft (Air Tractor Inc., Olney, TX, USA) [28]. Further, the spray deposition and drift from the low drift nozzle were characterized for aerial spray application of different application altitudes [31]. Using airborne remote sensing the cotton injury was assessed by spray drift from aerially applied glyphosate by spray deposition and drift sampling and measurement [32]. The second involves concentration of residues within air masses under stable atmospheric conditions. For the latter case, this mass of air with entrained chemical can drift long distances. In this aspect, Fritz et al. monitored and documented atmospheric conditions over a six-month period at two locations in Texas [22]. The measured meteorological data was used to relate low-level temperature inversion with respect to time of day, duration, and other associated meteorological conditions. Thomson et al. collected data to quantify atmospheric stability throughout the day and showed consistent patterns of atmospheric stability occurring primarily between the hours of 18:00 p.m. and 06:00 a.m. during summer months on clear days at the Mississippi Delta latitudes [6,7]. Further, results verified correctness of the three-degree rule in Arkansas when spraying on clear days in the summer months.

Huang et al. determined atmospheric stability at different time intervals for determination of aerial application timing to avoid the drift caused by temperature inversion [8].

Many controlled studies have used wind tunnels to assess the droplet spectra and thus suitability of nozzles and spray mixtures for aerial and ground application [33,34]. Gil et al. developed a test bench that could examine effects of wind to complement drift quantification methods in the field or laboratory [35]. The aim was to provide a simple field method near a ground sprayer to obtain meaningful near-field drift data as a function of weather conditions. In their study, it was found that drift potential was not significantly influenced when trials were conducted at average wind velocities less than 1 m/s. Wind speed and direction greatly influenced spatial distribution of spray sample recovered along the length of the test bench.

For field studies on localized spray drift, wind speed and direction have been documented to be primary factors in downwind spray drift of crop protection materials, along with height of application when using both aerial [28,31] and ground spray platforms [36]. Fritz et al. also indicated through field experiments that wind speed was the dominant meteorological factor in transport of aerially applied sprays [37]. Wind effects have been incorporated into simulations of downwind spray drift [24], which have been verified for accuracy through field trials [24,38–40].

Owing to the difficulty in properly characterizing wind in drift studies, the study presented herein compares two statistical methods for acquiring weather data to observe their effect on statistically modeled downwind concentration of aerially applied spray tracer. A sensitivity analysis on selected variables, including wind speed, was performed so that their effects on modeled concentration could further be compared using both weather acquisition methods.

2. Materials and Methods

2.1. Field Experiment

Spray tests were conducted over a 15-ha open field planted in Bermuda grass at the USDA ARS, Stoneville MS, with downwind drift samplers placed at 100, 135, 200, and 320 m. Water mixed with Syl-Tac® adjuvant at standard mix of 0.25% *v/v* was sprayed rate of 18.7 L/ha from an Air Tractor 402B aircraft (Air Tractor, Olney, TX, USA) through 56 CP-09 straight stream (SS) nozzles (Transland LLC, Wichita Falls, TX, USA) at a release height of 3.7 m, following the same procedures as indicated by Thomson et al. [5]. Rubidium Chloride (RbCl) tracer was mixed in the tank and sampled and collected on mylar sheets and used for analyzing spray swath patterns and off-target drift. To assure consistent height of application, height of each pass was recorded using a Lasertech ULS laser (Laser Technologies, Inc., Centennial, CO, USA). The CP-09 SS nozzles were set at a 30-degree angle and operating pressure was 303 kPa corresponding to an approximate per-nozzle flow rate of 2.4 L/min. Swath width was 18.3 m and aircraft speed was 56 m/s. Nozzle parameters were set to closely match spray characteristics for a TeeJet® D6-46 hollow cone tip [41]. Volume Median Diameter (VMD) (where half of the volume of spray contains droplets larger than the VMD) was modeled to be 218 μm using the USDA ARS AATRU models [42]. This droplet size indicates a large percentage of small droplets, and the experiment was purposely set up to induce drift at long downwind distances. At this VMD, 41% of the spray volume indicated droplet diameter of 200 μm or less, and this is also consistent with the nozzle setup used by Thomson et al. [41].

Spray samplers were oriented on the day of the experiment so that wind direction was nearly perpendicular to the direction of aircraft travel. To collect enough sample at the specified downwind distances, four passes of the aircraft were made for each run. After each run, Mylar sheets were removed from stands (Figure 1) using clean rubber gloves and toothpicks on the side of each sampling sheet to direct Mylar into its own zip-lock bag. The samples were returned to the edge of the field and immediately placed into ice chests where they were protected from light. Runs were repeated sixteen times. In the laboratory, each sheet was shaken for 20 min within a rinse solution of 1% HNO_3 (nitric acid solution), which was also used as a calibration blank on the AAnalyist 600 Atomic Absorption Spectrometer (Perkin Elmer, Waltham, MA, USA). Results obtained from the spectrometer for each sheet were determined in concentration units of ng/L.

Figure 1. Field sampling stand. Mylar sheets are held on the left bracket.

Weather data were recorded for each pass at 2-sec intervals using Kestrel 4500 weather trackers (Nielsen-Kellerman, Boothwyn, PA USA); one was set to log weather data at the beginning of each run, and one was read manually the instant spraying commenced. Using dual weather trackers facilitated testing different methods of characterizing weather effects. During the experiment, the air temperature ranged from 25 to 30 °C; relative humidity ranged from 42 to 75%; wind speed ranged from 3.3 to 5.5 m s^{-1}. The first 4 runs were flown in an afternoon and testing continued the next day, significantly past time ranges that would indicate stable atmosphere. Wind direction was from N to NE, perpendicular to the flight-line.

Mean weather data recorded during each test run is shown in Table 1 in which all weather variables were entered into the model. Depending on how weather data were averaged for the comparisons presented herein, the actual values shown under each heading will vary. Based on air temperatures at two heights and windspeed data recorded at an adjacent tall tower [6,7], the average atmospheric stability ratio was determined as -0.36 °Cs^2/m^2 indicating the desirable unstable atmospheric conditions. Thus, there was no danger of entrainment of spray in an inversion layer, which would produce inaccurate results for this spray drift study.

Table 1. Meteorological data obtained during the spray tests.

Test Run *	Mean Air Temp (°C)	Mean Relative Humidity (%)	Mean Solar Irradiance (kW/m^2)	Mean Wind Velocity (m/s)	Mean Wind Direction ** (Degrees)
1	29	44	0.6	4.8	2
2	29	43	0.5	5.0	−6
3	29	42	0.4	4.8	3
4	29	43	0.4	4.4	−3
5	25	75	0.7	5.5	−9
6	25	74	0.6	3.3	−9
7	27	64	1.0	4.0	−6
8	28	55	0.9	3.6	8
9	29	57	0.9	3.9	−17
10	29	55	1.0	4.3	−15
11	30	54	0.9	3.9	−19
12	30	56	0.8	3.3	−27
13	30	51	0.9	3.8	−22
14	30	51	0.4	4.1	−11
15	30	52	0.3	4.0	−32
16	29	52	0.3	3.9	−42

* Data for each run represented are an average of four flights/run. ** Wind Direction is relative to the sampling line.

2.2. Statistical Analysis

Two methods of accounting for weather were compared. The first one averaged each of the five weather variables at the start of each pass within a single run. This method is indicated hereafter as Weather Method 1. The second method used a ten-minute average of each weather variable within which all four passes were made. This method is indicated hereafter at Weather Method 2.

PROC Mixed in SAS 9.4 provided a model-based solution for the chosen effects and tested for significance of input variables on modeled downwind residue concentration. The five weather variables and sampling distance were input to the model and the response variable was residue concentration at each sampling location. Effective distance to a sampler was adjusted for each run to account for changes in wind direction. A natural-log (ln) transform was applied to the dependent variable 'concentration' and downwind sampling distance. Using the ln transform gave the best model fit to actual residue data at each sampling location, as verified with SAS runs of PROC GLM. Covariance parameter estimates and non-significant variables showing little effect were progressively removed from the model as their presence influenced results. This provided robust models by which sensitivity analysis using two weather characterization methods could be achieved.

3. Results

Distance to the downwind sampler had a significant effect on the level of spray concentration at significance level of $p = 0.01$ for both weather methods. The solar radiation intensity was also significant at $p = 0.01$. Interaction of wind speed and wind direction was significant for Weather Method 1, as was the interaction of downwind distance and wind direction at significance level of $p = 0.05$. Weather Method 2 showed no such interaction effect of wind speed and wind direction (significance level of $p = 0.8677$), but also showed an interaction effect of downwind distance and wind direction at significance level of $p = 0.05$.

Sensitivity analysis was performed on two significant variables—wind speed and solar radiation—to see how they behaved using the two weather methods. Minimum and maximum values for these variables over duration of the experiment were used.

The effect of solar radiation on modeled downwind concentration was pronounced and comparable between the two methods, with Weather Method 2 showing a slightly greater effect. Both methods indicated a greater effect of solar radiation differences with downwind distance, as this was probably due to the longer residence time for a droplet in the air, further promoting evaporation.

Use of Weather Method 1 shows an unusual result for wind speed (Table 2). As wind speed increased, modeled sample concentration decreased markedly. Contrary to this, in-field residue data across all runs showed a weakly increasing trend in downwind concentration with increasing wind speed (Figure 2). This is trend is consistent with what had been observed in a propeller wash drift study [43]. Weather method 2 shows no such discrepancy, with an expected increase in sampled concentration with wind speed (Table 3). Values for modeled ln(concentration) were also in the range of values observed in field data across all runs. Mean ln(concentration) across all field runs was 0.331 ng/L, and this falls squarely within modeled ranges illustrated in columns 2 and 3 of Table 2. Values of modeled ln(concentration) using Weather Method 1 (Table 2) were consistently higher than this field-derived result.

Table 2. Modeled ln (concentration) at minimum and maximum values of variables wind speed and solar radiation for Weather Method 1. Value of the other variable was held to minimum value to illustrate the effect of changes in the variable of interest.

Downwind Distance (m)	Modeled ln(conc) at Minimum Value of Solar (ng/L)	Modeled ln(conc) at Maximum Value of Solar (ng/L)	% Difference between Min and Max Values	Modeled ln(conc) at Minimum Wind Speed (ng/L)	Modeled ln(conc) at Maximum Wind Speed (ng/L)	% Difference between Min and Max Values
100	1.049	0.654	−37.7	1.049	0.351	−66.5
135	0.999	0.605	−39.4	0.999	0.301	−69.9
200	0.926	0.533	−42.4	0.926	0.228	−75.4
320	0.833	0.439	−47.3	0.833	0.134	−83.9

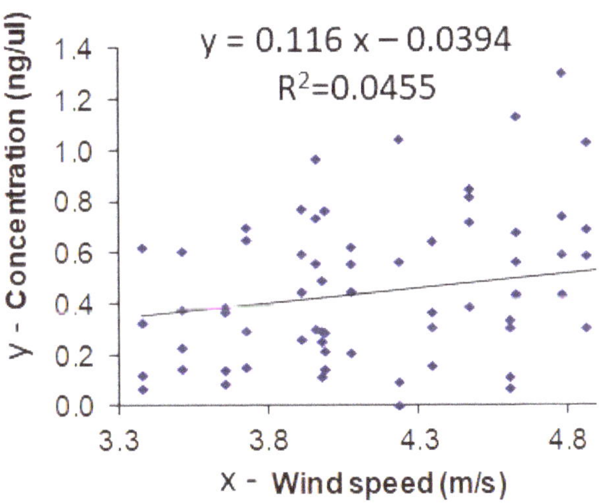

Figure 2. Sampled concentration of RbCl as a function of wind speed.

It is clear from statistical model solutions that Weather Method 2 (using weather averages over the first ten minutes of a run) yielded more reasonable results than Weather Method 1 with respect to the effect of wind speed. This points to the importance of careful characterization of weather when performing drift studies. Proper characterization may be more difficult when performing a drift study such as that described herein, that relies on several passes per run (four in this case). Although it is not immediately clear why Weather

Method 2 performed better, it could be because Weather Method 1 used an average of four discrete weather events. It is very possible for wind speed and direction to change abruptly, and ten-minute weather averages (Weather Method 2) prevented discrete events from markedly influencing results. Ten minutes was the approximate time for all four passes to be completed for each run.

Table 3. Modeled ln (concentration) at minimum and maximum values of variables wind speed and solar radiation for Weather Method 2. Value of the other variable was held to minimum value to illustrate the effect of changes in the variable of interest.

Downwind Distance (m)	Modeled ln(conc) at Minimum Value of Solar (ng/L)	Modeled ln(conc) at Maximum Value of Solar (ng/L)	% Difference between Min and Max Values	Modeled ln(conc) at Minimum Wind Speed (ng/L)	Modeled ln(conc) at Maximum Wind Speed (ng/L)	% Difference between Min and Max Values
100	0.688	0.398	−42.2	0.688	0.723	5.1
135	0.643	0.352	−45.3	0.643	0.677	5.3
200	0.575	0.285	−50.4	0.575	0.610	6.1
320	0.489	0.198	−59.5	0.489	0.524	7.2

Statistical significance was not greatly influenced by weather characterization method, although the interaction of wind direction and wind speed was only significant at the 0.05 level for Weather Method 1. Wind speed by itself was not significant for either method, but the wind speed/wind direction interaction prompted further investigation. Wind speed results from Figure 2 (and modeled for the second weather case, (Table 3) indeed indicate only a slight effect on sampled concentration of residue.

4. Conclusions

Based on results, the following conclusions can be made regarding treatment effects on downwind sampled concentration of RbCl:

- Use of Weather Method 2, that implemented weather averages over the length of a run, resulted in model-based estimates for RbCl concentration that compared well with field data averages.
- Model results using Weather Method 2 showed only slight sensitivity to changes in wind speed, and this was more pronounced further downwind. The degree of this effect was also consistent with field results.
- The effects of changes in solar radiation were comparable between the two methods.
- Use of Weather Method 1 resulted in unexpected (inverse) relationship of RbCl concentration with respect to increases in wind speed via sensitivity analysis and would not be recommended for use in a statistical model of downwind spray drift.
- Statistically modeled downwind sampled concentration is highly sensitive to characterization of wind, and care should be taken in its measurement for realistic drift assessment.
- The experiment procedure and statistical analytical methods can be adopted for UAV spray application of plant protection materials to characterize spray deposition and off-target drift.

5. Further Discussion

The procedure and statistical methods used in this study indicating comparative drift results hold for application of plant protection materials using general agricultural aviation aircraft. Precision agriculture is now advancing to the stage of smart agriculture, which can include targeted spraying [1]. In addition to conventional manned agricultural aircraft, Unmanned Aerial Vehicles (UAVs) have been developed and adapted for targeted spray application of plant protection materials. In the future, UAVs could possibly be developed to match the tank capacities of conventional manned agricultural aircraft. UAV-based systems typically have smaller tank capacities between 10 to 60 L, and this is a more popular setting for agricultural plant protection. UAV robot swarms could be used to cover large areas.

UAV spray application began in the 1980s when crop dusting was conducted using a spray system developed on a Yamaha UAV in Japan [44]. In the US, an experiment was conducted to determine the effectiveness of using an off-the-shelf Yamaha RMAX UAV for dispersing pesticides to reduce human disease due to insects [45]. Huang et al. developed a multirotor UAV spraying system for low-volume aerial spray application [46]. Now, the technology has been developed and applied worldwide [47]. Similar to conventional aerial spray application, UAV-based spray application faces the challenge of off-target drift [48,49]. Nozzle atomization characteristics, spray deposition, and off-target drift are topics of the research to maximize spray deposition within the crop canopy and minimize off-target losses in the field and environment. For this purpose, UAV-based spraying systems need to adopt parameters from the system investigation and evaluation. The parameters include:

- Pesticide parameters—the values of the parameters depend on spray volumes, total spray amount of the system to determine spray flight operation parameter specification, coefficient of variation to measure the uniformity of the total system spray application, density of droplets, and effective swath width. Spray volumes can be high volume, medium volume, low volume, very low volume, and ultra-low volume. Among them very low (5~50 L/hm^2) and ultra-low (<5 L/hm^2) volumes are for UAV spray application, which can be referred to select appropriate spray nozzles.
- Flight parameters, including flight speed, flight altitude, and flight position accuracy.
- Weather parameters including air temperature, relative humidity, precipitation, wind speed, wind direction, and atmospheric stability.

Off-target drift of UAV spray application can be categorized as vapor drift and airborne drift. Vapor drift mainly occurs in the spray of herbicide such as dicamba (3, 6-dichloro-2-methoxybenzoic acid), able to control several annual and perennial broadleaf weeds. Drift of dicamba is mainly vaporized and it has deleterious effects on broadleaf plants, especially non-dicamba-tolerant (NDT) row crops. Zhang et al. assessed field NDT soybean damage from dicamba by hyperspectral remote sensing using machine learning algorithms [50].

Airborne drift occurs in the spray of herbicides such as glyphosate (primarily glyphosate [N-(phosphonomethyl) glycine) either alone or with other herbicides to manage a broad spectrum of weeds. To determine the natural downwind drift, a tracer (such as RbCl as described in this study) can be mixed in the spray solution and deposition can be determined using mylar (or other types) of spray samplers placed downwind from the spray event. Water sensitive papers can also be placed downwind or clipped to plant leaves (depending on research objectives). Droplet size and density can be calculated to determine distribution and relative quantity spray deposition. Use of water sensitive papers does not require a tracer. Droplet density or tracer concentration can be related the downwind distance to assess spray drift potential [28,31].

In addition to natural spray drift, off-target spray movement can also be exacerbated by spray application under stable atmospheric conditions. The main objective of the aircraft pilot is simply to avoid spraying under stable atmospheric conditions. Guidelines for avoidance of these conditions have been developed through research studies [6,8].

The study presented herein was an integrative investigation of parameter effects. For spray application involving single- or multi-rotor UAVs, downwash air turbulence can have a great effect on spray characteristics. Also, altitude, speed and position of a UAV are susceptible to wind disturbance. This is a much different issue than that caused by propwash effects from general aviation aircraft [42]. So, how is turbulence and wind disturbance characterized for the spray deposition and off-target drift? Results of research studies can be integrated to provide useful UAV parameter specifications for application of plant protection materials to improve efficacy and decrease off-target drift.

Author Contributions: Conceptualization, S.J.T. and Y.H.; methodology, S.J.T.; validation, S.J.T.; formal analysis, S.J.T. and Y.H.; investigation, S.J.T. and Y.H.; resources, Y.H.; data curation, S.J.T.; writing—original draft preparation, S.J.T.; writing—Y.H. All authors have read and agreed to the published version of the manuscript.

Funding: This research received no external funding.

Data Availability Statement: Data availability depends on the request with the rules and regulation of the agency with the disclaimer as "Mention of a trade name, proprietary product, or specific equipment does not constitute a guarantee or warranty by the U.S. Department of Agriculture and does not imply approval of the product to the exclusion of others that may be available".

Conflicts of Interest: The authors declare no conflict of interest.

References

1. Huang, Y.; Zhang, Q. *Agricultural Cybernetics*; Springer Nature: Berlin, Germany, 2021.
2. Ganesh, A.S. A top-down approach. In *The Hindu*; Retrieved 2022-03-11; ISSN International Centre: Paris, France, 2014; ISSN 0971-751X.
3. Johnson, M.A. *McCook Field 1917–1927*; Landfall Press: Dayton, OH, USA, 2002; pp. 190–191.
4. Huang, Y. Agricultural aviation perspective on precision agriculture in the Mississippi Delta. *Smart Agric.* **2019**, *1*, 12–30.
5. Thomson, S.J.; Huang, Y.; Hanks, J.E.; Martin, D.E. Improving Flow Response of a Variable-rate Aerial Application System by Interactive Refinement. *Comput. Electron. Agric.* **2010**, *73*, 99–104. [CrossRef]
6. Thomson, S.J.; Huang, Y.; Fritz, B.K. Temporal indications of atmospheric stability affecting off-target spray drift in the midsouth U S. In *Proceedings of the American Society of Agricultural and Biological Engineers International (ASABE)*; ASABE: St. Joseph, MI, USA, 2011; 8p.
7. Thomson, S.J.; Huang, Y.; Fritz, B.K. Atmospheric stability intervals influencing the potential for off-target movement of spray in aerial application. *Int. J. Agric. Sci. Technol.* **2017**, *5*, 1–17. [CrossRef]
8. Huang, Y.; Thomson, S.J. Atmospheric stability determination at different time intervals for determination of aerial application timing. *J. Biosyst. Eng.* **2016**, *41*, 337–341. [CrossRef]
9. Huang, Y.; Fisher, D.K.; Silva, M.; Thomson, S.J. A real-time web tool for aerial application to avoid off-target movement of spray induced by stable atmospheric conditions in the Mississippi Delta. *Appl. Eng. Agric.* **2019**, *35*, 31–38. [CrossRef]
10. Huang, Y.; Fisher, D.K. An open-sourced web application for aerial applicators to avoid spray drift caused by temperature inversion. *Appl. Eng. Agric.* **2021**, *37*, 77–87. [CrossRef]
11. EPA (Environmental Protection Agency). Pesticide Registration (PR) Notice 2001-x Draft: Spray and Dust Drift Label Statements for Pesticide Products. Available online: https://www.epa.gov/pesticide-registration/pesticide-registration-notices-year (accessed on 20 December 2022).
12. Cooper, C.D.; Alley, F.C. *Air Pollution Control: A Design Approach*, 2nd ed.; Waveland Press, Inc.: Prospect Heights, IL, USA, 1994.
13. Yates, W.E.; Akesson, N.B.; Coutts, H.H. Evaluation of drift residues from aerial applications. *Trans. ASAE* **1966**, *9*, 389–393. [CrossRef]
14. Yates, W.E.; Akesson, N.B.; Coutts, H.H. Drift hazards related to ultra-low-volume and diluted sprays applied by agricultural aircraft. *Trans. ASAE* **1967**, *10*, 628–632. [CrossRef]
15. Pasquill, F.; Smith, F.B. *Atmospheric Diffusion: A Study of the Dispersion of Windborne Material from Industrial and Other Sources*, 3rd ed.; Ellis Horwood Limited: Chichester, UK, 1983.
16. Bird, S.L. A compilation of aerial spray drift field study data for low-flight agricultural application of pesticides. In *Chelsea, Mich.: Environmental Fate of Agrochemicals: A Modern Perspective*; Leng, M.L., Loevey, E.M., Zubkoff, P.L., Eds.; Lewis Publishers: Chelsea, MI, USA, 1995.
17. Ganzlemeier, H.; Rautmann, D.; Spangenberg, R.; Streloke, M.; Herrmann, M.; Wenzelburger, H.; Walte, H. *Studies on the Spray Drift of Plant Protection Products*; Blackwell Wissenschafts-Verlag GmbH: Berlin, Germany, 1995.
18. Arvidsson, T. *Spray Drift as Influenced by Meteorological and Technical Factors. A Methodological Study*; Acta Universitatis Agriculturae Sueciae, Agraria 71; Swedish University of Agricultural Sciences: Uppsala, Sweden, 1997.
19. Smith, D.B.; Bode, L.E.; Gerard, P.D. Predicting ground boom spray drift. *Trans. ASAE* **2000**, *43*, 547–553. [CrossRef]
20. Hewitt, A.J.; Johnson, D.; Fish, J.D.; Hermansky, C.G.; Valcore, D.L. The development of the spray drift task force database on pesticide movements for aerial agricultural spray applications. *Environ. Toxicol. Chem.* **2001**, *21*, 648–658. [CrossRef]
21. Maber, J.; Dewar, P.; Praat, J.P.; Hewitt, A.J. Real time spray drift prediction. *Acta Hortic.* **2001**, *566*, 493–498. [CrossRef]
22. Fritz, B.K.; Hoffmann, W.C.; Lan, Y.; Thomson, S.J.; Huang, Y. Low-level atmospheric temperature inversions: Characteristics and impacts on aerial applications. *Int. Agric. Eng. J.* **2008**, *X*, PM-08001.
23. Turner, D.B. *Workbook of Atmospheric Dispersion Estimates: An Introduction to Dispersion Modeling*, 2nd ed.; CRC Press: Boca Raton, FL, USA, 1994.
24. Teske, M.E.; Bird, S.L.; Esterly, D.M.; Curbishley, T.B.; Ray, S.L.; Perry, S.G. AgDRIFT®: A model for estimating near-field spray drift from aerial applications. *Environ. Toxicol. Chem.* **2002**, *21*, 659–671. [CrossRef] [PubMed]

25. Teske, M.E.; Thistle, H.W.; Ice, G.G. Technical advances in modeling aerially applied sprays. *Trans. ASAE* **2003**, *46*, 985–996. [CrossRef]
26. EPA (Environmental Protection Agency). *Meteorological Monitoring Guidance for Regulatory Modeling Applications*; EPA-454/R-99-005; USEPA Office of Air Quality Planning and Standards: Research Triangle Park, NC, USA, 2000.
27. *ASAE S561.1*; Procedures for Measuring Drift Deposits from Ground, Orchard, and Aerial Sprayers. ASABE: St. Joseph, MI, USA, 2004.
28. Huang, Y.; Thomson, S.J. Characterization of in-swath spray deposition for CP-11TT flat-fan nozzles used in low volume aerial application of crop production and protection materials. *Trans. ASABE* **2011**, *54*, 1973–1979. [CrossRef]
29. Huang, Y.; Zhan, W.; Fritz, B.K.; Thomson, S.J. Optimizing selection of controllable variables to minimize downwind drift from aerially applied sprays. *Appl. Eng. Agric.* **2012**, *28*, 307–314. [CrossRef]
30. Felsot, A.S.; Unsworth, J.B.; Linders, J.B.H.J.; Roberts, G.; Rautman, D.; Harris, C.; Carazo, E. Agrochemical spray drift; assessment and mitigation—A review. *J. Environ. Sci. Health Part B Pestic. Food Contam. Agric. Wastes* **2011**, *46*, 1–23. [CrossRef]
31. Huang, Y.; Thomson, S.J. Characterization of spray deposition and drift from a low drift nozzle for aerial application at different application altitudes. *Int. J. Agric. Biol. Eng.* **2011**, *4*, 1–6.
32. Huang, Y.; Thomson, S.J.; Ortiz, B.V.; Reddy, K.N.; Ding, W.; Zablotowicz, R.M.; Bright, J.R. Airborne remote sensing assessment of the damage to cotton caused by spray drift from aerially applied glyphosate through spray deposition measurements. *Biosyst. Eng.* **2010**, *107*, 212–220. [CrossRef]
33. Hoffmann, W.C.; Fritz, B.K.; Thornburg, J.W.; Bagley, W.E.; Birchfield, N.B.; Ellenberger, J. Spray drift reduction evaluations of spray nozzles using a standardized testing protocol. *J. ASTM Int.* **2010**, *7*, JAI102820. [CrossRef]
34. Ferguson, C.; O'Donnell, C.C.; Chauhan, B.S.; Adkins, S.W.; Kruger, G.R.; Wang, R.; Ferreira, P.H.U.; Hewitt, A.J. Determining the uniformity and consistency of droplet size across spray drift reducing nozzles in a wind tunnel. *Crop Prot.* **2015**, *76*, 1–6. [CrossRef]
35. Gil, E.; Gallarta, M.; Balsari, P.; Marucco, P.; Almajanoc, M.P.; Llop, J. Influence of wind velocity and wind direction on measurements of spray drift potential of boom sprayers using drift test bench. *Agric. For. Meteorol.* **2015**, *202*, 94–101. [CrossRef]
36. Fritz, B.K. Meteorological effects on deposition and drift of aerially applied sprays. *Trans. ASABE* **2006**, *49*, 1295–1301. [CrossRef]
37. Arvidsson, T.; Bergstrom, L.; Kreuger, J. Spray drift as influenced by meteorological and technical factors. *Pest Manag. Sci.* **2009**, *67*, 586–598. [CrossRef] [PubMed]
38. Bird, S.L.; Perry, S.G.; Ray, S.L.; Teske, M.E. Evaluation of the AgDISP aerial spray algorithms in the AgDRIFT model. *Environ. Toxicol. Chem.* **2002**, *21*, 672–681. [CrossRef]
39. Woods, N.; Craig, I.P.; Dorr, G.; Young, B. Spray drift of pesticides arising from aerial application in cotton. *J. Environ. Qual.* **2001**, *30*, 697–701. [CrossRef] [PubMed]
40. Thistle, H.W.; Teske, M.E.; Richardson, B.; Strand, T.M. Model physics and collection efficiency in estimates of pesticide spray drift model performance. *Trans. ASABE* **2020**, *63*, 1939–1945. [CrossRef]
41. Thomson, S.J.; Womac, A.; Mulrooney, J. Reducing pesticide drift by considering propeller rotation effects from aerial application near buffer zones. *Sustain. Agric. Res.* **2013**, *2*, 41–51. [CrossRef]
42. CPNozzles. Calculation Tools. Available online: https://cpnozzles.com/calculation-tools/ (accessed on 20 December 2022).
43. Thomson, S.J.; Young, L.D.; Bright, J.R.; Foster, P.N.; Poythress, D.D. *Effects of Spray Release Height and Nozzle/Atomizer Configuration on Penetration of Spray in a Soybean Canopy—Preliminary Results*; Technical Paper AA07-008; National Agricultural Aviation Association (NAAA): Washington, DC, USA, 2007.
44. Sato, A. *The RMAX Helicopter UAV*; Public Report; Aeronautic Operations; Yamaha Motor Co., Ltd.: Shizuoka, Japan, 2003.
45. Miller, J.W. *Report on the Development and Operation of an UAV for an Experiment on Unmanned Application of Pesticides*; AFRL, USAF: Youngstown, OH, USA, 2005.
46. Huang, Y.; Hoffmann, W.C.; Lan, Y.; Wu, W.; Fritz, B.K. Development of a spray system for an unmanned aerial vehicle platform. *Appl. Eng. Agric.* **2009**, *25*, 803–809. [CrossRef]
47. Chen, P.; Douzals, J.P.; Lan, Y.; Cotteux, E.; Delpuech, X.; Pouxviel, G.; Zhan, Y. Characteristics of unmanned aerial spraying systems and related spray drift: A review. *Front. Plant Sci.* **2022**, *13*, 2726. [CrossRef]
48. Wang, C.; Herbst, A.; Zeng, A.; Wongsuk, S.; Qiao, B.; Qi, P.; Bonds, J.; Overbeck, V.; Yang, Y.; Gao, W.; et al. Assessment of spray deposition, drift and mass balance from unmanned aerial vehicle sprayer using an artificial vineyard. *Sci. Total Environ.* **2021**, *777*, 146181. [CrossRef] [PubMed]
49. Biglia, A.; Grella, M.; Bloise, N.; Comba, L.; Mozzanini, E.; Sopegno, A.; Pittarello, M.; Dicembrini, E.; Eloi Alcatrão, L.; Guglieri, G.; et al. UAV-spray application in vineyards: Flight modes and spray system adjustment effects on canopy deposit, coverage, and off-target losses. *Sci. Total Environ.* **2022**, *845*, 157292. [CrossRef] [PubMed]
50. Zhang, J.; Huang, Y.; Reddy, K.N.; Wang, B. Assessing crop damage from dicamba on non-dicamba-tolerant soybean by hyperspectral imaging through machine learning. *Pest Manag. Sci.* **2019**, *75*, 3260–3272. [CrossRef] [PubMed]

agronomy

MDPI

Article

Study on Spray Deposition and Drift Characteristics of UAV Agricultural Sprayer for Application of Insecticide in Redgram Crop (*Cajanus cajan L. Millsp.*)

Yallappa Dengeru [1], Kavitha Ramasamy [1,*], Surendrakumar Allimuthu [1], Suthakar Balakrishnan [1], Ayyasamy Paramasivam Mohan Kumar [1], Balaji Kannan [2] and Kalarani Muthusami Karuppasami [3]

1 Department of Farm Machinery and Power Engineering, Agricultural Engineering College and Research Institute, Tamil Nadu Agricultural University, Coimbatore 641003, India
2 Department of Physical Science and Information Technology, Agricultural Engineering College and Research Institute, Tamil Nadu Agricultural University, Coimbatore 641003, India
3 Directorate of Crop Management, Tamil Nadu Agricultural University, Coimbatore 641003, India
* Correspondence: kavitha@tnau.ac.in

Citation: Dengeru, Y.; Ramasamy, K.; Allimuthu, S.; Balakrishnan, S.; Kumar, A.P.M.; Kannan, B.; Karuppasami, K.M. Study on Spray Deposition and Drift Characteristics of UAV Agricultural Sprayer for Application of Insecticide in Redgram Crop (*Cajanus cajan L. Millsp.*). *Agronomy* **2022**, *12*, 3196. https://doi.org/10.3390/agronomy12123196

Academic Editors: Xiongkui He, Fuzeng Yang and Baijing Qiu

Received: 19 September 2022
Accepted: 19 October 2022
Published: 16 December 2022

Abstract: Insecticide applications are typically being carried out with traditional manual spraying equipment in redgram, which leads to inadequate control of insects due to higher crop height. The modern deployment of tractor-drawn spray machines causes serious damage to the crop. In this connection, unmanned aerial vehicle (UAV) spray technology has great potential for precise insecticide application in redgram crops. One of the important machine parameters influencing droplet deposition and drift characteristics in UAV sprayers is downwash airflow generated by a multi-rotor propeller. A field experiment was carried out at the redgram research field (N11.01, E76.92), Tamil Nadu Agricultural University, Coimbatore, Tamil Nadu, during 2021–2022 to study the spray drift and deposition characteristics of an autonomous UAV sprayer. The Imidacloprid (a.i. 17.8SL) insecticide mixed with water in a ratio of 1 mL per liter was sprayed with a UAV sprayer. Water-sensitive paper samples were kept at upper, middle, and bottom positions on the leaves, and data were analyzed for the spray droplet size, deposition rate, droplet density, and area coverage both in target and non-target areas using Spray Deposit Scanner software. UAV spray droplet deposition rate (2.93 ± 0.17, 2.01 ± 0.08, and 2.21 ± 0.162.38 μL cm^{-2}), droplet density (47 ± 4.04, 53 ± 3.61, and 52 ± 8.74 droplets cm^{-2}), and area coverage (15.72 ± 0.39, 16.60 ± 0.71, and 14.99 ± 0.39%) were highest in the upper layer as compared to the middle layer (droplet deposition rate: 1.21 ± 0.08, 1.07 ± 0.03, and 0.77 ± 0.02 μL cm^{-2}; droplet density: 42 ± 2.52, 43 ± 8.50, and 38 ± 2.52 droplets cm^{-2}; area coverage: 10.95 ± 0.81, 11.22 ± 0.56, and 8.57 ± 0.44%) and bottom layer (droplet deposition rate: 0.41 ± 0.06, 0.35 ± 0.03, and 0.33 ± 0.03 μL cm^{-2}; droplet density: 22 ± 4.36, 17 ± 3.51, and 19 ± 4.51 droplets cm^{-2}; area coverage: 2.78 ± 0.29, 2.95 ± 0.45, and 2.46 ± 0.20%, respectively). In the spray drift test, there was a higher droplet deposition rate (1.63 ± 0.09, 1.93 ± 0.05, and 1.82 ± 0.06 μL cm^{-2}), area coverage (14.40 ± 0.07, 17.54 ± 0.36, and 16.42 ± 0.30%), and droplet density (46 ± 3.61, 54 ± 2.08, and 45 ± 3.21 No's cm^{-2}) in the target area as compared to the non-target area (droplet deposition rate: 0.88 ± 0.02, 0.46 ± 0.03, 0.22 ± 0.05, and 0.00 μL cm^{-2}; droplet density: 23 ± 1.53, 11 ± 2.08, 6 ± 1.53, and 0.00 droplets cm^{-2}; area coverage: 7.58 ± 0.34, 4.41 ± 0.19, 2.16 ± 0.05, and 0.00%, respectively), which may have been due to the downwash airflow produced by the multi-rotor propeller of the UAV sprayer. Finally, the UAV-based spraying technology results showed that the downwash air produced by the six-rotor propeller improved the penetrability of insecticide to crop leaves and led to a higher droplet deposition rate, droplet density, area coverage, and droplet penetrability on the upper layer, middle layer, and bottom layer of the plants.

Keywords: droplet size; deposition rate; downwash airflow; flat fan nozzle; redgram leaf; UAV sprayer

1. Introduction

The pigeon pea or redgram *(Cajanus cajan L. Millsp.)* is the second most significant pulse crop and a rich source of protein. It is mainly intercropped with pulses, cereals, millets, and oilseeds. Conventional methods of insecticide and pesticide spray application lead to excessive application of chemicals, lower spray uniformity, deposition, and coverage, resulting in a higher cost of pesticide as well as environmental pollution, as well as increased drudgery and reduced area coverage [1]. Application of insecticide is being carried out with traditional spraying equipment on the redgram crop, which leads to inadequate control of pests due to higher crop height. On the other hand, tractor-drawn spray machines cause serious damage to the crop. Unmanned aerial vehicle (UAV) spray technology has a great potential for precise insecticide application in red gram crops.

UAV spray technology is one that can be controlled by a pilot on a ground station autonomously using pre-programmed flight plans [2–4]. The UAV is attached with spray tanks, which can store the insecticide and pesticide and spray them over the crops. These fly at the proper height, helping the pesticide penetrate perfectly into the crop [5–7]. The most essential benefit of using a multi-rotor UAV for chemical spraying is that, due to its unique rotor structure and principle of motion, it generates a powerful downwash airflow during flight operation, changing the crop disturbance and improving liquid penetration [8,9]. It has been reported that liquid sprayed using UAVs has quite a good deposition impact on the bottom part of crops [10–13].

Many investigators have carried out research on the spray effects of UAV spray technology. For insecticides sprayed by UAV sprayers, droplet size is one of the most important factors affecting droplet drift and deposition rate [14,15]. Wind speed is one of the meteorological conditions that might cause spray drift [16,17]. The UAV machine operation parameters, viz., flight height, travel speed, payload, and configuration of the UAV model, have a great impact on the distribution and penetration of droplets [18]. The height and forward speed of UAV flight parameters have a great impact on the spray droplet distributions [19,20]. The downwash airflow velocity produced by the rotors can create a strong velocity distribution for plants during spraying, helping spray droplets to atomize much further with enhanced deposition onto the crop surface. As a result, spray droplet velocity has positive effects on spray swath, deposition, and drift [21–23]. Shengde et al. [24] used a wind velocity sensor to evaluate the effect of rotor wind field on the uniformity and penetration of droplet deposition in three spatial directions: X, Y, and Z. Spray droplet drift is decreased, and the impact of droplet deposition rate is increased when chemicals are applied with a UAV sprayer [25,26]. Spray drift from unmanned aerial vehicles (UAVs) was studied at various pressures (2, 3, 4, and 5 bar). Liu et al. [27] studied spray drift from unmanned aerial vehicles (UAVs) at various pressures (2, 3, 4, and 5 bar). The drift typically dropped, increasing the distance downwind from the UAV test platform. Berner and Chojnacki [28] conducted the tests on the relationship between the speed of the UAV rotor and spray deposition. Fritz et al. [29] tested the effects of spray rate and droplet size on droplet deposition on wheat and found that lower spray rates with larger droplet sizes resulted in greater deposits. It has been reported that liquid sprayed using UAV has quite a good deposition impact on the bottom part of crops [10,30]. When the flying height increases, the vertical velocity of the UAV downwash flow near the crop canopy will decrease [31]. The rotor wind field, flight environment (temperature, humidity wind direction, and wind power), equipment performance, and pesticide characteristics (viscosity, density) also have an effect on the spraying effect [32–34]. Xinyu et al. [35] evaluated UAV spray and deposition in a rice field using Z-3 UAV operating parameters that could control 90% of drift in 8 m. Yang et al. [36] studied the influence of UAV rotor downwash airflow spray width using the XV–2 model. The results showed that UAV flight height mainly influenced the spray width, and the effective spray width was 10.0 m at a flight height of 6.0 m and had better effect on droplet deposition at 2.0 m height of spray. When droplets below 200 m are subjected to a 5 km/h crosswind, they can drastically deviate from their intended path [37]. Choi et al. [38] conducted tests using a six-rotor UAV with three heights and two speeds of operational parameters. Lan et al. [39] conducted research on the impact of the UAV downwash on droplet deposition distribution using a DJI T16 UAV, and it was found that with an increase in flight height, the change of the downwash wind field led to a gradual decrease in droplet deposition in the effective spray area and were deposited more uniformly [24,27]. For decades, empirical and mechanistic models have been developed to predict pesticide deposition and drift from aerial pesticide application [40–42]. In insect control operations, the quality of insecticide application is important. The proper dosage should be applied uniformly, the toxicant should reach the target, and droplet size and density on the target should be correct [43]. The aim of this investigation was to explore the spray droplet deposition rate, droplet size, droplet density, area coverage, uniformity distribution, and drift characteristics in the redgram crop sprayed by a UAV sprayer. It also provides technical reference and guidance for proper and safe aerial spraying in agricultural production.

2. Materials and Methods

2.1. Equipment

The UAV used in the present investigation was an E610P six-rotor electric (M/s. EFT Electronic Technology Co., Ltd., Hefei City, China), as shown in Figure 1. The spraying system of the UAV sprayer mainly consists of BLDC motors, lithium polymer (LiPo) batteries, a flight controller, an RC receiver, a GNSS RTK GPS unit, a pesticide tank, a pump, nozzles, and a supporting frame. The UAV sprayer has two LiPo batteries of 6 cells each, with a capacity of 16,000 mAh to supply the necessary current required for the propulsion system. A 24 V BLDC motor coupled with a pump was used to pressurize spray liquid and then to atomize it into fine spray droplets. This UAV sprayer has four numbers of 2020A-132 series flat fan nozzles (M/s. Ningbo Licheng Agricultural Spray Technology Co., Ltd., Zhejiang, China), which are mounted and screwed below the BLDC motor base plate. This UAV model has the functions of GPS route planning and breakpoint return, which can complete aerial spraying operations autonomously. The main performance indicators are shown in Table 1.

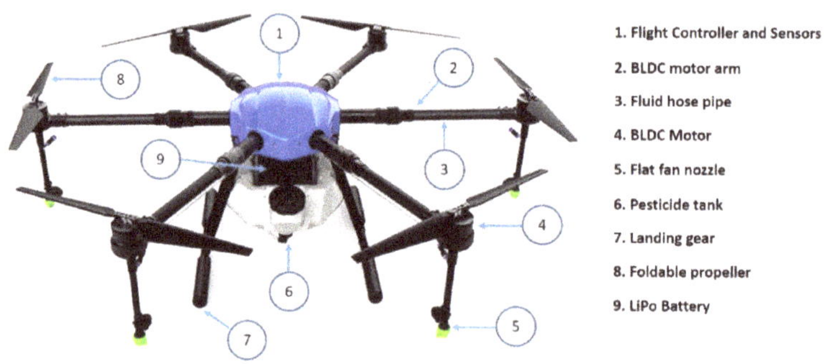

Figure 1. Six-rotor electric autonomous UAV sprayer.

2.2. Field Plot

The experiment was carried out at the redgram research field (N11.01, E76.92), Department of Pulses, Tamil Nadu Agricultural University, Coimbatore, Tamil Nadu state, India. The details of crop parameters are mentioned below in Table 2. The redgram crop parameters were average row to row, plant to plant spacing, and average crop height measured during the UAV spraying operation. Figure 2 depicts the application of insecticide in the redgram crop using a UAV agricultural sprayer.

Table 1. Specifications of UAV agricultural sprayer.

Main Parameter	Norms and Numerical Value
Type	Hexacopter
Item model	E610P
Unfold fuselage size, (L × W × H), mm	2000 × 1800 × 670
Folding size, (L × W × H), mm	950 × 850 × 670
Power source	12S 160,000 mAh LiPo Battery
Payload capacity, L	10
Self-weight, kg	6.9
Take-off weight, kg	26
Flight height, m	1–20
Forward travel speed, ms^{-1}	0–8
Type of spray nozzle	Flat fan shape (2020A-132 series)
Number of nozzles	4
Discharge rate, $l\ m^{-1}$	0–3.2
Swath width of spray, m	3–5
Liquid pressure, $kg\ cm^{-2}$	3.4
Remote controller distance, km	1.5
No-load flight time, min	25
Charging time, min	90

Table 2. Details of crop parameters.

Crop Parameters	
Crop	Redgram
Variety	Co8
Plot size (L × W), mm	70 × 40
Height of crop, m	1.9 to 2.3
Stage of crop	>110 DAS
Row to row spacing, m	0.70 to 0.74
Plant-to-plant spacing, m	0.15 to 0.21

Figure 2. UAV agricultural spray operation in redgram crop.

2.3. Spray sampling Scheme

2.3.1. Sampling of Spray Deposition

Water-sensitive papers (produced by AAMS, Maldegem, Belgium) are a kind of spray card specifically designed for visualizing deposit patterns. It has a dye-coated surface that produces visible stains from aqueous droplets, and we used a 50 × 50 mm WSP to collect the amount of spray droplets deposition on the surface of the leaves in three locations [44,45]. At each sampling point, the WSP was clamped with a double-ended clip and kept on the leaves at three different heights of the plants, viz., 40 cm, 110 cm, and 210 cm from the ground level. The arrangement and locations of WSP samples on redgram leaf are shown in Figure 3a,b. The field layout with the location of WSP spray deposition samples is shown in Figure 4.

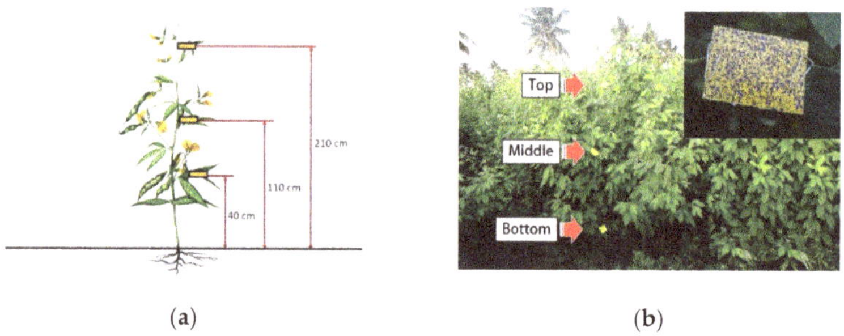

Figure 3. Layout and position of water sensitive paper for the collection of spray droplet deposition characteristics: (**a**) layout of WSP samples in the upper, middle, and bottom at 210, 110, and 40 mm height from the ground, respectively, and (**b**) the position of WSP samples on redgram leaf in the field.

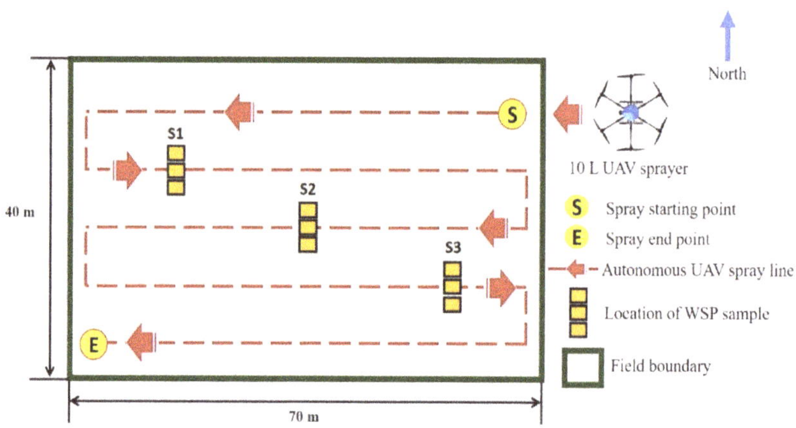

Figure 4. Schematic diagram for the location of WSP spray.

2.3.2. Sampling of Spray Drift

The UAV agricultural spray drift experiment was conducted from 09:50 a.m. to 10:30 a.m. on 8 December 2021 at the redgram research field (N11.01, E76.92), Tamil Nadu Agricultural University, Coimbatore, Tamil Nadu state, India. Weather conditions, viz., wind velocity, air temperature, and relative humidity, were recorded during the spray drift experiment in the redgram field [46]. A WSP with a size of 50 × 50 mm was placed on flat iron plate attachments to square iron rods at all collection points for the measurement of spray droplet deposition rate in the canopy. The height of WSP samples were adjusted to the head of the redgram crop canopy by flexible clamps [25], as shown in Figure 5a,b.

A transmitter (SKYDROID, T10 2.4 Hz 10CH FHSS) with a screen recorded the flight deviation, travel speed, and altitude of each flight using ground station equipment. The total length of the target sample line was close to three meters, which is greater than the effective spraying width calibrated during the laboratory test. The water-sensitive paper collection cards were mounted below the UAV autonomous spray line fixed at 0 m, and with consideration of the natural wind direction, single-line WSP collection cards were placed and spaced at −1.0, 0.0, and −2.0 m on the upside wind flow direction of the UAV path line, whereas another set of five WSPs collection cards were placed at 1.0, 2.0, 3.0, 4.0, and 5.0 m on the down side of the wind flow [12,25]. The schematic experimental layout of WSP drift sample locations is shown in Figure 6.

(a) (b)

Figure 5. Placement of WSP sample for spray drift: (**a**) layout position of WSP sample on iron stand at 210 mm distance from the ground level, and (**b**) position of the WSP sample iron stand in the redgram field crop.

2.4. Selection and Feeding of UAV Spray Operational Parameters

The nozzle 2020A-132 series flat fan nozzles (M/s. Ningbo Licheng Agricultural Spray Technology Co., Ltd., Zhejiang, China) were selected for the test [47]. When the UAV is spraying, the four nozzles behind the fuselage are turned on for spraying. The spray flow rate of four nozzles were tested using a handheld portable-type sensor-based digital nozzle tester (AAMS, Maldegem, Belgium). For measuring liquid operating pressure, a digital liquid pressure gauge was connected to the output of a water spray hose pipe, and the other end was connected to the nozzles of the lateral hose pipe. The maximum flow rate of single nozzles was measured at 0.8 l m^{-1} at 100% spray motor speed, and the total nozzle flow rate and liquid pressure, including the four nozzles, were 3.2 l m^{-1} and 3.4 kg cm^{-2}, respectively. The height of spray of 1.6 m (vertical distance between the crop canopy and the tip of the drone sprayer nozzle) was set for the complete spray operation. The height of spray and crop height layout is shown in Figure 7. The main sucking insects were leafhoppers and white flies in the redgram crop. Imidacloprid insecticide was mixed with water in a ratio of 1 mL per liter (as per the redgram crop package of practice published by the Tamil Nadu Agricultural University, Coimbatore, India). The diluted chemical was sprayed, and operational parameters during the spraying time are presented in Table 3.

Table 3. UAV sprayer operational parameters.

Operational Parameters	Norms and Numerical Values
Forward speed, ms^{-1}	3
Height of spray, m	1.6
Swatch width of spray, m	3.8
Discharge rate, l m^{-1}	3.2
Liquid pressure, kg cm^{-2}	3.4

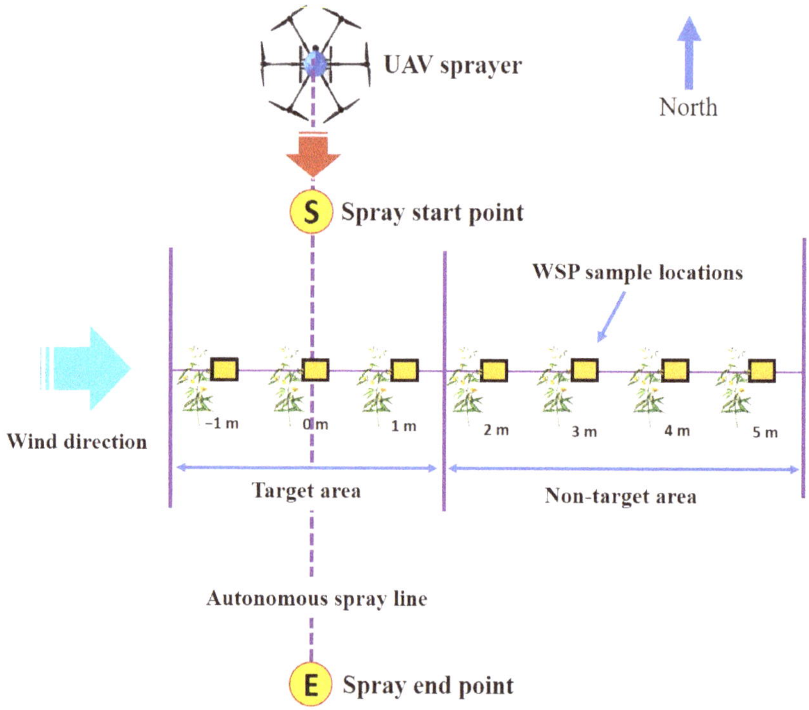

Figure 6. Schematic experimental layout of WSP drift sample locations.

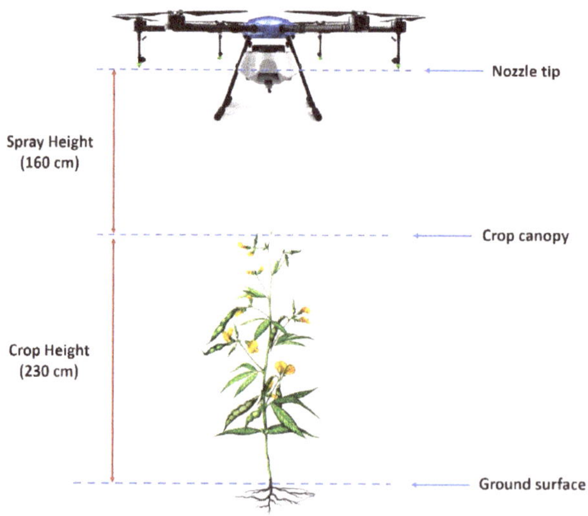

Figure 7. Schematic experimental layout of height of spray and crop height.

2.4.1. Autonomous Spraying System

The Agri Assistant mobile app (JIYI K++V2, V1.5.1) was used for UAV autonomous spraying operations in the redgram field. It also provides the live status of the UAV spray operational parameters, viz., field GPS location and satellite connection strength, field area, time required to spray a given area, battery voltage warning, spray motor speed, height of flight, and spray swath width. Initially, the GPS co-ordinates of the redgram field boundary such as latitude, longitude, and altitude were entered. The operational parameters of the UAV sprayer, viz., flight forward speed, spray height, spray swath width, and nozzle flow rate were chosen as input operation modes for the spray test. Normal operation conditions, such as a flight speed of 3 ms^{-1} and a flight height of 1.6 m above the crop canopy with a swath width of 3.8 m, were used to ensure the validity and authenticity of the test results [25]. After providing the field boundary and spray operational parameters to the Agri Assistant app, the UAV sprayer starts in auto take-off mode and begins its spraying operation by clicking the auto take-off option on the Agri Assistant mobile device at the ground station. The UAV autonomous spraying operation of the Agri Assistant app screen is shown in Figure 8. The layout of predefined and actual autonomous spray path map layout is shown in Figure 9.

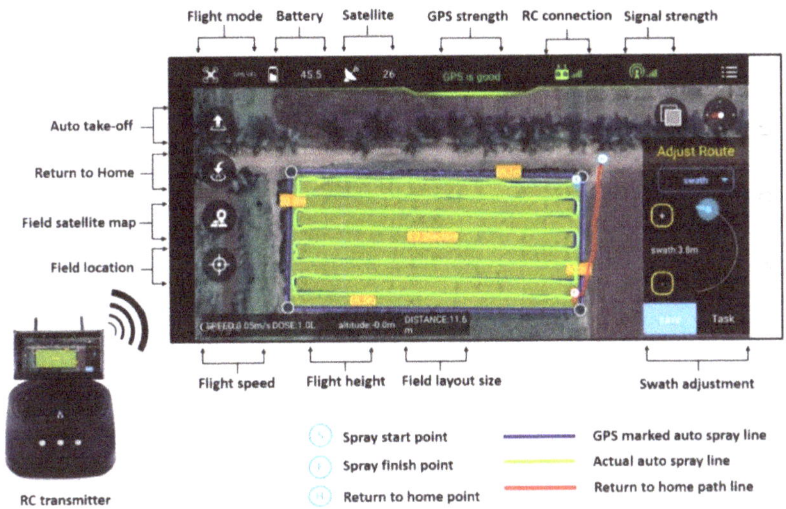

Figure 8. Screen view of the Agri Assistant mobile app for UAV autonomous spraying operations.

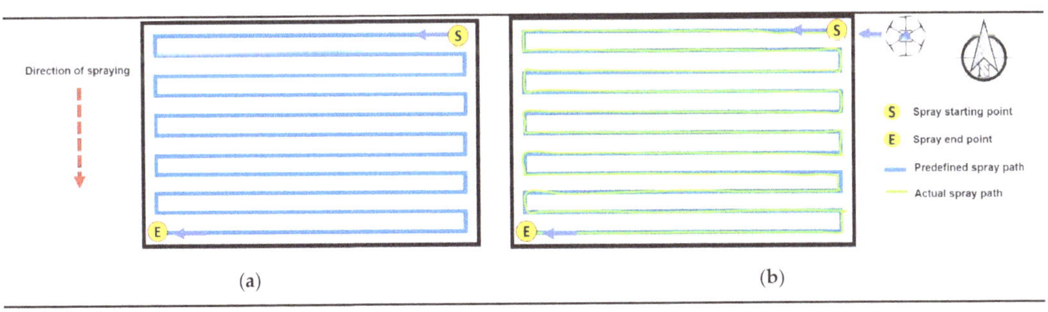

Figure 9. Field maps of UAV spray operation with autonomous mode. (**a**) Predefined autonomous spray path map. (**b**) Actual autonomous spray path map.

2.4.2. Recording of Meteorological Parameters

During UAV spraying operations, the different meteorological parameters such as wind velocity, air temperature, humidity, and rainfall meteorological parameters were recorded in order to avoid the ill effects of climate on the performance of the spraying operation [16]. A portable anemometer (LUTRON AM 4202, Vane type, rang: 0.4~30.0 ms^{-1}) was mounted on a square iron pipe at 2.0 m above the crop canopy to measure the wind velocity every 60 s [36]. Weather conditions, including wind speed, air temperature, and relative humidity, were recorded and are presented in Table 4.

Table 4. Meteorological data during the UAV spray test in the redgram field.

Date	**8 December 2021**	
Time	08:30 a.m. to 09:45 a.m.	
Location	Redgram research field (N11.01, E76.92), TNAU, Coimbatore, Tamil Nadu state, India	
Environmental parameters	Air temperature, ° C	22.4 to 26.6°
	Relative humidity, %	56.2 to 61.1
	Wind velocity, ms^{-1}	0.92 to 1.24 1.93 to 2.47 (drift experiment)
	Rainfall, mm	0

2.5. Collection of WSP and Spraying Effectiveness Analysis

After every spraying test, WSPs were collected and placed in marked envelopes one by one according to the serial number, then transferred to the laboratory for further study. The deposited amount and coverage density of the droplets at upper, middle, and bottom locations were analyzed as suggested by Zhu et al. [48].

2.5.1. Deposit Scan Software

WSPs were analyzed by using a Micro Droplet Analayzer and Macro Droplet Analyzer instruments (developed by LABLINE—DMS 101, India), and then images were processed with Deposit Scan software (developed by USDA, Wooster, OH, USA) [49]. The deposition rate (μL cm^{-2}); Dv0.1, Dv0.5, and Dv0.9 (μm); deposition density (No's cm^{-2}); and droplet penetrability (%) were studied. Mean deposition, mean deposition density, and coefficient of variation (CV) were also calculated. Dv0.1 is the droplet diameter (μm), wherein 10% of the spray volume contained in droplets was smaller than this value. Similarly, Dv0.5 and Dv0.9 are droplet diameters, wherein 50% and 90% of the spray volumes contained in the droplets were smaller than these values, respectively. The Dv0.5, also known as the volume median diameter (VMD), is the droplet size median at which the accumulation of all droplets, from small to large, equals 50% of the total volume of the droplets, which is a critical index to measure the size of the droplet [47]. Figure 10 depicts the step-by-step procedure for droplet size analysis using stereo micro and macro scope with Deposit scan software. The image processing procedure used in the study was similar to what Martin [50] reported previously. DepositScan software for finding droplet deposition rate (μL cm^{-2}) as per the work of Zhu et al. [48] is shown in Figure 11.

2.5.2. Droplet Penetrability (%)

In order to characterize the droplet penetrability between the various collection points in the experiment, this study measured the coefficient of variation (CV) of droplet deposition density at each collection point. The droplet penetrability was measured by the CV of the amount of droplet deposition on the upper, middle, and bottom layers of each collection point [44]. The smaller the CV value, the more uniform the droplet deposition and the better the penetration [51]. It was calculated by the following equation:

$$Coefficient\ of\ Variation\ (CV) = \frac{SD}{X} \times 100 \quad (1)$$

$$Mean(X) = \frac{\sum X_i}{N} \quad (2)$$

$$Standard\ deviation\ (SD) = \sqrt{\frac{\sum_1^N (X_i - X)^2}{N-1}} \quad (3)$$

where

X is the deposition value of every sampling point (μL cm^{-2});
X_i is the average deposition value of every sampling point (μL cm^{-2});
N is the number of sampling points.

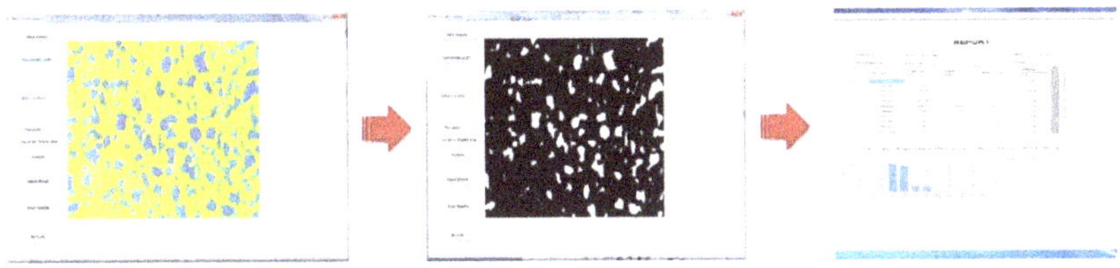

(a) (b) (c)

Figure 10. Procedure flow chart for droplet size analysis using stereo micro and macro scope: (**a**) image view of spray droplet sample under microscope; (**b**) conversion of the image to black and white and with adjustment of contrast and threshold; and (**c**) output results of volume mean diameter (VMD), droplet density (DD), and percentage of volume coverage.

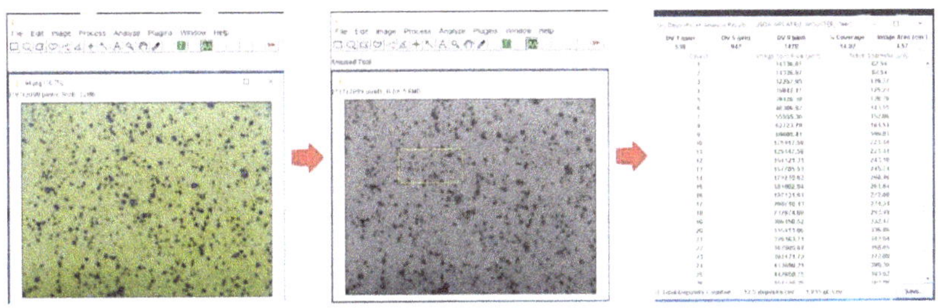

(a) (b) (c)

Figure 11. Flow chart for the procedure for droplet deposition rate under DepositScan software: (**a**) import of water-sensitive paper (WSP) image to Deposit Scan software, (**b**) conversion of the imported color images to black and white under 8 bit with the sample area marked, and (**c**) selection of the green AA Tool for analysis.

2.5.3. Effective Spray Width

The effective swath width is the distance between the points on either side of a single swath, wherein the rate of spray deposit equals one-half of the effective application rate. Effective spray width was determined in a manner that provided the most uniform overall application rate. The effective spray width test was performed according to the method described in the ASAE Standard [47].

2.5.4. Application Rate

The application rate was calculated as per the ASABE standard [52]. The mean value of discharge rate, travel speed, and effective spray width were measured, and application rate was calculated with the formula below:

$$Application\ rate\ (R) = \frac{Q \times K}{S \times W} \quad (4)$$

where

R = Application rate, l ha^{-1};
Q = Output rate, l min^{-1};
K = Constant, 600;
S = Travel speed, km h^{-1};
W = Effective spray width, m^3.

3. Results and Discussion

3.1. Distribution and Analysis of Spray Droplet Deposition Characteristics

WSPs were placed on the leaf at 210 cm, 110 cm, and 40 cm above the ground level. The sampling locations were divided into upper, middle, and bottom layers. The average droplet deposition of each sample obtained by WSP analysis using DepositScan software is presented in Table 5.

Table 5. Characteristics of droplet deposition for each layer at each location.

WSP Position	Location of WSP	Spray Droplet Size (µm)			Droplet Density (No's cm^{-2})
		Dv0.1	Dv0.5	Dv0.9	
Upper	U1	200 ± 9.45	423 ± 5.03	508 ± 5.69	47 ± 4.04
	U2	223 ± 5.29	472 ± 2.65	622 ± 3.00	53 ± 3.61
	U3	236 ± 4.36	492 ± 5.57	724 ± 6.51	52 ± 8.74
Middle	M1	239 ± 2.65	424 ± 2.08	537 ± 6.11	42 ± 2.52
	M2	241 ± 6.24	484 ± 3.00	686 ± 4.16	43 ± 8.50
	M3	134 ± 3.06	304 ± 3.06	656 ± 4.58	38 ± 2.52
Bottom	B1	219 ± 4.00	372 ± 4.51	754 ± 3.51	22 ± 4.36
	B2	213 ± 3.61	355 ± 4.00	635 ± 4.04	17 ± 3.51
	B3	282 ± 9.54	394 ± 4.58	742 ± 3.61	19 ± 4.51

It is observed from Figure 12 that the droplet deposition rate decreased from the upper to the bottom layer, and droplet deposition rate in the upper (2.93 ± 0.17, 2.11 ± 0.08, and 2.36 ± 0.16 µL cm^{-2}), middle (1.22 ± 0.08, 1.07 ± 0.03, and 0.76 ± 0.02 µL cm^{-2}), and bottom layers (0.43 ± 0.06, 0.35 ± 0.03, and 0.32 ± 0.03 µL cm^{-2}) were found in locations 1, 2, and 3, respectively. The coverage per unit area decreased from the upper layer to the bottom layer. Coverage per unit area in the upper (15.72 ± 0.49, 16.60 ± 0.71, and 14.94 ± 0.39%), middle (10.95 ± 0.81, 11.22 ± 0.56, and 8.57 ± 0.44%), and bottom layers (2.78 ± 0.29, 2.95 ± 0.45, and 2.46 ± 0.20%) was found in locations 1, 2, and 3, respectively. As shown in Table 5, the droplet deposition density on the upper (47 ± 4.04, 53 ± 3.61, and 52 ± 8.74 droplets cm^{-2}), middle (42 ± 2.52, 43 ± 8.50, 38 ± 2.52 droplets cm^{-2}), and lower layers (22 ± 4.36, 17 ± 3.51, and 19 ± 4.51 droplets cm^{-2}) was found in locations 1, 2, and 3, respectively. Overall, the droplet deposition density was the highest in the upper layer and the lowest in the middle and bottom layers. Yang et al. [53] found that the RPAAS rotor's downwash airflow generated a pressure difference between the top and lower surfaces of the leaf, resulting in torque and allowing spray droplets to penetrate up to fourfold to the bottom surface of the leaf. The droplet penetrability is another typical index to evaluate the deposition effect. In smaller CV, more uniform distribution of the spray droplet was observed. The best droplet penetrability was assessed on the basis of the coefficient of variation. The droplet penetrability was best in the upper layer, with the CV of deposition density reaching 6.34% compared to the middle and bottom layers at 6.45% and 13.02%, respectively. The findings of this research are in line with previous research and hence confirms that downwash and wind turbulence caused by RPAAS aircraft rotor blades aid in droplet deposition and canopy penetration [11,22,54–57].

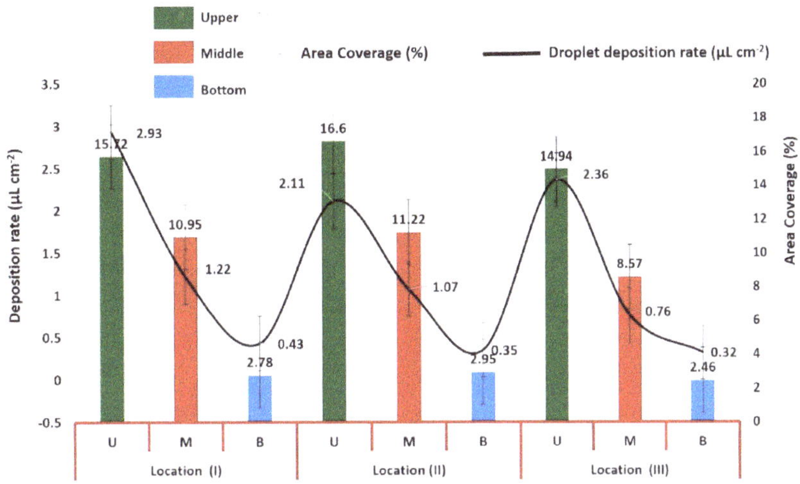

Figure 12. Droplet deposition rate and area coverage.

3.2. Effective Spray width and Application Rate of Chemical Usage

The average effective spray width was found to be 3100 mm at a height of spray of 1200 mm from the above crop canopy. It was observed that the swath width increased by increasing the height of spray and operating pressure. In addition, when the flight height was 1.0 to 3.0 m, the effective spray width was 3.1 to 3.8 m. This is in line with the earlier study [51].

The average 3 ms^{-1} travel speed was maintained during autonomous spray operation, and the actual field capacity of the UAV sprayer was found to be 2.84 ha h^{-1}.

The theoretical application rate was calculated as per the procedure explained in 2.5.4, and it was found to be 57.34 l ha^{-1}. The actual application rate was found to be 54.0 l ha^{-1} (≈15.0 L diluted chemical was consumed for covering 0.28 ha area of redgram field). This diluted imidacloprid (a.i. 17.8SL) chemical application rate was matched as per the recommendation of the redgram crop package of practice published by Tamil Nadu Agricultural University, Coimbatore, India. UAV spray droplet deposition drift results in target and non-target area zones.

For UAV spray drift, the sampling area was divided into two zones on the basis of the direction of the natural wind. The samples in the target area zone were placed at −1.0, 0.0, and 1.0 m distances, and the samples in the non-target area zone were placed at 2.0, 3.0, 4.0, and 5.0 m distances, as presented in Table 6.

Table 6. Amount of spray drift droplet deposition.

Sample Zone	Drift Sample Distance (m)	Droplet Density (No's cm^{-2})
Target area	−1.0	46 ± 3.61
	0.0	54 ± 2.08
	1.0	45 ± 3.21
Non-target area	2.0	23 ± 1.53
	3.0	11 ± 2.08
	4.0	6 ± 1.53
	5.0	0.00

Figure 13 shows that the large amount of droplets were deposited at sampling points of −1.0, 0.0, and 1.0 m distance. Samples in the target area were found at 1.63 ± 0.09, 1.93 ± 0.05, and 1.82 ± 0.06 μL cm^{-2}, respectively, indicating that the droplets mainly deposited below the UAV sprayer flight route. This may have been due to the downwash air produced by the propeller of the UAV sprayer [55–57]. Li et al. [21] reported that the downwash air velocity had effects on droplet deposition and standing crop. A few droplets were also collected on WSPs in the non-target area zone because the droplets drifted with the wind to the downwind side. The droplet deposition in the non-target area samples, viz., 2.0, 3.0, 4.0, and 5.0 m, were found to be 0.88 ± 0.02, 0.46 ± 0.03, 0.22 ± 0.05, and 0.00 μL cm^{-2}, respectively. From Figure 13, it is observed that the amount of droplet deposition decreased as lateral distance increased from the central line of flight path and was almost nil at 5.0 m from the UAV flight center line. The results support the findings of Bird et al. [58], who concluded that, under identical stability situations, increased wind speeds are likely to increase off-target deposition. Spraying droplets of UAV is affected by many factors, and the droplet size is one of them [17,59]. The ideal droplet size is 50–300 μm. If the droplet size was less than 50 μm, the droplets would be easy to drift. The effect of height, velocity, and nozzle flow rate on droplet deposition dispersion during drone operation is proportional to crop canopy height [60]. If the droplet size is greater than 300 μm, the droplets would find it difficult to penetrate the crop canopy and adhere to the target. As indicated by Bergeron et al. [37], our findings show that droplet size distribution may have a major impact on spray drift.

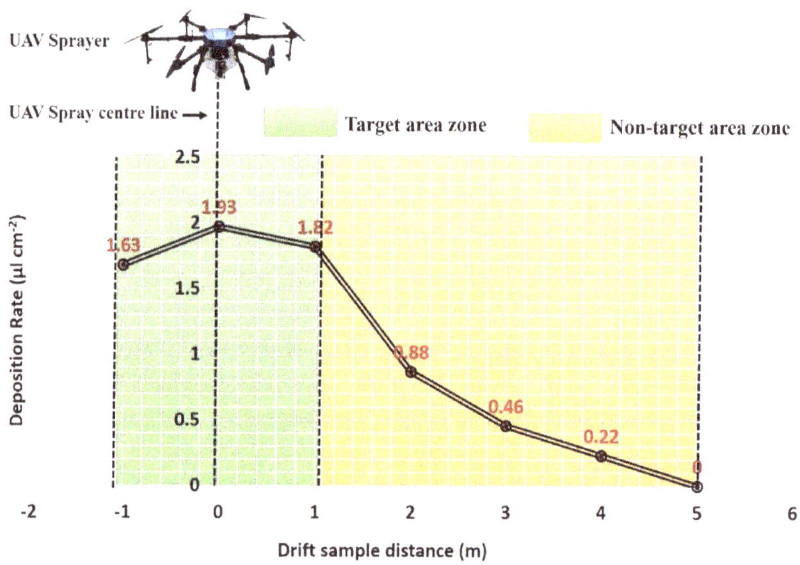

Figure 13. Droplet deposition at different drift sampling distances.

Table 6 and Figure 14 show that the large amount of droplet density and coverage area were found at sampling points of −1.0, 0.0, and 1.0 m distance. Samples of droplet density in the target area (−1.0 m to 1.0 m) were found at 46 ± 3.61, 54 ± 2.08, and 45 ± 3.21 No's cm^{-2}, and 23 ± 1.53, 11 ± 2.08, 6 ± 1.53, and 0.00 No's cm^{-2} droplet density were found in the non-target area (2.0 m to 5.0 m). Similarly, more area coverage (%) was found in the target area at 14.40 ± 0.07, 17.54 ± 0.36, and 16.42 ± 0.30% compared to the non-target area at 7.58 ± 0.34, 4.41 ± 0.19, 2.16 ± 0.05, and 0.00%, indicating that the droplets were mainly found more so below the UAV sprayer flight route (target area). This may have been due to the downwash air produced by the propeller of the UAV sprayer [11,54,60]. Yang [10] reported that the downwash air velocity had effects on area coverage and droplet density on crop leaf. Few droplets were also collected on WSPs in the non-target area zone because the droplets drifted with the wind to the downwind side. According to other studies [55–57,61], the reason for the difference in droplet deposition results could have been that droplets with smaller sizes are more likely to drift outside the target area due to the influence of the environmental wind, whereas droplets with larger sizes, which are less affected by the environmental wind due to the mass of the droplet, are more likely to be deposited in the target area.

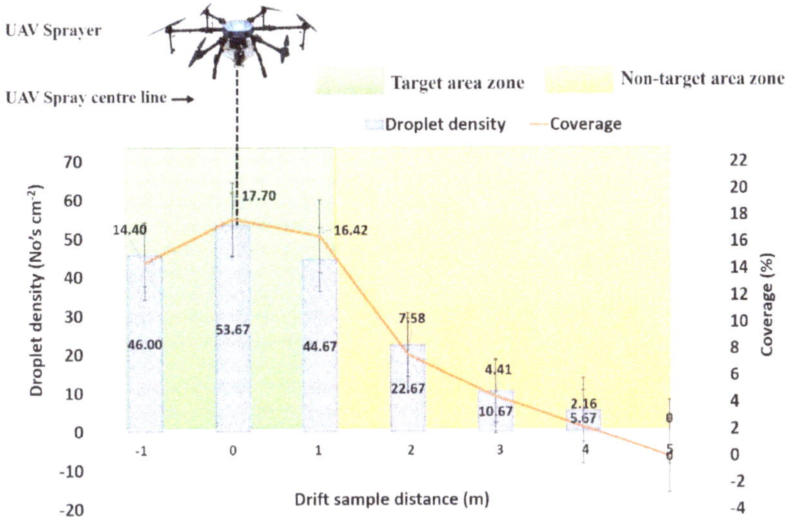

Figure 14. Droplet density and coverage at different drift sampling distances.

From Figure 14 and Table 4, it is observed that the amount of droplet density and area coverage were decreased as lateral distance increased from the central line of the flight path. Area coverage and droplet density were almost nil at 5.0 m from the UAV flight center line. These findings were comparable to those of Fritz [17], indicating that higher wind speed resulted in greater downwind ground deposition. The findings revealed that wind speed was the most important meteorological element in the transport and destiny of aerially applied sprays.

4. Conclusions

In this study, a multi-rotor UAV sprayer with a payload capacity of 10 liters was used in fully autonomous mode to apply insecticide to a redgram crop. The average deposition rate and area coverage in the upper layer were higher (2.47 μL cm^{-2}) when compared to the middle layer (1.02 μL cm^{-2}) and bottom layers (0.37 μL cm^{-2}) of redgram leaf. Overall, the droplet deposition rate (2.47 μL cm^{-2}), droplet density (51.00 droplets cm^{-2}), and area coverage (15.77%) were highest in the upper layer as compared to middle layers (0.75 μL cm^{-2}, 19.33 droplets cm^{-2}, 10.25%, respectively) and bottom layers (0.37 μL cm^{-2}, 19.00 droplets cm^{-2}, 2.73%, respectively). The droplet penetrability was best in the upper layer, with the CV of deposition density reaching 6.34% when compared to the middle and bottom layers (6.45% and 13.02%, respectively).

Spray droplet is one of the most critical aspects affecting the droplet deposition rate, area coverage, and droplet density for insecticide spraying by UAV in the redgram field. There was higher droplet deposition rate (1.79 μL cm^{-2}), area coverage (16.17%)/ and droplet density (48.00 droplets cm^{-2}) in the target area as compared to the non-target area (deposition rate: 0.39 μL cm^{-2}, coverage: 3.54% and droplet density: 10.00 droplets cm^{-2} respectively), which may have been due to the downwash air produced by the propeller of the UAV sprayer. The downwash air velocity had effects on droplet deposition and the standing crop. Droplet deposition rate decreased as the lateral distance from the center of the UAV spray line increased (from 1.93 ± 0.05 μL cm^{-2} to 0.22 ± 0.05 μL cm^{-2}), and similarly, the spray droplet drift distance was reduced with the increase in droplet size, which showed that the increase in spray droplet size can effectively minimize the spray droplet drift. The UAV spray method is quite effective, not only in terms of reach but also in mitigating health risks faced by farmers who walk through the fields with handheld sprayers, exposing themselves to toxic chemicals. Moreover, it will help to overcome the shortage of labor.

Author Contributions: Conceptualization, Y.D., K.R. and A.P.M.K.; methodology, Y.D., K.R., S.A. and S.B.; software, B.K.; validation, Y.D. and K.R.; formal analysis, Y.D., K.R. and S.B.; investigation, Y.D., K.R., S.A., S.B., A.P.M.K. and B.K.; resources, Y.D., S.B. and A.P.M.K.; data curation, Y.D., S.B. and A.P.M.K.; writing—original draft preparation, Y.D., K.R., S.B. and A.P.M.K.; writing—review and editing, K.R., S.A., S.B., B.K., A.P.M.K. and K.M.K.; visualization, Y.D., K.R., S.A. and S.B.; supervision, K.R., S.A. and K.M.K.; project administration, K.R., S.A. and K.M.K.; funding acquisition, K.R., S.A. and K.M.K. All authors have read and agreed to the published version of the manuscript.

Funding: This research was funded by Indian Council of Agricultural Research (ICAR), All India Co-ordinated Research Project (AICRP) on Farm Implements and Machinery (FIM), TNAU Coimbatore, Tamil Nadu.

Institutional Review Board Statement: Not applicable.

Informed Consent Statement: Not applicable.

Data Availability Statement: Not applicable.

Acknowledgments: The authors wish to acknowledge the financial assistance provided by the Indian Council of Agricultural Research (ICAR), All India Coordinated Research Project (AICRP) on Farm Implements and Machinery (FIM) and Tamil Nadu Agricultural University, Coimbatore, Tamil Nadu (India) for providing necessary research facilities for conducting the experiment.

Conflicts of Interest: The authors declare no conflict of interest.

References

1. Wang, G.; Han, Y.; Li, X.; Andaloro, J.; Chen, P.; Hoffmann, W.C.; Han, X.; Chen, S.; Lan, Y. Field evaluation of spray drift and environmental impact using an agricultural unmanned aerial vehicle (UAV) sprayer. *Sci. Total Environ.* **2020**, *737*, 139793. [CrossRef]
2. Huang, H.; Deng, J.; Lan, Y.; Yang, A.; Deng, X.; Zhang, L. A fully convolutional network for weed mapping of unmanned aerial vehicle (UAV) imagery. *PLoS ONE* **2018**, *13*, e0196302. [CrossRef]
3. Yao, W.; Wang, X.; Lan, Y.; Jin, J. Effect of UAV prewetting application during the flowering period of cotton on pesticide droplet deposition. *Front. Agric. Sci. Eng.* **2018**, *5*, 455–461. [CrossRef]
4. Xiongkui, H.; Bonds, J.; Herbst, A.; Langenakens, J. Recent development of unmanned aerial vehicle for plant protection in East Asia. *Int. J. Agric. Biol. Eng.* **2017**, *10*, 18–30.
5. Kirk, I.W. Aerial spray drift from different formulations of glyphosate. *Trans. ASAE* **2000**, *43*, 555. [CrossRef]
6. Ru, Y.; Zhou, H.; Jia, Z.; Wu, X.; Fan, Q. Design and application of electrostatic spraying system. *J. Nanjing For. Univ. (Nat. Sci. Ed.)* **2011**, *35*, 91–94.
7. Huang, Y.; Hoffmann, W.C.; Lan, Y.; Wu, W.; Fritz, B.K. Development of a spray system for an unmanned aerial vehicle platform. *Trans. ASABE* **2009**, *25*, 803–809. [CrossRef]
8. Bozdogan, N.Y. Assessment of buffer zone for aquatic organisms in pesticide application. *Int. J. Agric. Biol. Eng.* **2016**, *9*, 227–234.
9. Yang, S.H.; Zheng, Y.J.; Liu, X.X. Research status and trends of downwash airflow of spray UAVs in agriculture. *Int J Precis Agric Aviat* **2019**, *2*, 1–8. [CrossRef]
10. Yang, S. Spray Droplet Deposition and Distribution Inside Crop Canopy and Control Efficiency Applied by Unmanned Aerial Vehicle. Ph.D. Thesis, Chinese Academy of Agricultural Sciences, Beijing, China, 2014; pp. 1–44. (In Chinese with English Abstract)
11. Qing, T.; Ruirui, Z.; Liping, C.; Min, X.; Tongchuan, Y.; Bin, Z. Droplets movement and deposition of an eight-rotor agricultural UAV in downwash flow field. *Int. J. Agric. Biol. Eng.* **2017**, *10*, 47–56. [CrossRef]
12. Wang, J.; Lan, Y.; Zhang, H.; Zhang, Y.; Wen, S.; Yao, W.; Deng, J. Drift and deposition of pesticide applied by UAV on pineapple plants under different meteorological conditions. *Int. J. Agric. Biol. Eng.* **2018**, *11*, 5–12. [CrossRef]
13. He, X.; Zeng, A.; He, J. Effect of wind velocity from orchard sprayer on droplet deposit and distribution. *Nongye Gongcheng Xuebao (Trans. Chin. Soc. Agric. Eng.)* **2002**, *18*, 75–77.

14. Liu, X.J.; Zhou, H.P.; Zheng, J.Q. Research advances of the technologies for spray drift control of pesticide application. *Trans. CSAE* **2005**, *21*, 186–190.
15. Wang, X.; He, X.; Song, J.; Wang, Z.; Wang, C.; Wang, S.; Wu, R.; Meng, Y. Drift potential of UAV with adjuvants in aerial applications. *Int. J. Agric. Biol. Eng.* **2018**, *11*, 54–58. [CrossRef]
16. Smith, D.B.; Bode, L.E.; Gerard, P.D. Predicting ground boom spray drift. *Trans. ASAE* **2000**, *43*, 547. [CrossRef]
17. Fritz, B.K. Meteorological effects on deposition and drift of aerially applied sprays. *Trans. ASABE* **2006**, *49*, 1295–1301. [CrossRef]
18. Zhang, J.; He, X.; Song, J.; Zeng, A.; Liu, Y.; Li, X. Influence of spraying parameters of unmanned aircraft on droplets deposition. *Nongye Jixie Xuebao=Trans. Chin. Soc. Agric. Mach.* **2012**, *43*, 94–96.
19. An, J.; Xiang, W.; Han, Z.; Xiao, K.; Wang, Z.; Wang, X.; Li, Y. Validation of the Institute of Atmospheric Physics emergency response model with the meteorological towers measurements and SF6 diffusion and pool fire experiments. *Atmos. Environ.* **2013**, *81*, 60–67. [CrossRef]
20. Gao, Y.Y. Study on distribution of pesticide droplets in gramineous crop canopy and control effect sprayed by unmanned aerial vehicle. *Northeast Agric. Univ. Harbin, China* **2013**.
21. Li, J.; Lan, Y.; Shi, Y. Research progress on airflow characteristics and field pesticide application system of rotary-wing UAV. *Trans. Chin. Soc. Agric. Eng.* **2018**, *34*, 104–118.
22. Zhang, H.; Zheng, J.; Zhou, H.; Dorr, G.J. Droplet deposition distribution and off-target drift during pesticide spraying operation. *Nongye Jixie Xuebao/Trans. Chin. Soc. Agric. Mach.* **2017**, *48*, 114–122.
23. He, Y.; Wu, J.J.; Fang, H.; Zheng, Q.S.; Xiao, S.P.; Cen, H.Y. Research on deposition effect of droplets based on plant protection unmanned aerial vehicle: A review. *J. Zhejiang Univ. (Agric. Life Sci.)* **2018**, *44*, 392–398. (In Chinese)
24. Shengde, C.; Lan, Y.; Jiyu, L.; Zhiyan, Z.; Aimin, L.; Yuedong, M. Effect of wind field below unmanned helicopter on droplet deposition distribution of aerial spraying. *Int. J. Agric. Biol. Eng.* **2017**, *10*, 67–77.
25. Chen, S.; Lan, Y.; Zhou, Z.; Ouyang, F.; Wang, G.; Huang, X.; Cheng, S. Effect of droplet size parameters on droplet deposition and drift of aerial spraying by using plant protection UAV. *Agronomy* **2020**, *10*, 195. [CrossRef]
26. Hoffmann, W.C.; Hewitt, A.J. Comparison of three imaging systems for water-sensitive papers. *Appl. Eng. Agric.* **2005**, *21*, 961–964. [CrossRef]
27. Liu, Q.; Chen, S.; Wang, G.; Lan, Y. Drift Evaluation of a Quadrotor Unmanned Aerial Vehicle (UAV) Sprayer: Effect of Liquid Pressure and Wind Speed on Drift Potential Based on Wind Tunnel Test. *Appl. Sci.* **2021**, *11*, 7258. [CrossRef]
28. Berner, B.; Chojnacki, J. Use of drones in crop protection. In Proceedings of the IX International Scientific Symposium "Farm Machinery and Processes Management in Sustainable Agriculture", Lublin, Poland, 22–24 November 2017.
29. Fritz, B.K.; Kirk, I.W.; Hoffmann, W.C.; Martin, D.E.; Hofman, V.L.; Hollingsworth, C.; McMullen, M.; Halley, S. Aerial application methods for increasing spray deposition on wheat heads. *Appl. Eng. Agric.* **2006**, *22*, 357–364. [CrossRef]
30. Lou, Z.; Xin, F.; Han, X.; Lan, Y.; Duan, T.; Fu, W. Effect of unmanned aerial vehicle flight height on droplet distribution, drift and control of cotton aphids and spider mites. *Agronomy* **2018**, *8*, 187. [CrossRef]
31. Dixon, D.; Boon, S.; Silins, U. Watershed-scale controls on snow accumulation in a small montane watershed, southwestern Alberta, Canada. *Hydrol. Process.* **2014**, *28*, 1294–1306. [CrossRef]
32. Salcedo, R.; Garcera, C.; Granell, R.; Molto, E.; Chueca, P. Description of the airflow produced by an air-assisted sprayer during pesticide applications to citrus. *Span. J. Agric. Res.* **2015**, *13*, e0208. [CrossRef]
33. Ling, W.; Du, C.; Mengchao, Z.; Yu, W.; Ze, Y.; Shumao, W. CFD simulation of low-attitude droplets deposition characteristics for UAV based on multi-feature fusion. *IFAC-Pap.* **2018**, *51*, 648–653. [CrossRef]
34. Li, J.; Shi, Y.; Lan, Y.; Guo, S. Vertical distribution and vortex structure of rotor wind field under the influence of rice canopy. *Comput. Electron. Agric.* **2019**, *159*, 140–146. [CrossRef]
35. Xinyu, X.; Kang, T.; Weicai, Q.; Yubin, L.; Huihui, Z. Drift and deposition of ultra-low altitude and low volume application in paddy field. *Int. J. Agric. Biol. Eng.* **2014**, *7*, 23–28.
36. Yang, F.; Xue, X.; Cai, C.; Sun, Z.; Zhou, Q. Numerical simulation and analysis on spray drift movement of multirotor plant protection unmanned aerial vehicle. *Energies* **2018**, *11*, 2399. [CrossRef]
37. Bergeron, V. Designing intelligent fluids for controlling spray applications. *Comptes Rendus Phys.* **2003**, *4*, 211–219. [CrossRef]
38. Choi, D.S.; Ma, K.C.; Kim, H.J.; Lee, J.H.; Oh, S.A.; Kim, S.G. Control standards of three major insect pests of Chinese cabbage (Brassica campestris) using drones for pesticide application. *Korean J. Appl. Entomol.* **2018**, *57*, 347–354.
39. Lan, Y.; Qian, S.; Chen, S.; Zhao, Y.; Deng, X.; Wang, G.; Qiu, X. Influence of the Downwash Wind Field of Plant Protection UAV on Droplet Deposition Distribution Characteristics at Different Flight Heights. *Agronomy* **2021**, *11*, 2399. [CrossRef]
40. Teske, M.E.; Bird, S.L.; Esterly, D.M.; Curbishley, T.B.; Ray, S.L.; Perry, S.G. AgDrift®: A model for estimating near-field spray drift from aerial applications. *Environ. Toxicol. Chem. Int. J.* **2002**, *21*, 659–671. [CrossRef]
41. Bilanin, A.J.; Teske, M.E.; Barry, J.W.; Ekblad, R.B. AGDISP: The aircraft spray dispersion model, code development and experimental validation. *Trans. ASAE* **1989**, *32*, 327–0334. [CrossRef]
42. Bache, D.H.; Sayer, W.J.D. Transport of aerial spray, I. A model of aerial dispersion. *Agric. Meteorol.* **1975**, *15*, 257–271. [CrossRef]
43. Miller, P.C.H.; Ellis, M.B. A review of spray generation, delivery to the target and how adjuvants influence the process. *Plant Prot. Q.* **1997**, *12*, 33–38.
44. Guo, S.; Li, J.; Yao, W.; Zhan, Y.; Li, Y.; Shi, Y. Distribution characteristics on droplet deposition of wind field vortex formed by multi-rotor UAV. *PLoS ONE* **2019**, *14*, 22–24. [CrossRef]

Small-sized spraying robots that replaced manual operations have been widely used in traditional orchards [10]. However, the capacity of a single spraying robot is extremely limited and the amount of pesticide is not sufficient to complete the entire spraying tasks. The distance between water sources and orchards in arid areas is usually far, so the task of transporting water back and forth is particularly onerous. Therefore, it is necessary to use a dosing robot to transport the water needed for the spraying robots back and forth. Further, the collaboration between the spraying robot and the dosing robot not only reduces labor but also greatly increases productivity. Collaborative navigation is one of the most important technologies for the collaborative task of spraying and dosing robots.

At this stage, most of the robot groups are based on global navigation satellite systems (GNSS) for collaborative navigation. With the promotion and application of satellite positioning technology, Jo et al. used the difference in GNSS data between outdoor robots to calculate the distance between robots and to realize the cooperative positioning of robot groups [11]. However, the high density of fruit trees planted in tulip orchards makes it a huge workload to manually measure the GNSS position information of each fruit tree and then use this information as a target point to assist the robot group in cooperative localization. Tim et al. developed a ground-based agricultural robot for high-throughput crop phenotyping. The robot navigated autonomously by using path tracking between GPS waypoints. However, as the season progresses, this capability was lost when the sorghum grew taller than the GPS antenna [12]. Liang Zhang et al. developed an integrated BDS/IMU automatic navigation system for orchard spraying robots [13]. The system used a combination of BDS and IMU loosely coupled with path planning based on the terrain features of the orchard. Although this navigation system can make the spraying robot track the preset path smoothly and stably, it cannot satisfy the cooperative navigation of spraying and dosing. Xu Aigong et al. proposed a cooperative navigation vehicle localization method based on the BDS/UWB combination [14]. Firstly, the combined observation equation of BDS/UWB was constructed, and then the position parameters were solved by using the extended Kalman filter (EKF). Finally, the vehicle positioning result of cooperative navigation was obtained. Although the positioning accuracy of this method is significantly better than that of a single BDS system, it is not applicable to traditional orchards, where navigation satellite signals are easily lost [15].

Currently, visual navigation and laser navigation are also widely used in orchards [16,17]. Although vision sensors are low-cost and information-rich, they are susceptible to light and cannot be steadily navigated in the orchard [18,19]. Laser navigation would be more stable in a traditional orchard, and LiDAR build maps are used by some for collaborative navigation [20–22]. Cheein et al. proposed an Extended information filter-simultaneous localization and mapping (EIF-SLAM) method using fruit trees as environmental references [23]. The method utilized olive tree trunk information recognized by the robot's on-board 2D LiDAR and monocular vision, and reconstructed the trunk information into a feature map. Its experimental results demonstrated that the map drawn by the robot by this method was consistent with the real scene. However, the trunk information could not be obtained when there were people or similar objects interfering next to the trunk, which led to mapping errors. To reduce the relative errors that occurred during laser or visual measurements, Gimenez et al. proposed a method for constructing maps based on solving optimization problems with nonlinear constraints [24]. The method measured the polar angle of fruit trees by robotic vehicle-mounted 2D LiDAR, and then optimized this measurement information with the absolute GPS position information of fruit trees in the four corners of the orchard. However, the method is not suitable for identifying fruit tree trunks that are obscured by canopy or branches and leaves. To identify lower fruit tree trunks, Shalal et al. used vehicle-mounted 2D LiDAR and vision to detect fruit tree trunks [25], and GPS information from multiple known points within the fruit tree rows was used to correct the detected trunk information and generate an orchard map. However, this map only contains the location information of the center point of the fruit trees in the orchard, and there are measurement errors in the vision sensors carried by different robots. Whether the orchard map constructed based on this method can provide cooperative positioning information for the robot group needs to be further experimented.

In this paper, a map-based construction of a collaborative navigation system for spraying-dosing robot groups was developed for the production needs of traditional orchard spraying and dosing sessions in arid regions. In this study, a 3D orchard point cloud map was constructed first, then paths and path points were set on the 2D projection map, and finally, the spraying and dosing robot groups achieved collaborative navigation on the map according to the developed collaborative navigation strategy and motion control algorithm. Synergizing the spraying and dosing robots in traditional orchards can not only alleviate the labor shortage but also improve production efficiency, and the research of the cooperative navigation system is an important foundation for the intelligent development of multiple robots in orchards.

This paper is organized as follows: Section 2 describes the system components and research methodology of the spraying–dosing robot group; Section 3 describes the specific experimental protocol and experimental results; Section 4 discusses the experimental results and Section 5 presents the conclusions of this article.

2. Materials and Methods

2.1. Structure and Design

2.1.1. Navigation System

The spraying–dosing robot group consists of a spraying robot and a dosing robot, and its hardware system is shown in Figure 1. The spraying robot consists of two main parts, the traction robot and the sprayer, which are connected by pins. The spraying robot and the dosing robot communicate through the wireless routers and receive each other's navigation information for cooperative navigation.

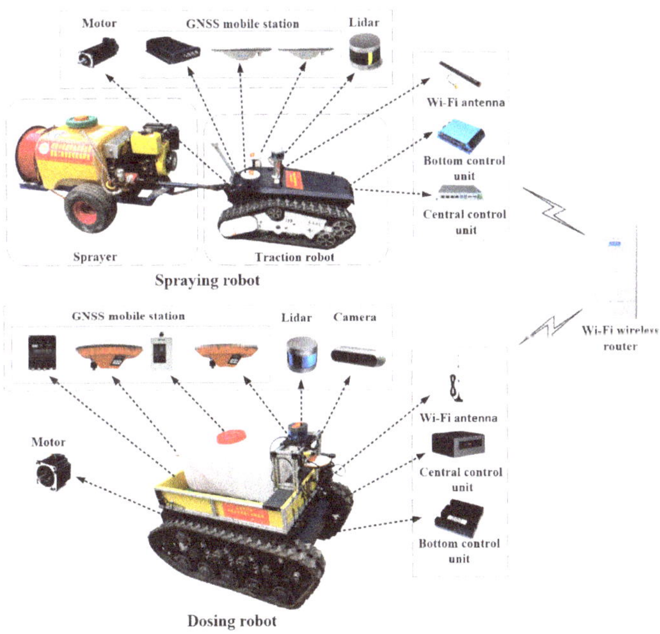

Figure 1. The hardware structure of the spraying–dosing robot cooperative navigation system.

The central control unit of the spraying robot adopts UNO-2484G computer, the bottom control unit includes a KYDBL2450-2E motor controller and two 48 V DC brushless motors, the centimeter-level accuracy RTK-GNSS in the positioning unit is composed of

M600 mobile station and T300 base station produced by Sinan, and the model of LIDAR is Velodyne VLP-16. The central control unit of the dosing robot adopts BOXER-8150AI computer, the bottom control unit includes a VSY100D60-2 motor controller and two 48 V DC brushless motors, the positioning unit adopts CGI-410 mobile station and i70 base station produced by Sinan, and the model of LIDAR is RS-LIDAR-16.

The software system of the spray–dosing robot is shown in Figure 2. The information interaction layer is used for the information interaction between the spraying robot and the dosing robot under the Ubuntu system; the information processing layer is the cooperative navigation program based on the Robot Operating System (ROS); the execution layer is based on the Ubuntu system's mobile platform control, which outputs speed information to the underlying servo controller and finally realizes the autonomous navigation of the spraying robot and the dosing robot.

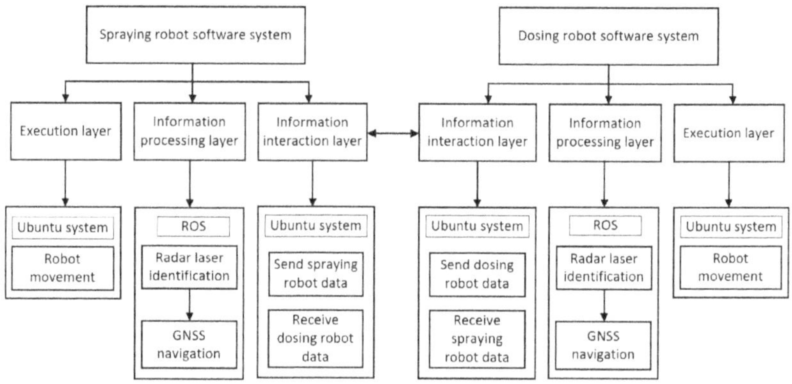

Figure 2. Software system structure of the spraying–dosing robot.

2.1.2. Communication System Design

A stable and reliable wireless communication system is a prerequisite for the collaborative navigation of the spraying–dosing robot group, under which the robot group can co-locate and navigate based on exchanged position and velocity information [26,27]. To save development time, the wireless communication system is based on Wi-Fi communication, using the existing IEEE 802.11 series of communication standards and the free public frequency band of 2.4 GHz. The data frame format for robot information interaction uses the communication data frame format developed by the team research [28,29].

The two robots communicate point-to-point in a local area network with the help of a wireless router, where the wireless router (ADV1) has a connection (Internet Protocol, IP) address of 192.168.62.1. The IP address of the spraying robot communication system is set to 192.168.62.2, the sub-virtual mask set to 24, and the gateway address is the same as the router address. Similarly, the IP address of the dosing robot is set to 192.168.62.20, and the rest is the same as the spray delivery robot. After the setting is completed, check whether the address of the wireless network card is modified successfully by opening the terminal.

2.2. Map Construction

In this section, based on the structure and design of the robot, we construct a point cloud map of the orchard that provides a priori environmental information for collaborative navigation.

Firstly, we remotely control the body of the spraying robot to drive continuously through the orchard while recording point cloud information. Secondly, we process the collected point cloud data and perform point cloud stitching by finding the neighboring points in two adjacent frames. To reduce the amount of data during point cloud stitching, we refer to the lightweight and ground-optimized LiDAR mapping algorithm (LeGO-LOAM), which well balances the accuracy and efficiency of building maps [30–32]. Thirdly, we use the adjacent points in the point cloud to calculate the attitude estimate between two adjacent frames and generate LiDAR odometry information. Finally, we optimized the LiDAR odometry information using real-time GNSS pose information as a constraint and constructed a point cloud map of the orchard, as shown in Figure 3.

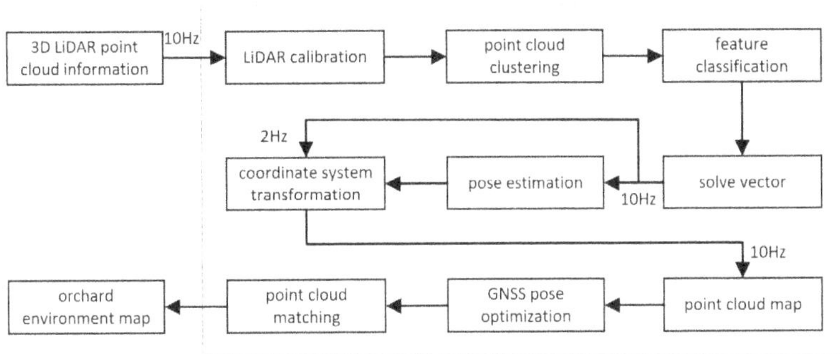

Figure 3. Construction process of orchard point cloud map.

2.2.1. Feature Classification

We establish the local coordinate system XYZ with the center point of the spraying robot body as the origin (LiDAR is at the center point of the spraying robot body). Since the LiDAR was calibrated before being used, the calibrated point cloud image is projected directly onto the XOY plane frame by frame here. In the current frame of the point cloud p^t, the distance from a single point p_i^t to the LiDAR is the value of that point. As shown in Figure 4, any point with horizontal scan angle α_i and vertical scan angle β_i in the XYZ spatial coordinate system has coordinates of x_i^t, y_i^t and z_i^t, respectively. The local coordinate system XYZ is rotated around X-axis, Y-axis and Z-axis by the angles θ_{roll}, θ_{pitch} and θ_{yaw}, respectively.

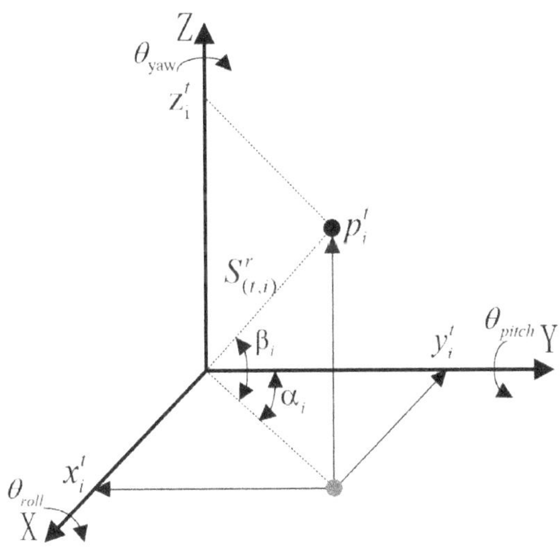

Figure 4. Point cloud reprojection.

In order to distinguish the projected points on the XOY plane matrix and to improve the speed of searching the point cloud, we label the individual points p^t in the point cloud matrix according to Equation (1). The main markers are the row number index number

r_i, column number index number c_i and the value $v^r_{(t,i)}$ of the plane in which the point is located. Each point is unique according to the index and value.

$$\begin{aligned} r_i &= \frac{\left(\arctan\left(\frac{Z_i^t}{\sqrt{X_i^{t2}+Y_i^{t2}}}\right)\times\frac{180}{\pi}+15\right)}{2} \\ c_i &= \frac{\left(\arctan\left(\frac{X_i^t}{Y_{R(t)}^t}\right)\times\frac{180}{\pi}-90\right)}{0.2}+\frac{360}{2} \\ v^r_{(t,i)} &= \sqrt{X_i^{t2}+Y_i^{t2}+Z_i^{t2}} \end{aligned} \tag{1}$$

In Equation (1), X_i^t, Y_i^t, Z_i^t are the horizontal, vertical and height coordinates of any point x in space in the point cloud of the current frame.

After marking the points in a single frame utilizing reprojection, the ground points and non-ground points p^t can be quickly distinguished. The principle is that the points smaller than 1° of the LiDAR vertical scanning angle are ground points and the rest are non-ground points. For non-ground points, the breadth-first search (BFS) algorithm is used to search the surrounding neighborhoods adjacent to any point p_i^t according to the row index number, as shown in Figure 5. If the row number of any point p_i^t is r_i and the column number is c_i, the row number of the neighboring points is r_{i-1} or r_{i+1}, and the column number of the neighboring points is c_{i-1} or c_{i+1}. According to Equation (2), determining whether p_i^t and its neighboring points are in the same plane when the angle between two points is less than 60° is considered as the same plane, and vice versa is a different plane [19]. After the neighboring points of point p_i^t are traversed one by one, then search the adjacent points of neighborhood, such as point p_j^t, until all single-frame point clouds are traversed.

$$\delta = \arctan\frac{v^r_{(t,j)}\sin\gamma}{v^r_{(t,i)}-v^r_{(t,j)}\cos\gamma} j \in (r\pm 1, c\pm 1) \tag{2}$$

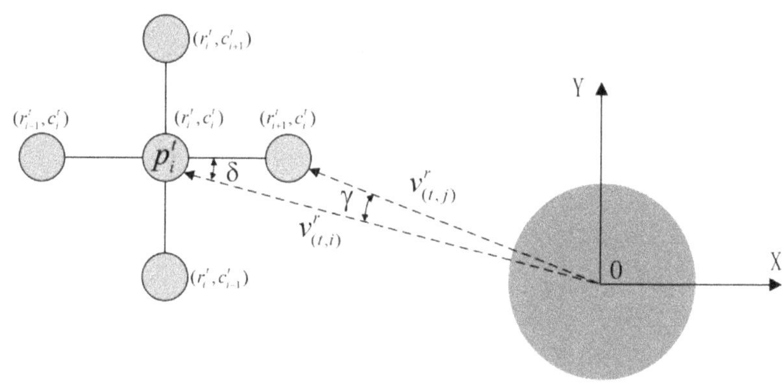

Figure 5. Point cloud search.

In Equation (2), δ is the angle between any point and the adjacent point, $v^r_{(t,j)}$ is the Euclidean distance between the adjacent point and the LiDAR, $v^r_{(t,i)}$ is the Euclidean distance between any point and the LiDAR, and γ is the angular difference between any point and the adjacent point.

In the orchard environment, there will be a lot of weeds, rugged ground, leaves blown by the wind, etc. This will result in scattered points in the point cloud acquired in the current frame, and these scattered points will have a large impact on the accuracy of feature classification. In order to obtain stable and reliable point cloud features, when the number of point clouds in the plane of non-ground point cloud p_i^t is less than 30, these discrete point clouds are removed directly. After point cloud clustering, only large objects such as tree trunks, ground, signage, and other feature point clouds are left, as shown in Figure 6. Figure 6a shows the original point cloud, and Figure 6b shows the point cloud after clustering. It can be observed that the number of point clouds after clustering is dramatically reduced compared with the original point cloud. Only stable point cloud features are retained, such as green for ground point clouds and black for non-ground object features.

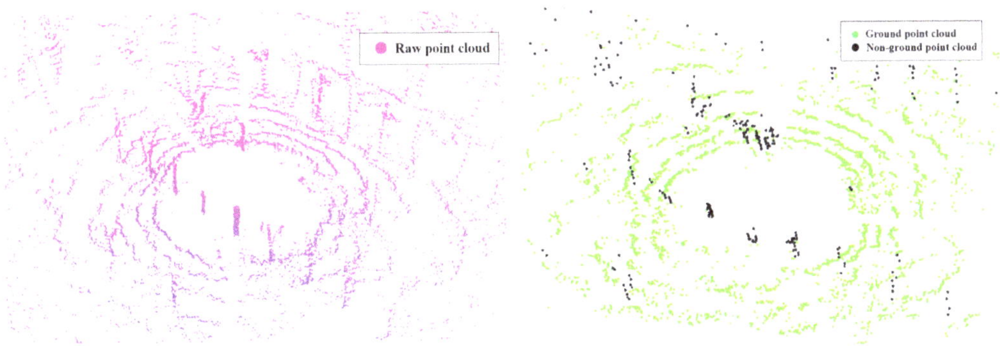

(**a**) (**b**)

Figure 6. Results of point cloud clustering: (**a**) Raw point cloud; (**b**) the point cloud after clustering.

Then, the clustered point cloud planar features are divided again into 6 regions equally according to the LiDAR horizontal scanning angle (0∼360°), and the same row index number c in each region is regarded as a subset. Drawing on the method of BFS search, the five points adjacent to the left and right of the point cloud particles p_i^t in the same subset are taken for comparison each time. The smoothness S_i of each point cloud particle in the same subset is calculated by substituting the point cloud particles into Equation (3) according to the distance between the neighboring points p_j^t and p_i^t. When the smoothness S_i of the point cloud particles is greater than the set threshold S_{th} (set to 5×10^{-3}) then it is an edge feature, and conversely, S_i is less than the set threshold then it is a surface feature.

$$S_i = \frac{1}{|P| \cdot \|r_i\|} \| \sum_{j \in S\ j \neq i} (p_j^r - p_i^r) \| \quad (3)$$

In Equation (3), P is a continuous subset of 5 adjacent points on the projection plane of the current frame.

The non-ground point cloud with the maximum smoothness $S_{\max}$ in each subset within a single aliquot is set as the edge feature $n\mathbb{F}_e^t$, and the minimum smoothness S_{min} is set as the surface feature $n\mathbb{F}_s^t$. The F_e^t and F_s^t of the 6 aliquots are the point cloud collections of edge and plane features of the current frame point cloud. As shown in Figure 7a, the red points in the figure are the point clouds of edge features, and the green points are the point clouds of surface features in the current frame. The set of non-ground surface features with maximum smoothness in the aliquot is the set of non-ground edge features nF_e^t, and the set of ground surface features with minimum smoothness in each aliquot is the set of ground edge features nF_s^t. The F_e^t and F_s^t of 6 aliquots are the sets of non-ground edge feature and ground surface feature point clouds for the current frame, respectively. Here, F_e^t belongs to $\mathbb{F}_e^t$ and F_s^t to $\mathbb{F}_s^t$. The overlapping point clouds can be removed by smoothing calculation. As shown in Figure 7b, the yellow points are the point clouds with non-ground edge features for the current frame, and the blue points are the point clouds with distinctive ground surface features.

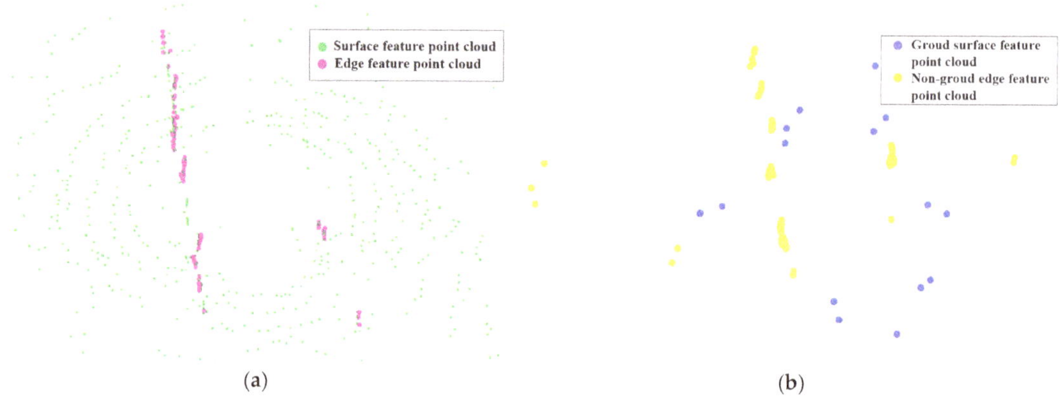

Figure 7. Point cloud feature classification: (**a**) Surface feature point cloud and edge feature point cloud; (**b**) non-ground edge feature point cloud and ground plane feature point cloud.

After downsampling the point cloud by feature classification, the surface features and edge features of the current frame point cloud can be obtained, and this operation can greatly reduce the number of single-frame point clouds, where each point cloud feature set contains the point cloud, as shown in Equation (4).

$$\begin{aligned} \mathbb{F}_s^t &= \left\{ [p_1^t, \ldots, p_i^t, \ldots, p_{n_1}^t]^T \; 1 \leq i \leq n_1 \right\} \\ \mathbb{F}_e^t &= \left\{ [p_1^t, \ldots, p_i^t, \ldots, p_{n_2}^t]^T \; 1 \leq i \leq n_2 \right\} \\ F_s^t &= \left\{ [p_1^t, \ldots, p_i^t, \ldots, p_{n_3}^t]^T \; 1 \leq i \leq n_3 \right\} \\ F_e^t &= \left\{ [p_1^t, \ldots, p_i^t, \ldots, p_{n_4}^t]^T \; 1 \leq i \leq n_4 \right\} \end{aligned} \tag{4}$$

In Equation (4), n_1, n_2, n_3, n_4 are the number of point clouds in the current frame for $\mathbb{F}_s^t$, $\mathbb{F}_e^t$, F_s^t, and F_e^t, respectively.

2.2.2. Posture Estimation

We perform correlation calculation on the surface feature point clouds to obtain the point cloud particles with the same surface features in the point clouds of the two adjacent frames. Then, the surface feature Iterative Closest Points (ICP) matching is performed by the same point cloud particles (neighboring points) to obtain the corresponding plane rotation matrix. Under the constraint of this rotation matrix, we perform correlation calculation on the edge feature point clouds to obtain the point cloud particles with the same edge features in the adjacent two frames. Immediately after that, we perform ICP matching of the edge features by the neighboring points to obtain the corresponding edge rotation matrix relationship. The pose estimation of a single-frame point cloud can be obtained by unifying the two matrices, which is the pose estimation of the LiDAR sensor relative to the previous frame. The steps of point cloud pose estimation are shown in Figure 8. The principle is to feature match the point clouds of two adjacent frames in LiDAR to obtain the conversion relationship from a single point to an edge point cloud or a surface point cloud. Then, the coordinate system of the original point cloud is transformed according to the transformation relationship to obtain the point cloud pose estimation.

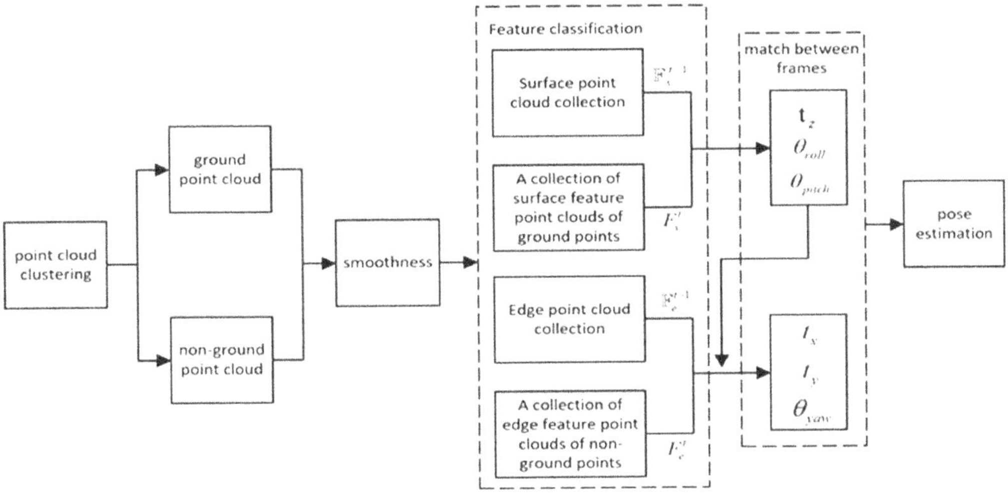

Figure 8. The point cloud pose estimation.

The covariance is used to determine the correlation between two adjacent point clouds in the above equation, as shown in Equation (5). We start the calculation from the second frame by default, and the first frame is only used to record the time information. If the correlation coefficient $\rho_{\mathbb{F}_s^{t-1}F_s^t}$ between the points in the current frame $\mathbb{F}_s^t$ and the points in the previous frame $\mathbb{F}_s^{t-1}$ is greater than 0, the point cloud particles in the surface features of the two frames are correlated. At this time, the correlated points in $\mathbb{F}_s^{t-1}$ are recorded as ground points and saved at this time, otherwise, they are regarded as non-ground points and discarded directly. Similarly, if the correlation coefficient $\rho_{\mathbb{F}_e^{t-1}F_e^t}$ between the points in the current frame F_e^t and the points in the previous frame $\mathbb{F}_e^{t-1}$ is greater than 0, the point cloud particles in the edge features of the two frames are correlated. Then, we record the relevant points in $\mathbb{F}_e^{t-1}$ as the same edge and save them according to the point markers.

$$\begin{aligned}\rho_{\mathbb{F}_s^{t-1}F_s^t} &= \frac{\mathrm{cov}\left(\mathbb{F}_s^{t-1}, F_s^t\right)}{\sigma_{\mathbb{F}_s^{t-1}} \cdot \sigma_{F_s^t}} \\ \rho_{\mathbb{F}_e^{t-1}F_e^t} &= \frac{\mathrm{cov}\left(\mathbb{F}_e^{t-1}, F_e^t\right)}{\sigma_{\mathbb{F}_e^{t-1}} \cdot \sigma_{F_e^t}}\end{aligned} \quad (5)$$

In Equation (5), $\rho_{\mathbb{F}_s^{t-1}F_s^t}$ is the correlation coefficient between two adjacent frames of the point cloud, which p_i^t is in the same plane as p_i^{t-1}, $\rho_{\mathbb{F}_e^{t-1}F_e^t}$ is the correlation coefficient where p_i^t and p_i^{t-1} are on the same side in two adjacent frames of the point cloud, $\mathrm{cov}\left(\mathbb{F}_s^{t-1}, F_s^t\right)$ is the covariance between p_i^t and p_i^{t-1} in the surface feature point cloud of two adjacent frames, and $\mathrm{cov}\left(\mathbb{F}_e^{t-1}, F_e^t\right)$ is the covariance between p_i^t and p_i^{t-1} in the edge feature point cloud of two adjacent frames.

The point cloud particles with correlation coefficients $\rho_{\mathbb{F}_s^{t-1}F_s^t} > 0$ and $\rho_{\mathbb{F}_e^{t-1}F_e^t} > 0$ are substituted into Equation (6) to find the surface feature vector λ_{jlm} corresponding to the feature maxima A_s of the current frame surface feature point cloud p_i^t and the previous frame surface feature points p_j^{t-1}, p_l^{t-1} and p_m^{t-1}. Similarly, the edge feature vector λ_{jl} corresponding to the feature maxima A_e of the current frame edge feature point p_i^t and the previous frame edge feature points p_j^{t-1} and p_l^{t-1} can also be found.

$$
\begin{aligned}
A_s - \lambda_{jlm}\hat{\Sigma}_{(\mathbb{F}_s^{t-1}, F_s^t)} x_{p_i^t} &= 0 \\
A_e - \lambda_{jl}\hat{\Sigma}_{(\mathbb{F}_e^{t-1}, F_e^t)} x_{p_i^t} &= 0,
\end{aligned} \quad (6)
$$

In Equation (6), $x_{p_i^t}$ is the value of the current frame feature point p_i^t in the local coordinate system XYZ, $\hat{\Sigma}_{(\mathbb{F}_s^{t-1}, F_s^t)}$ is the surface feature-related particle swarm covariance matrix, and $\hat{\Sigma}_{(\mathbb{F}_e^{t-1}, F_e^t)}$ is the edge feature-related particle group covariance matrix.

According to the area of space quadrilateral method, the distance d_ε from the current frame surface feature point p_i^t to the previous frame plane feature vector λ_{jlm} can be found. According to the area of the parallelogram method, the distance d_H from any point in the current frame edge feature point cloud to the previous frame edge feature vector λ_{jl} can be found. In Figure 9, $\vec{s}_{jl}$ is the direction vector of p_j^t and p_l^t in the previous frame edge feature point clouds, and $\vec{s}_{jm}$ is the direction vector of p_j^t and p_m^t.

$$
\begin{aligned}
d_\varepsilon &= \frac{\lambda_{jl}}{\left|\vec{s}_{jl}\right|} \\
d_H &= \frac{\lambda_{jlm}}{\left|\vec{s}_{jl} \times \vec{s}_{jm}\right|},
\end{aligned} \quad (7)
$$

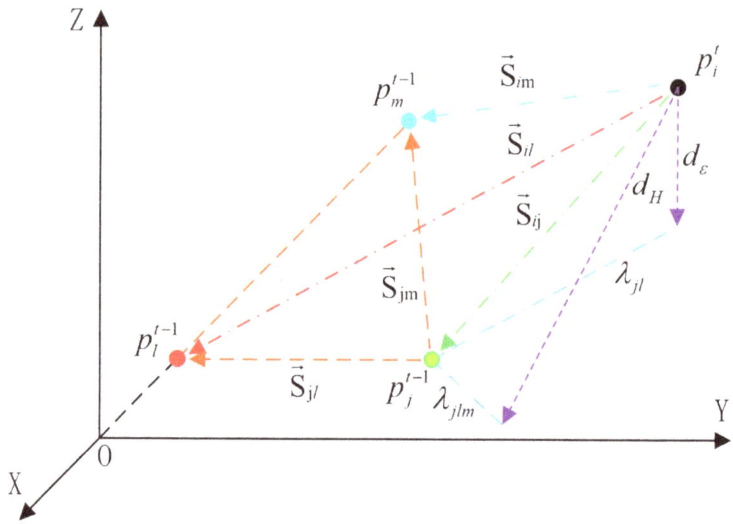

Figure 9. Principle of eigenvector solving based on area method.

The smaller the values of d_ε and d_H, the shorter the distance between the current frame feature point p_i^t and the previous frame surface feature $\mathbb{F}_s^{t-1}$ or edge feature point cloud $\mathbb{F}_e^{t-1}$, as shown in Figure 10. When point p_i^t is perpendicular to the previous frame feature $\mathbb{F}_s^{t-1}$, the distance $\min(d_H)$ between feature point p_i^t and the corresponding point p_i^{t-1} of the previous frame feature point cloud is the shortest. This vector value is used to estimate the point cloud pose between two frames in the fastest way. The initial frame point cloud is used as the starting point to establish the map coordinate system as $X_M Y_M Z_M$. The distance D between the initial frame point p_i^0 and p_i^{t-1} is calculated, and this distance is used to obtain the pose estimate of the current frame point cloud relative to the map coordinate system.

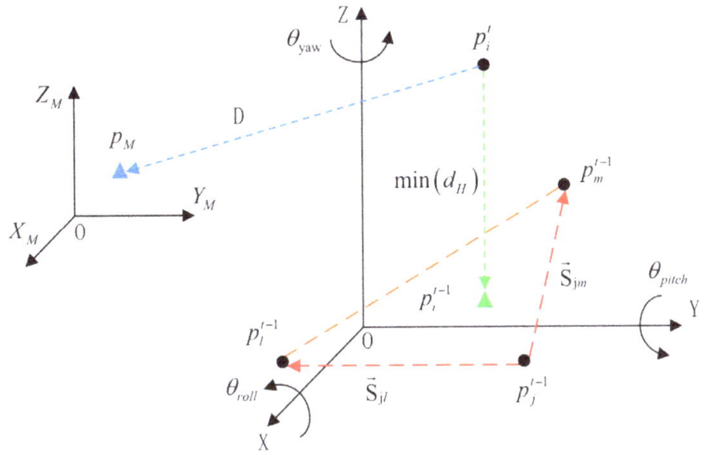

Figure 10. Principle of point cloud pose estimation.

The coordinate system rotation is the inverse rotation of the vector. Referring to Euler's formula according to the rule of clockwise rotation, the relationship between the current frame point cloud p_i^t in XYZ coordinate system and the corresponding point p_M in $X_M Y_M Z_M$ coordinate system can be obtained, as shown in Equation (8).

$$p_M = Q(p_M, p_i^t) = R_y R_x R_z p_i^t + T = R p_i^t + T \tag{8}$$

In Equation (8), Q is the transformation function of the current frame point p_i^t in XYZ coordinate system, and the point p_M in $X_M Y_M Z_M$ coordinate system, R is the rotation matrix of p_i^t and p_M, and T is the translation matrix of p_i^t verses p_M.

There is a certain error between p_i^t obtained by Equation (8) and the true value, as shown in Equation (9).

$$f(X) = D(p_M, M) = D(R p_i^t + T, M), \tag{9}$$

In Equation (9), $f(X)$ is the error function and M is the true value of p_i^t in $X_M Y_M Z_M$ coordinate system.

The initial value $f(X)$ is substituted into Equation (10) and iterated until the conversion error is minimized (10^{-6}) or the number of iterations is satisfied (set to 500). The pose estimate $\xi_i[t_x, t_y, t_z, \theta_{roll}, \theta_{ptich}, \theta_{yaw}]$ of the local single-frame point cloud in the point cloud map coordinate system can be obtained, and the single-frame point cloud is converted into the point cloud map after substituting it into Equation (8).

$$\begin{aligned} &\mathrm{J}^T \mathrm{J} \Delta x_{lm} = \mathrm{J}^T f(X) \\ &\mathrm{J} = \left[\begin{array}{cccccc} \frac{\partial d_H}{\partial \theta_{roll}} & \frac{\partial d_H}{\partial \theta_{ptich}} & \frac{\partial d_\varepsilon}{\partial \theta_{yaw}} & \frac{\partial d_\varepsilon}{\partial t_x} & \frac{\partial d_\varepsilon}{\partial t_y} & \frac{\partial d_H}{\partial t_z} \end{array} \right]^T, \end{aligned} \tag{10}$$

In Equation (10), Δx_{lm} is the iteration step length and J is the Jacobi matrix of single-frame feature point observables.

2.2.3. Map Optimization

LiDAR odometry information will drift when the robot turns too many times or the rotation angle is too large, which leads to an increase in the error of the point cloud map. In order to control the drift of the LiDAR odometry and improve the accuracy of the point cloud map, this section adopts a GNSS optimization method based on graph theory to optimize the map, as shown in Figure 11. We used the GNSS information with fixed frequency (2 Hz) to constrain the LiDAR pose estimation several times and obtain an orchard environment map based on GNSS information. In Figure 11, the solid black line is the robot's trajectory, the green dots represent the LiDAR point clouds of the current frame with continuously updated position estimates, the red pentagrams represent GNSS information (X_G, Y_G, Z_G), and the red dots are the robot's positions aligned with GNSS frequency with position point coordinates of $\left(X_m, Y_m, Z_m, \theta_{roll}^m, \theta_{pitch}^m, \theta_{yaw}^m\right)$, where X_m, Y_m, Z_m are the robots' coordinates in the cooperative coordinate system and $\theta_{roll}^m, \theta_{pitch}^m, \theta_{yaw}^m$ are the position points' cross roll, pitch and yaw angle information.

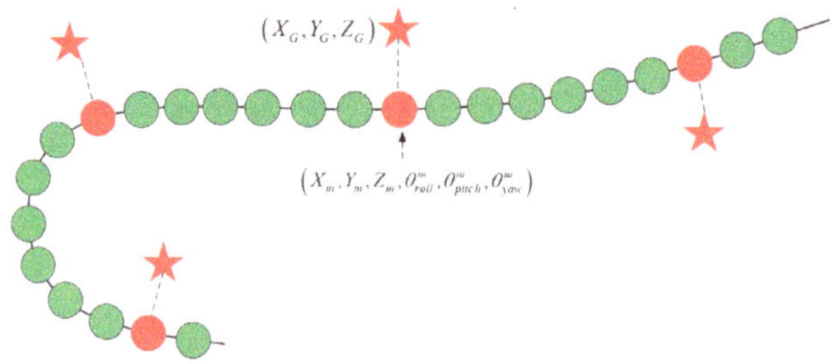

Figure 11. Principle of map optimization based on GNSS information.

According to Equation (10), the covariance matrix Σ_{ri} of the point cloud position pose ξ_i for each current frame can be obtained. After substituting Σ_{ri} and the covariance matrix Σ_{mj} of the current GNSS position data into Equation (11), the position of the robot can be obtained after integrating the data from the beginning to the present.

$$A_i = \Sigma_{i=1}^{n} \Sigma_{ri}^{-1} + \Sigma_{mj}^{-1}\; i \in [1, 2, 3, \ldots, n], \tag{11}$$

In Equation (11), A_i is the Point cloud pose and GNSS position data integration accuracy estimation matrix.

$$\begin{aligned} A_{ri} &= A_i^{-1} \Sigma_{ri}^{-1} \\ A_{mj} &= A_i^{-1} \Sigma_{mj}^{-1} \\ \xi_{mi} &= A_{ri} T_r + A_{mj} m_j, \end{aligned} \tag{12}$$

In Equation (11), T_r is the robot's current position parameters and ξ_{mi} the GNSS-corrected geoid position parameters.

According to Equation (12), the positional parameters ξ_{mi} can be obtained and a point cloud map optimized based on GNSS information can be generated, as shown in Figure 12.

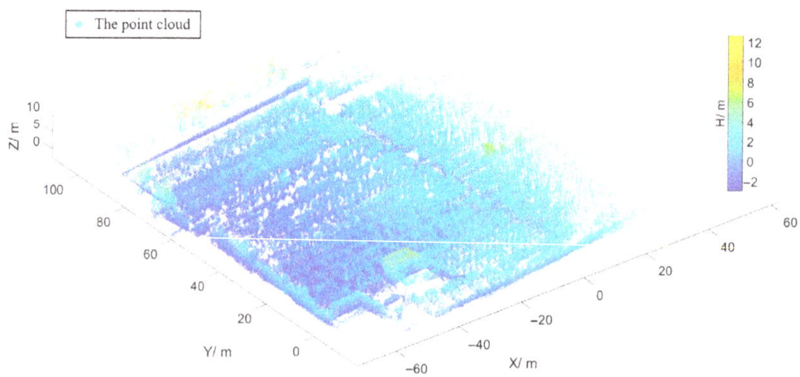

Figure 12. Orchard point cloud map.

2.3. Motion Control

To achieve the best path tracking results for the spraying robot and the dosing robot in their respective navigation methods, pure pursuit control based on the trajectory motion model is used for the spraying robot, and D-A control based on RTK-GNSS is used for the dosing robot.

2.3.1. Pure Pursuit Control

Based on the research of our team members, we adopt the pure pursuit algorithm based on robot kinematic model for the spraying robot [28]. The pure pursuit algorithm is a widely applicable geometric method for low-speed movement control with the advantages of few parameters, high predictability, and accurate linear tracking capability in narrow scenarios [33]. The principle of this algorithm is to change the lateral deviation of the robot's current position point from the linear trajectory by the magnitude of the pre-sight distance to achieve a good tracking effect.

2.3.2. D-A Control

The dosing robot takes the target point as the tracking object and adopts the Distance-Angle (D-A) tracking control method [34]. Firstly, input the longitude and latitude information of the target point measured by RTK-GNSS at static time, and then calculate the direction difference and distance between the current point and the target point. It is judged with the set threshold value to adjust the direction and movement strategy of the next driving of the dosing robot. In addition, to prevent collision between two robots during cooperative movement, each robot adopts the collision strategy of the LiDAR detection of obstacles in the direction of movement and the emergency stop.

2.4. Collaborative Navigation Strategy

Based on the construction of a 3D orchard point cloud map, a master–slave command-based collaborative navigation strategy was further developed, as shown in Figure 13. The cooperative navigation consists of two parts, which are the navigation of the spraying robot in the orchard and the navigation of the dosing robot on the road. The spraying robot as the master robot sends the co-navigation command to the dosing robot and publishes the information about the co-dosing point. The dosing robot receives and responds to the cooperative request issued by the spraying robot and turns on the fixed-point navigation function. The dosing robot first goes to the water source to fill the water, then moves to the collaborative dosing point issued by the spraying robot after filling the water, while feeding the navigation status information to the spraying robot. After the dosing robot arrives at the collaborative dosing point, it doses the spraying robot within the waiting time. After the dosing task is finished, the spraying robot continues to perform the spraying task. The dosing robot returns to the water source and waits for a new cooperative navigation command with the next cooperative dosing point.

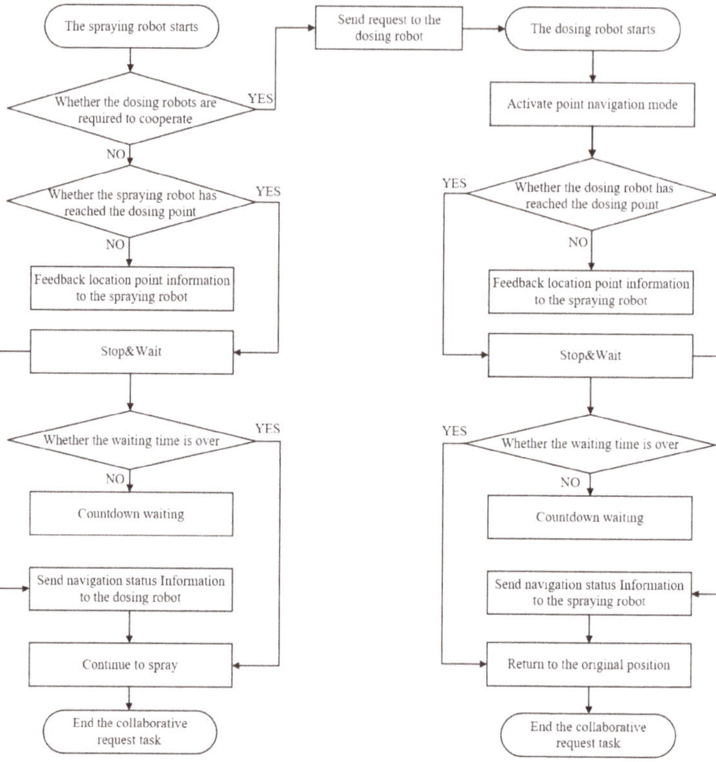

Figure 13. Spraying–dosing robot cooperative navigation strategy.

3. Results

To test the performance of the cooperative navigation system of the spraying–dosing robot group, the communication system and navigation control were experimented on separately. The experiment site was selected from the National Persimmon Germplasm Resource Park of Northwest Agriculture and Forestry University, in which the persimmon orchards were all traditional orchards with fruit trees spaced about 2.4 m apart in rows, as shown in Figure 14.

Figure 14. Traditional orchard environment.

The experimental orchard environment in which the spraying–dosing robot group work consists of a garage, an orchard, a house, a road and a water source. Figure 15a shows a schematic diagram of the orchard environment and Figure 15b shows a physical diagram of the orchard environment.

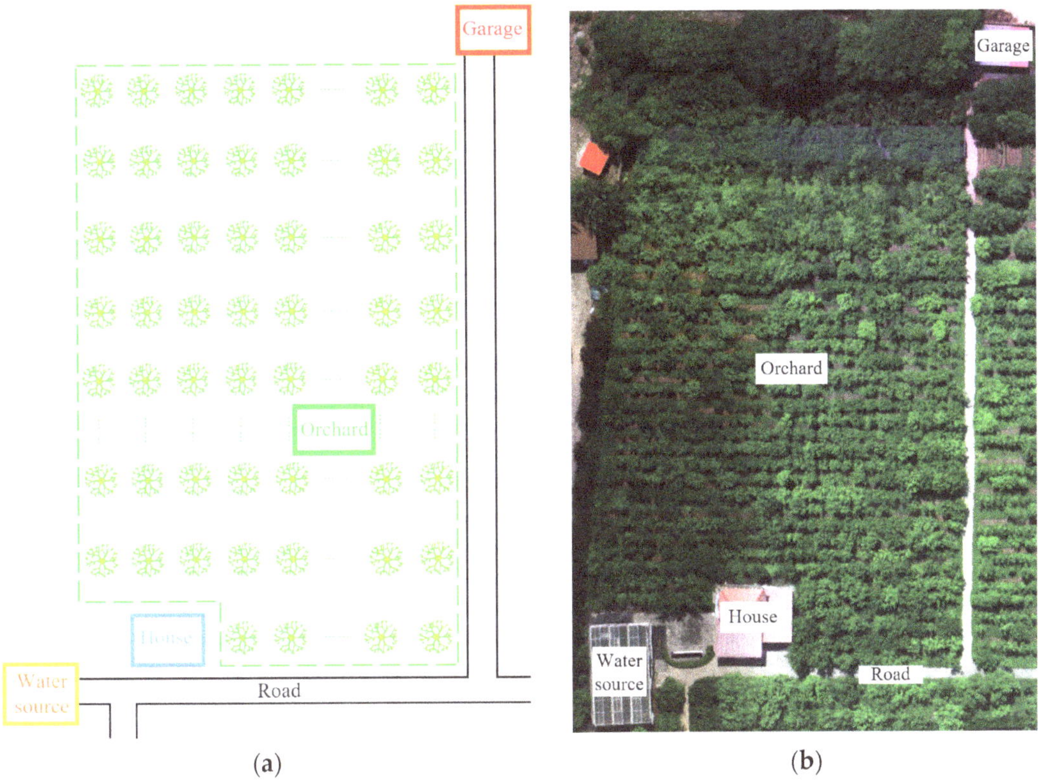

Figure 15. Map of the orchard environment: (**a**) Diagram of the orchard environment; (**b**) physical picture of the orchard environment.

3.1. Communication Experiment

To verify the reliability of the information interaction between the spraying robot and the dosing robot, the communication system of the cooperative navigation system was experimented upon. The dosing robot was docked at point D on the water source, and the spraying robot was docked at points A, B and C in the orchard, as shown in Figure 16. To ensure the minimum communication distance of the robot group, the wireless router was deployed in the middle of the orchard. The height of the wireless router deployment is affected by the orchard environment. If the deployment position is too high, the robot group and the wireless signal are easily affected by the low branches of the fruit trees, resulting in its poor passability; if the deployment position is too low, it is easy to aggravate the influence of weeds on signal propagation (including reflection, bypass, etc.) [29]. Therefore, to ensure the wireless signal strength of the communication system, the wireless router was deployed at the fork of the main trunk of the fruit tree.

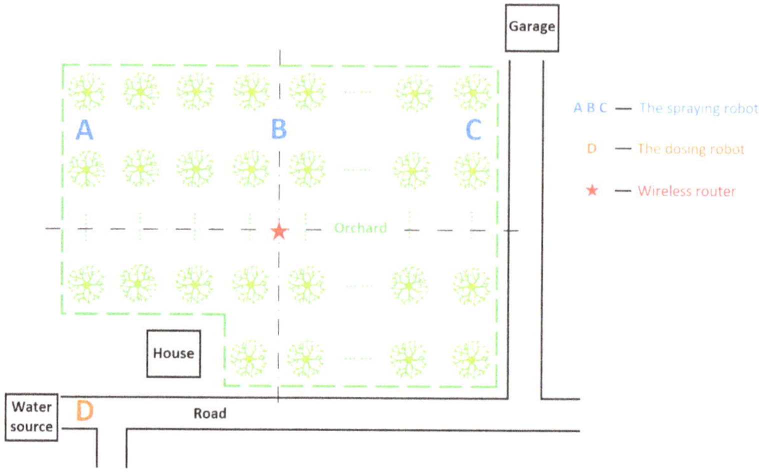

Figure 16. Communication experiment schematic.

Before starting the experiment, the wireless communication router ADV1 was turned on to set up a LAN for the two robots, and then the dosing robot was connected to the spraying robot via SSH commands and IP addresses. The experiment was conducted by sending 500 packets from the spraying robot to the dosing robot. Then, according to the location between the two robots, the IP addresses of the robots connected from the spraying robot to the dosing robot were checked and the number of packets received by the dosing robot after communication was checked, and the packet loss rate was calculated. The experiment was repeated three times independently and the average packet loss rate at each experiment point was calculated.

The rate of communication data loss between the spraying robot and the dosing robot in the experiment orchard environment is shown in Table 1.

Table 1. Packet loss in robotic communication systems.

Location	Distance/m	Group Number	Received Packets/PCS	Packet Loss/%	Average Packet Loss/%
A	84.4	1	494	1.2	1.7
		2	490	2	
		3	491	1.8	
B	88.1	1	491	1.8	2.1
		2	488	2.4	
		3	489	2.2	
C	91.3	1	485	3.0	3.4
		2	483	3.4	
		3	481	3.8	

The average values of the packet loss rate of the dosing robot at A, B and C are 1.7%, 2.1% and 3.4%, respectively.

3.2. Cooperative Navigation Control Experiment

To test the effect of motion control on the cooperative navigation system, the experiment of cooperative navigation control is carried out for the spraying–dosing robot group. Since the spraying robot and the dosing robot chose different models of RTK-GNSS base stations in the positioning module, to ensure the consistency of the positioning information, it is necessary to match the positioning points of the two base stations. Firstly, the base stations of the two robots were respectively set up on a fixed tripod. Secondly, the base station of the spraying robot was used to measure and record the position point information of the base station of the dosing robot. Finally, the position point information was input into the base station of the dosing robot.

The 3D point cloud map needs to be projected into a 2D point cloud map before the experiment. Then, the navigation preset paths and points of the spraying–dosing robot group are set on the 2D point cloud map. In Figure 17, the red pentagram is the navigation starting point of the spraying robot, the green line is the preset path of the spraying robot, the yellow line is the preset path of the dosing robot, and the blue dot is the cooperative dosing point. The spraying robot travels through the orchard in the direction of the red arrow, and the dosing robot goes from the garage to the water source, where it finishes filling up and heads to the collaborative dosing point in the orchard.

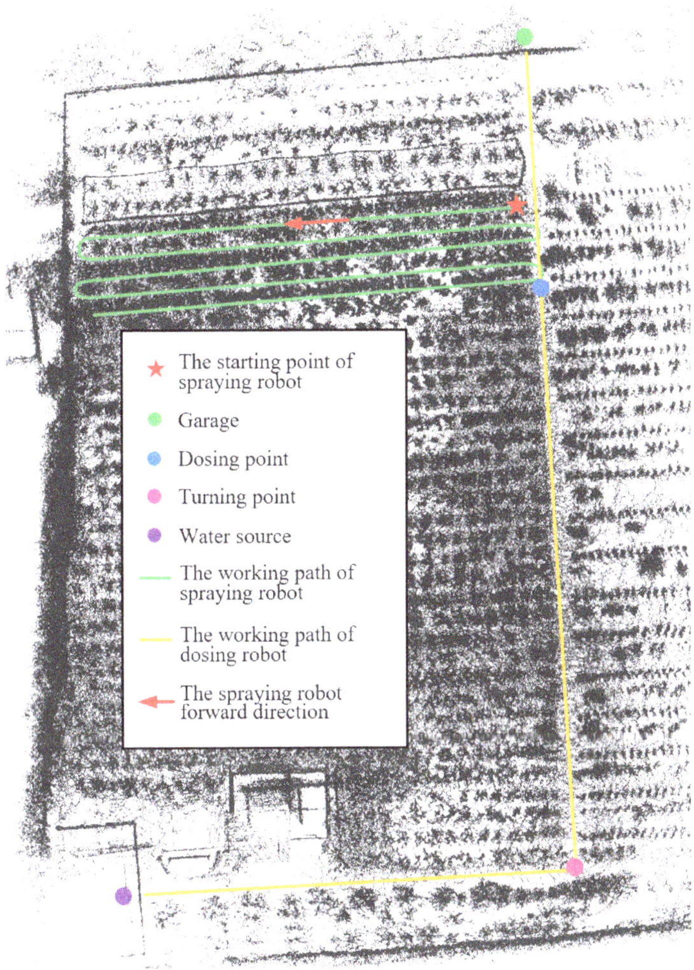

Figure 17. Pre-set path for spraying–dosing robot group.

In Figure 18, a preset point is set at the midpoint of each end of the row, and the two points are connected as a preset path for spraying between the rows. The spraying robot adopts smooth steering when turning at the head of the ground, so two path points are needed at the head of the ground to assist in turning [35].

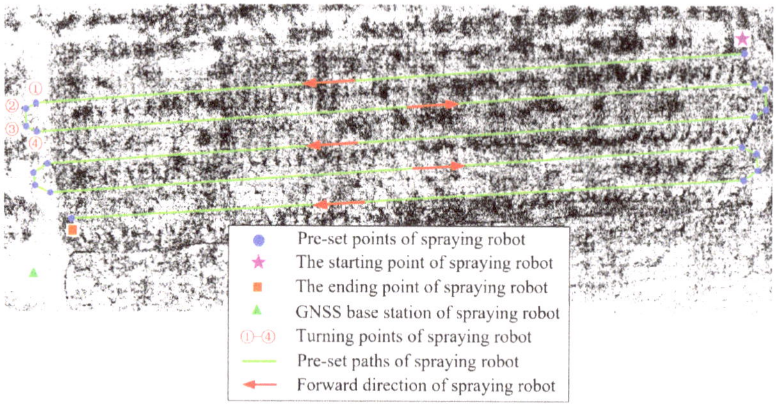

Figure 18. Pre-set paths for spraying robot.

Based on the results of several previous orchard experiments, the walking speed of both the spraying robot and the dosing robot was set to 0.75 m/s in this experiment [7,28]. During the experiment, the spraying–dosing robot group were moved along the preset paths on the map, and the GNSS coordinate information of the driving trajectory was recorded at the same time. The experiment was repeated three times, in which the spraying robot was tested for continuous operation between five rows. The experimental scenarios are shown in Figure 19.

Figure 19. Collaborative navigation trials in orchards.

To verify the path-tracking effect of the spraying–dosing robot group, the motion trajectory of the spraying–dosing robot group during operation is plotted in Figure 20. When the spraying robot sprays in between rows, the actual motion trajectory is roughly the same as the preset path between rows, but there are also some trajectories with deviations. When the spraying robot turns at the head of the ground, the actual trajectory deviates more from the preset path at the head of the ground. The deviation of the trajectory of the movement of the dosing robot to the reservoir and the trajectory of the return to the orchard is smaller.

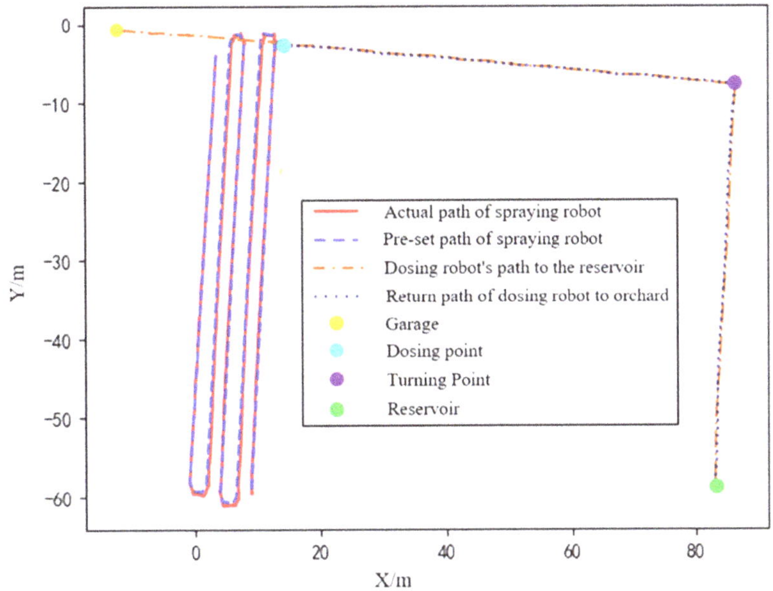

Figure 20. Movement trajectory of spraying–dosing robot group.

To analyze the performance of the cooperative navigation system more accurately, the lateral deviation and heading deviation of the spraying robot and the dosing robot from the known trajectory were calculated, respectively, as shown in Figure 21a–d.

The lateral deviation of the spraying robot was within 15 cm and the heading deviation was about 10° when spraying between rows in the orchard. However, the deviation of the spraying robot changed significantly during the head turn, the maximum value of the lateral deviation was less than 52 cm and the maximum value of the heading deviation was less than 45°. Since the spraying robot does not carry out a spraying operation when the ground head turns and no obstacles are blocking the ground head, the deviation value can meet the demand of cooperative ground head navigation. The maximum lateral deviation of the dosing robot was 30.1 cm, the average lateral deviation was 5.7 cm, the maximum heading deviation was less than 90°, and the minimum heading deviation was more than 10°. The maximum and average values of lateral deviation of the dosing robot were 29.7 cm and 5.7 cm, respectively, and the lateral deviation during linear motion was about 5 cm.

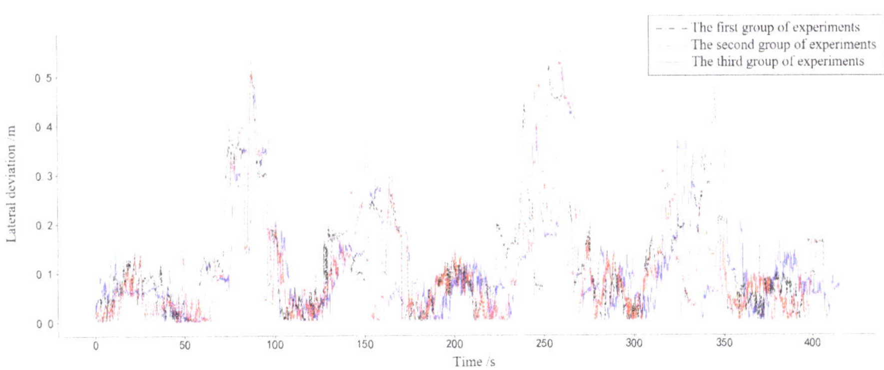

(a)

Figure 21. *Cont.*

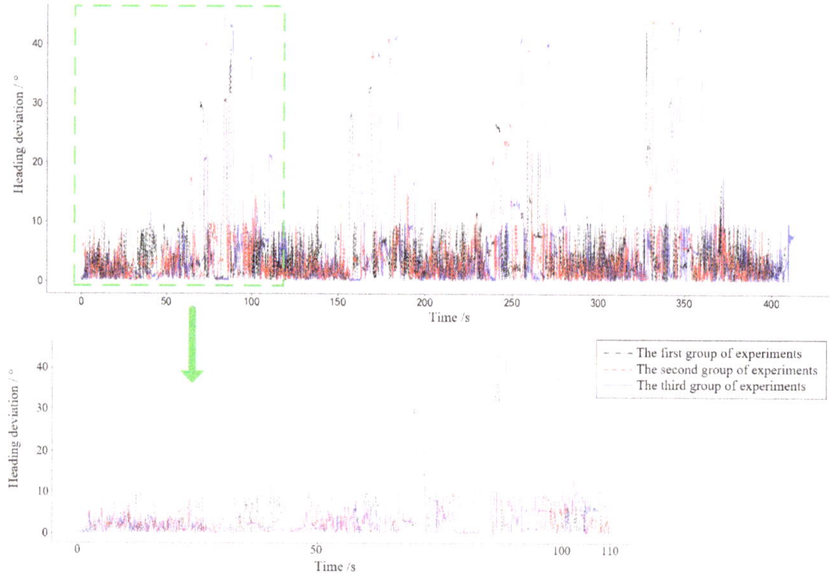

(**b**)

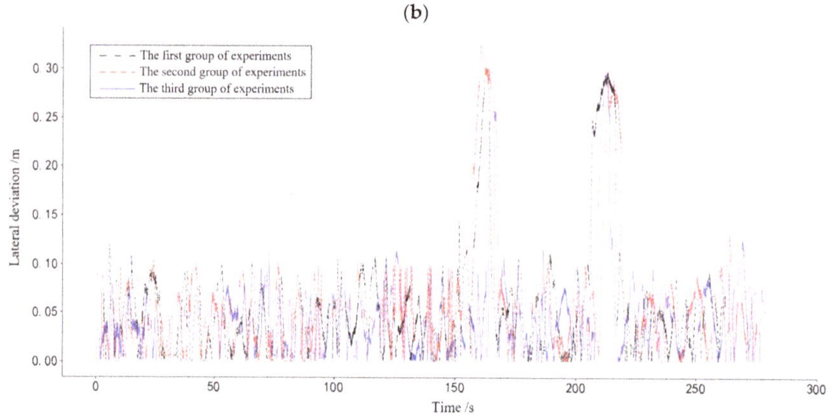

(**c**)

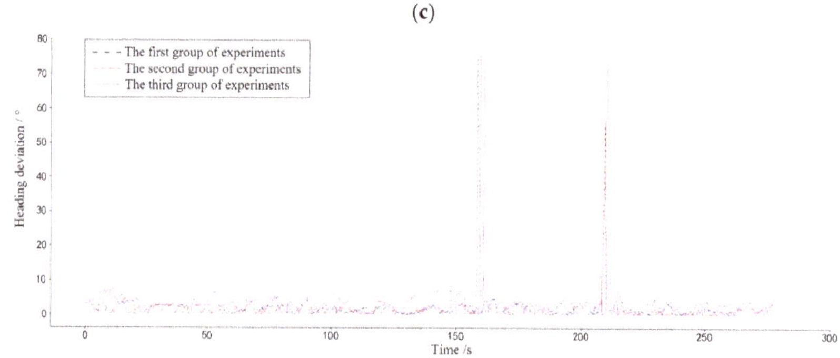

(**d**)

Figure 21. Results of cooperative navigation control experiments with spraying–dosing robot group. (**a**) Lateral error of spraying robot; (**b**) heading error of spraying robot; (**c**) lateral error of dosing robot; (**d**) heading error of dosing robot.

4. Discussion

The above experimental results demonstrated that the developed map-based construction of the spraying–dosing robot group cooperative navigation system can meet the navigation requirements of traditional orchard spraying and dosing tasks.

(1) In the experimental results of the communication system, the maximum communication distance of the spraying–dosing robot group was 91.3 m, and the maximum value of the packet loss rate was 3.8%, which was less than 5%. The experimental results demonstrated that the spraying–dosing robot group can achieve point-to-point messaging in the traditional orchard. Although the packet loss rate of communication between robot groups increased with the increase in communication distance, the packet loss rates were all less than 5%. Therefore, the Wi-Fi-based spraying–dosing robot communication system can meet the communication needs of the cooperative operation in orchards.

(2) The experimental results of cooperative navigation control demonstrated that the average lateral deviation of the spraying robot was 6.7 cm when it traveled at 0.75 m/s between orchard rows and 19.8 cm when it made a turn in the ground, while the average lateral deviation of the dosing robot was 5.7 cm. This is because the canopy above the robot's path is less shaded than the spraying robot. In addition, the working road of the dosing robot is a hard concrete road, while the road in the orchard where the spraying robot works is soft, and there may be slippage during the driving process. The actual motion trajectory of the dosing robot was approximately the same as the preset path, with an average lateral deviation of 5.7 cm, indicating that the D-A control method can ensure that the dosing robot followed the path point in a relatively stable manner. The average lateral deviation was 6.4 cm, but the lateral deviation and heading deviation increased to 51.7 cm and 43.1°, respectively, during the turn. To reduce the lateral deviation and heading deviation of the spraying robot when turning on the ground, the pure pursuit algorithm needs to be optimized in the future.

The spraying robot as a master robot in this article demonstrated a 17% reduction in maximum lateral deviation compared to the spraying robot based on the fuzzy control algorithm [28]. The dosing robot as a slave robot has reduced the maximum lateral deviation by 25.2% compared to the orchard transport robot based on following navigation [7].

5. Conclusions

In this study, a map-based construction of a collaborative navigation system for spraying–dosing robot groups was developed and field experiments were conducted in orchards, and the following are the specific findings of this study.

(1) The results of the navigation communication experiments demonstrated that the packet loss rate during communication was less than 5%, which proved that the communication system constructed in this paper met the communication needs of traditional orchards.

(2) The results of the navigation control experiments showed that the maximum value of the absolute lateral error in the row for the spraying robot was 24.9 cm, and the maximum value of the absolute lateral error for the dosing robot was 29.7 cm. The results proved that the cooperative map-based navigation system developed in this paper met the navigation accuracy requirements for spraying and dosing operations.

(3) In future work, the cooperative navigation strategy needs to be optimized and the pure pursuit control algorithm needs to be improved to reduce the deviation during the ground head turn.

Author Contributions: Conceptualization, F.Y.; methodology, W.M.; software, J.Q.; validation, J.Q., W.W. and W.M.; formal analysis, M.Y.; investigation, H.L.; resources, F.Y.; data curation, W.W.; writing—original draft preparation, J.Q.; writing—review, F.Y. and J.Q.; writing—editing, J.Q.; visualization, Z.R. and S.S.; supervision, F.Y.; project administration, F.Y.; funding acquisition, F.Y. All authors have read and agreed to the published version of the manuscript.

Funding: This research was funded by Shaanxi Province Key Research and Development - Key Industry Innovation Chain Agricultural Field Project (Program No. S2022-YF-ZDCXL-ZDLNY-0128) and the Major Science and Technology Project of Shaanxi Province of China (Program No. 2020zdzx03-04-01).

Data Availability Statement: Not applicable.

Acknowledgments: We also thank the critical comments and suggestions from the anonymous reviewers for improving the manuscript.

Conflicts of Interest: The authors declare no conflict of interest.

References

1. Vasconez, J.P.; Kantor, G.A.; Cheein, F.A.A. Human–robot interaction in agriculture, A survey and current challenges. *Biosyst. Eng.* **2019**, *179*, 35–48. [CrossRef]
2. Vásconez, J.P.; Cheein, F.A.A. Workload and production assessment in the avocado harvesting process using human-robot collaborative strategies. *Biosyst. Eng.* **2022**, *223*, 56–77. [CrossRef]
3. Vasconez, J.P.; Guevara, L.; Cheein, F.A. Social robot navigation based on HRI non-verbal communication: A case study on avocado harvesting. In Proceedings of the 34th ACM/SIGAPP Symposium on Applied Computing, Limassol, Cyprus, 8–12 April 2019; pp. 957–960.
4. Tourrette, T.; Deremetz, M.; Naud, O.; Lenain, R.; Laneurit, J.; De Rudnicki, V. Close coordination of mobile robots using radio beacons, A new concept aimed at smart spraying in agriculture. In Proceedings of the 2018 IEEE/RSJ International Conference on Intelligent Robots and Systems (IROS), Madrid, Spain, 1–5 October 2018; pp. 7727–7734.
5. National Bureau of Statistics of China. *China Statistical Yearbook;* National Bureau of Statistics of China: Beijing, China, 2021.
6. Zheng, Y.; Jiang, S.; Chen, B.; Lü, H.; Wan, C.; Kang, F. Review on technology and equipment of mechanization in hilly orchard. *Trans. Chin. Soc. Agric.* **2020**, *51*, 1–20.
7. Liu, Z.; Wang, X.; Ren, Z.; Mao, W.; Yang, F. Crawler tractor navigation path tracking control algorithm based on virtual radarmodel. *Trans. Chin. Soc. Agric. Mach.* **2021**, *52*, 375–385.
8. Gu, C.; Wang, X.; Wang, X.; Yang, F.; Zhai, C. Research progress on variable-rate spraying technology in orchards. *Appl. Eng. Agric.* **2020**, *36*, 927–942. [CrossRef]
9. Zhai, C.Y.; Zhao, C.J.; Ning, W.; John, L.; Wang, X.; Paul, W.; Zhang, H.H. Research progress on precision control methods of air-assisted spraying in orchards. *Trans. Chin. Soc. Agric. Eng.* **2018**, *34*, 1–15.
10. Liu, L.; Liu, Y.; He, X.; Liu, W. Precision Variable-Rate Spraying Robot by Using Single 3D LIDAR in Orchards. *Agronomy* **2022**, *12*, 2509. [CrossRef]
11. Jo, K.; Lee, J.; Kim, J. Cooperative multi-robot localization using differential position data. In Proceedings of the 2007 IEEE/ASME International Conference on Advanced Intelligent Mechatronics, Zurich, Switzerland, 4–7 September 2007.
12. Mueller-Sim, T.; Jenkins, M.; Abel, J.; Kantor, G. The Robotanist, A Ground-based Agricultural Robot for High-throughput Crop Phenotyping. In Proceedings of the IEEE International Conference on Robotics and Automation (ICRA), Singapore, 29 May–3 June 2017; pp. 3634–3639.
13. Zhang, L.; Zhu, X.; Huang, J.; Huang, J.; Xie, J.; Xiao, X.; Yin, G.; Wang, X.; Li, M.; Fang, K. BDS/IMU Integrated Auto-Navigation System of Orchard Spraying Robot. *Appl. Sci.* **2022**, *12*, 8173. [CrossRef]
14. Xu, A.; Cao, N.; Sui, X.; Wang, C.; Gao, S. Cooperative vehicle positioning method based on BDS/UWB. *Sci. Surv. Mapp.* **2020**, *45*, 1–8.
15. Li, M.; Imou, K.; Wakabayashi, K.; Yokoyama, S. Review of research on agricultural vehicle autonomous guidance. *Int. J. Agric. Biol. Eng.* **2009**, *2*, 1–16.
16. Radcliffe, J.; Cox, J.; Bulanon, D.M. Machine vision for orchard navigation. *Comput. Ind.* **2018**, *98*, 165–171. [CrossRef]
17. Zhang, S.; Guo, C.; Gao, Z.; Sugirbay, A.; Chen, J.; Chen, Y. Research on 2D Laser Automatic Navigation Control for Standardized Orchard. *Appl. Sci.* **2020**, *10*, 2763. [CrossRef]
18. Zhang, M.; Ji, Y.; Li, S.; Cao, R.; Xu, H.; Zhan, Z. Research Progress of Agricultural Machinery Navigation Technology. *Trans. Chin. Soc. Agric. Mach.* **2020**, *51*, 1–18.
19. Yan, C.; Xu, L.; Yuan, Q.; Ma, S.; Niu, C.; Zhao, S. Design and experiments of vineyard variable spraying control system based on binocular vision. *Trans. Chin. Soc. Agric. Eng.* **2021**, *37*, 13–22.
20. Liu, W.; He, X.; Liu, Y.; Wu, Z.; Yuan, C.; Liu, L.; Qi, P.; Li, T. Navigation method between rows for orchard based on 3D LiDAR. *Trans. Chin. Soc. Agric. Eng.* **2021**, *37*, 165–174.
21. Santos, L.C.; Aguiar, A.S.; Santos, F.N.; Valente, A.; Ventura, J.B.; Sousa, A.J. Navigation Stack for Robots Working in Steep Slope Vineyard. In *Advances. in Intelligent Systems and Computing, Proceedings of the Intelligent Systems and Applications, London, UK, 3–4 September 2020;* Arai, K., Kapoor, S., Bhatia, Eds.; Springer: Cham, Switzerland, 2021; Volume 1250, p. 1250.
22. Xue, X.; Xu, X.; Li, Z.; Hong, T.; Xie, J.; Chen, J.; Song, S. Design and test of variable spray model based on leaf wall area in orchards. *Trans. Chin. Soc. Agric. Eng.* **2020**, *36*, 16–22.

23. Cheein, F.A.; Steiner, G.; Paina, G.P.; Carelli, R. Optimized EIF-SLAM algorithm for precision agriculture mapping based on stems detection. *Comput. Electron. Agric.* **2011**, *78*, 195–207. [CrossRef]
24. Gimenez, J.; Herrera, D.; Tosetti, S.; Carelli, R. Optimization methodology to fruit grove mapping in precision agriculture. *Comput. Electron. Agric.* **2015**, *116*, 88–100. [CrossRef]
25. Shalal, N.; Low, T.; McCarthy, C.; Hancock, N. Orchard mapping and mobile robot localisation using on-board camera and laser scanner data fusion—Part A, Tree detection. *Comput. Electron. Agric.* **2015**, *119*, 254–266. [CrossRef]
26. Mao, W.; Liu, H.; Wang, D.; Yang, F.; Liu, Z. An Improved AODV Routing Protocol for Multi-Robot Communication in Orchard. *Smart Agric.* **2021**, *3*, 96.
27. Mao, W.; Liu, Z.; Liu, H.; Yang, F.; Wang, M. Research Progress on Synergistic Technologies of Agricultural Multi-Robots. *Appl. Sci.* **2021**, *11*, 1448. [CrossRef]
28. Mao, W.; Liu, H.; Hao, W.; Yang, F.; Liu, Z. Development of a Combined Orchard Harvesting Robot Navigation System. *Remote Sens.* **2022**, *14*, 675. [CrossRef]
29. Liu, Z.; Liu, H.; Mao, W.; Yang, F.; Wang, W.; Qin, J. Research on Wireless Signal Propagation Characteristics of Traditional Apple Orchard for Multi-robot. *Trans. Chin. Soc. Agric. Mach.* **2022**, *53*, 283–293.
30. Zhou, Z.; Cao, J.; Di, S. Overview of 3D Lidar SLAM algorithms. *Chin. J. Sci. Instrum.* **2021**, *42*, 13–27.
31. Shan, T.; Englot, B. Lego-loam, Lightweight and ground-optimized lidar odometry and mapping on variable terrain. In Proceedings of the 2018 IEEE/RSJ International Conference on Intelligent Robots and Systems (IROS), Madrid, Spain, 1–5 October 2018; pp. 4758–4765.
32. Shan, T.; Englot, B.; Meyers, D.; Wang, W.; Ratti, C.; Rus, D. Lio-sam: Tightly-coupled lidar inertial odometry via smoothing and mapping. In Proceedings of the 2020 IEEE/RSJ international conference on intelligent robots and systems (IROS), Las Vegas, NV, USA, 24 October–24 January 2020; pp. 5135–5142.
33. Chai, S.; Yao, L.; Xu, L.; Chen, Q.; Xu, T.; Yang, Y. Research on greenhouse agricultural machinery path tracking based on dynamic look ahead distance pure pursuit model. *J. Chin. Agric. Mech.* **2021**, *42*, 58–64.
34. Mao, W.; Liu, H.; Wang, X.; Yang, F.; Liu, Z.; Wang, Z. Design and experiment of a dual navigation mode orchard transport robot. *Trans. Chin. Soc. Agric. Mach.* **2022**, *53*, 27–39.
35. Zhou, J.; He, Y. Research Progress on Navigation Path Planning of Agricultural Machinery. *Trans. Chin. Soc. Agric. Mach.* **2021**, *52*, 1–14.

agronomy

MDPI

Article

Double-DQN-Based Path-Tracking Control Algorithm for Orchard Traction Spraying Robot

Zhigang Ren [1,2,3,4], Zhijie Liu [1,2,3,4], Minxin Yuan [1,2,3,4], Heng Liu [1,2,3,4], Wang Wang [1,2,3,4], Jifeng Qin [1,2,3,4] and Fuzeng Yang [1,2,3,4,*]

1 College of Mechanical and Electronic Engineering, Northwest A&F University, Yangling, Xianyang 712100, China
2 Apple Mechanized Research Base, Yangling, Xianyang 712100, China
3 Shanxi Key Laboratory of Apple, Yangling, Xianyang 712100, China
4 State Key Laboratory of Soil Erosion and Dryland Farming on Loess Plateau, Yangling, Xianyang 712100, China
* Correspondence: yangfzkm@nwafu.edu.cn

Abstract: The precise path-tracking control of tractors and trailers is the key to realizing agricultural automation. In order to improve the path-tracking control accuracy and driving stability of orchard traction spraying robots, this study proposed a navigation path-tracking control algorithm based on Double Deep Q-Network (Double DQN). Drawing on the driver's driving experience and referring to the principle of radar scanning and the principle of image recognition, a virtual radar model was constructed to generate a virtual radar map. The virtual radar map was used to describe the position relationship between the traction spraying robot and the planned path. Combined with the deep reinforcement learning method, all possible robot driving actions under the current virtual radar map were scored, and the best driving action was selected as the output of the network. In this study, a path-tracking algorithm was self-developed with a deep Q-network trained by driving the traction spraying robot in a simulated virtual environment. The algorithm was tested in both simulations and in a field to follow a typical 'U'-shaped path. The simulation results showed that the proposed algorithm was able to achieve accurate path-tracking control of the spraying trailer. The field tests showed that when the vehicle speed was 0.36 m/s and 0.75 m/s, the maximum lateral deviation of the algorithm was 0.233 m and 0.266 m, the average lateral deviation was 0.071 m and 0.076 m, and the standard deviation was 0.051 m and 0.057 m, respectively. Compared with the algorithm based on the virtual radar model, the maximum lateral deviation was reduced by 56.37% and 51.54%, the average lateral deviation was reduced by 7.8% and 5.0%, and the standard deviation was reduced by 20.31% and 8.1%, respectively. The results showed that the Double-DQN-based navigation path-tracking control algorithm for the traction spraying robot in the orchard had higher path-tracking accuracy and driving stability, which could meet the actual operational requirements of traditional orchards.

Keywords: conventional orchard; spraying trailer; navigation; path tracking; reinforcement learning

Citation: Ren, Z.; Liu, Z.; Yuan, M.; Liu, H.; Wang, W.; Qin, J.; Yang, F. Double-DQN-Based Path-Tracking Control Algorithm for Orchard Traction Spraying Robot. *Agronomy* **2022**, *12*, 2803. https://doi.org/10.3390/agronomy12112803

Academic Editor: Simon Pearson

Received: 25 September 2022
Accepted: 4 November 2022
Published: 10 November 2022

1. Introduction

China's orchards are currently divided into standard orchards and traditional orchards. The traditional orchards' planting area accounts for about 75% of all orchard area [1]. Pest and disease control work is an important part of orchard fruit tree management. With changes in season and climate, 8~15 spraying times per year occur, and this workload accounts for about 30% of the whole fruit tree management workload [2,3]. However, traditional orchards are densely planted, and their access roads are low and narrow, which makes it difficult for ordinary agricultural machinery to enter and operate [4]. Because the traction structure can effectively reduce the height of the whole machine to adapt to a low-channel working environment, it is widely used.

The goal of automatic navigation or guidance of agricultural machinery is to control the trajectory of the vehicle so that it maintains a constant distance from the adjacent driving line. Most scholars at home and abroad are committed to keeping the distance between the tractor and the planned path constant. In fact, different from self-propelled rice transplanters, fertilizer applicators and seeders, the goal of traction agricultural machinery navigation should be to maintain the operating points in the machine, such as the plough knife in the seeder, the flat shovel of the grader or the spraying trailer in the traction spraying robot, and keep a constant distance from the planned path. At present, domestic and foreign scholars have little research on the path-tracking control of the orchard traction spraying robot navigation, and the tractors are mainly wheeled tractors. As the steering principles of tracked vehicles and wheeled vehicles are different, the path-tracking algorithms of wheeled vehicles are difficult to apply directly to tracked vehicles [5,6]. However, accurate path tracking is the key for spraying robots to achieve precise spraying of agricultural crops [7]. Especially when working on a curved road, the driving track of a trailer is different from that of the tractor. Navigating the tractor alone can result in path gaps and overlaps, which can affect the effectiveness of spraying operations [8,9].

A traction agricultural machinery system is a non-holonomic system, which contains multiple nonlinear dynamic inputs and outputs, so it is not easy to realize their automatic navigation control. Modern control techniques such as Model Predictive Control (MPC) are now being applied to the control of traction agricultural machinery systems [10]. For example, Kayacan et al. combined nonlinear moving horizon estimation (NMHE) with a fast centralised NMPC approach to achieve accurate trajectory tracking of tractor-trailer farm machinery systems under varying soil conditions [11]. Yue et al. proposed a coordinated control method for tractor-trailer trajectory tracking control. The coordinated control consists of multiple levels of controllers, each consisting of a different algorithm [12]. Murillo et al. proposed a novel nonlinear mathematical model of an articulated tractor-trailer system that can be combined with the rolling horizon technique to improve the path-tracking performance of articulated systems [5]. Kayacan et al. proposed a nonlinear method of modelling the yaw and longitudinal dynamics of a tractor-trailer system to address the problem of limited performance of existing vehicle kinematic models for designing guidance systems. A complete nonlinear dynamics model and an accurate calculation of the sideslip are achieved [13]. However, all the above control methods rely on complex mathematical models and are not easy to apply in engineering.

Building accurate mathematical models is key to guaranteeing the effectiveness of the control, and the models applied in the controller can be either kinematic or kinetic [14,15]. For mechanical equipment with only a simple control device, the dynamic model is not applicable due to the lack of direct control force or torque. Therefore, this study studied the control method of the traction spraying robot based on the establishment of its kinematic model. In this regard, kinematic model-based controllers have been shown to be sufficiently accurate for vehicles operating at low acceleration [14,16,17].

To address these issues, this study attempted to use a set of positioning systems to obtain the pose information of the spraying trailer, created a kinematic model of the traction spraying robot to derive the pose information of the spraying trailer in real-time, and analyzed the path-tracking effect of the spraying trailer, which could effectively reduce the cost of equipment and facilitate its application. Based on this, we designed a Double-DQN-based robot navigation control system for accurate path tracking of the spraying trailer. The method was based on control criteria learned in a simulated environment (using a reward function) to optimize tracking performance. The method does not rely on complicated mathematical derivation and is easy to extend.

2. Materials and Methods

2.1. Hardware and Software Setup

In order to test the performance of the navigation path-tracking control algorithm of the orchard traction spraying robot based on Double-DQN, we built a test platform based on the representative 1KFL-30 crawler tractor widely used in China [6]. In order to make the tractor structure compact and reliable, the tractor was replaced with a Yuchai two-cylinder air-cooled engine. The spraying trailer was an orchard spraying machine developed by our team. The traction spraying robot is shown in Figure 1. The vehicle is small in size, and the crawler walking mechanism has good adaptability to various orchard types and road conditions. In addition, this diesel engine-based vehicle ensures high power, stability, ease of maintenance and low cost. The vehicle supports various tools to meet the requirements of various tasks (e.g., spraying, tilling, furrowing, fertilizer application, and transport). The main technical parameters of the traction spraying robot are shown in Table 1.

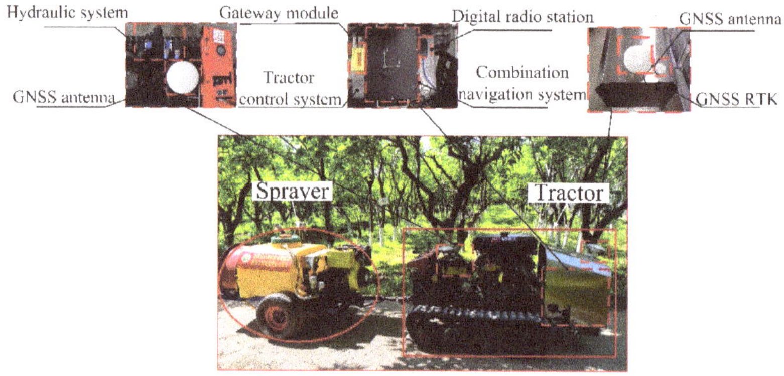

Figure 1. Orchard traction spraying robot.

Table 1. The main technical parameters of the orchard traction spraying robot.

Project	The Parameter Value
Dimensions (length × width × height) (mm)	3610 × 980 × 780
Maximum operating speed (km·h^{-1})	4.25
Spraying height (m)	2.6~3.2
Spraying width (m)	5~8
Pesticide box volume (L)	200
Track ground length (mm)	700
Gauge (mm)	700
Tractor supporting power (kW)	12
Sprayer supporting power (kW)	6.3

The structure of the navigation control system of the traction spraying robot is shown in Figure 2; it was mainly composed of navigation and positioning equipment, the host computer control system, a lower computer control system, a wireless remote control system and a wireless communication system. The navigation and positioning equipment was an XN422 combined navigation system (Xi'an North Jereh Optoelectronics Technology Co., Ltd., Xi'an, China). The equipment consisted of three parts: a reference station, a mobile station and a data link. It could provide position information with an accuracy of 0.03 m and heading information with an accuracy of 0.1° at a frequency of 100 Hz. The lower control system was deployed on the tractor. The lower control system transmitted the tractor pose information provided by the navigation and positioning equipment through R232, R422 serial communication, and CAN bus communication and received the navigation host computer control system or wireless remote control system command to control the robot. The host computer control system software was built based on Python3.6 and QT5.0 (The QT Company, Boston, CA, USA) framework. Its main functions include: (1) providing a visual interface, (2) path planning, (3) tractor autonomous learning and training, and (4) recording simulation and test data.

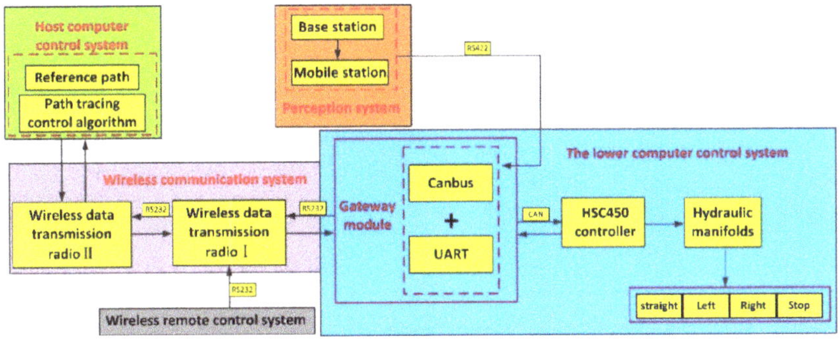

Figure 2. Structure of the navigation control system of the orchard traction spraying robot.

2.2. Kinematic Model of the Traction Spraying Robot

The tractor of the traction spraying robot included an electromagnetic hydraulic valve group, hydraulic cylinder and clutch. The onboard electronic control unit (ECU) outputted on-off signals by controlling the relay unit to adjust the state of the solenoid valve. Taking the forward motion state as an example, when the left track braked, the right track rotated, and the vehicle turned left. When the right track braked, the left track rotated, the vehicle turned right, and the turning radius was a fixed value. When the left and right tracks rotated at the same time, the vehicle moved forward. When the left and right tracks braked simultaneously, the vehicle stopped.

To simplify the kinematic analysis process, it was assumed that: (1) without considering the deformation, the tractor, the spraying trailer, and their connecting mechanism were rigid bodies, (2) the centroid of the tractor and spraying trailer coincided with the geometric center, (3) we could ignore the track section slip, and (4) the steering angular velocity and resistance coefficient of the crawler tractor remained unchanged during rotation [18].

The kinematic model of the traction spraying robot is shown in Figure 3. (x_i, y_i, θ_i) is defined as the positional information of each vehicle i = (0, 1), where $P_i(x_i, y_i)$ denotes the vehicle position coordinates, and θ_i denotes the heading of the vehicle (where I = 0 for the tracked tractor and i = 1 for the spraying trailer). L_0 denotes the distance from the tractor center of the mass coordinates to the hook-up point, L_1 denotes the distance from the spraying trailer center of the mass coordinates to the hook-up point, v_0 and v_1 denote, respectively, the instantaneous linear velocity of the center of mass of the tractor and the spraying trailer, and B is the tractor gauge width.

The instantaneous linear and angular velocities of the track speed and center of mass on the left and right sides of the tractor are related as in Equations (1) and (2).

$$v_0 = \frac{v_L + v_R}{2}, \tag{1}$$

$$\omega = \frac{v_R - v_L}{B}, \tag{2}$$

Based on the kinematic principles $\dot{x} = v\cos(\theta)$ and $\dot{\theta} = \omega$, Equations (1) and (2) are brought into it to yield the kinematic equation of the tracked tractor, as shown in Equation (3).

$$\begin{pmatrix} \dot{x}_0 \\ \dot{y}_0 \\ \dot{\theta}_0 \end{pmatrix} = \begin{pmatrix} \frac{\cos(\theta_0)}{2} & \frac{\cos(\theta_0)}{2} \\ \frac{\sin(\theta_0)}{2} & \frac{\sin(\theta_0)}{2} \\ \frac{1}{B} & -\frac{1}{B} \end{pmatrix} \begin{pmatrix} v_R \\ v_L \end{pmatrix}, \tag{3}$$

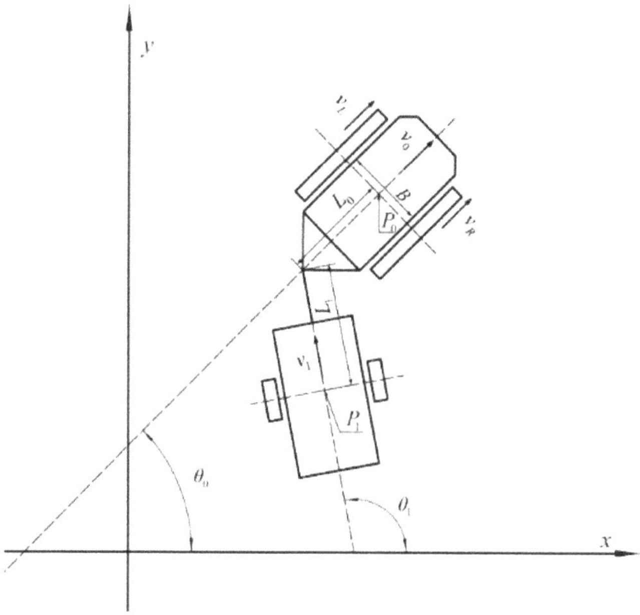

Figure 3. Kinematic model of the orchard traction spraying robot.

According to the kinematic principle, the expression for the position information of the spraying trailer is shown in Equation (4).

$$\begin{cases} \dot{x}_1 = v_1 \cos(\theta_1) \\ \dot{y}_1 = v_1 \sin(\theta_1) \end{cases}, \tag{4}$$

The relationship between the tractor position and the spraying trailer position can be derived from the tractor unit structure in Figure 3, as shown in Equation (5).

$$\begin{cases} x_1 = x_0 - (L_0 \cos\theta_0 + L_1 \cos\theta_1) \\ y_1 = y_0 - (L_0 \sin\theta_0 + L_1 \sin\theta_1) \end{cases}, \tag{5}$$

Deriving both sides of Equation (5) and substituting into Equation (1), Equation (6) is obtained as follows.

$$\begin{cases} v_1 \cos(\theta_1) = v_0 \cos(\theta_0) + L_0\dot{\theta}_0 \sin(\theta_0) + L_1\dot{\theta}_1 \sin(\theta_1) \\ v_1 \sin(\theta_1) = v_0 \sin(\theta_0) - L_0\dot{\theta}_0 \cos(\theta_0) - L_1\dot{\theta}_1 \cos(\theta_1) \end{cases}, \tag{6}$$

Multiplying the first formula of Equation (6) by $\sin(\theta_1)$ at both ends and the second formula by $\cos(\theta_1)$ at both ends, then subtracting the second formula from the first, gives Equation (7).

$$v_0(\cos(\theta_0)\sin(\theta_1) - \sin(\theta_0)\cos(\theta_1)) + L_0\dot{\theta}_0(\sin(\theta_0)\sin(\theta_1) + \cos(\theta_0)\cos(\theta_1)) + L_1\dot{\theta}_1 = 0 \tag{7}$$

Simplifying Equation (7) yields Equation (8):

$$\dot{\theta}_1 = \frac{1}{L_1}\left(v_0 \sin(\theta_0 - \theta_1) - L_0\dot{\theta}_0 \cos(\theta_0 - \theta_1)\right), \tag{8}$$

Similarly, multiplying the first formula of Equation (6) by $\cos(\theta_1)$ at both ends and the second formula by $\sin(\theta_1)$ at both ends, the two formulas are added together and simplified to give Equation (9):

$$v_1 = v_0 \cos(\theta_0 - \theta_1) + L_0\dot{\theta}_0 \sin(\theta_0 - \theta_1), \tag{9}$$

The tractor used in this study was a single-sided brake steering, which could only control the driving state of the tracks on the left and right sides. Based on the relationship between the instantaneous linear velocity of the center of mass of the tractor and the travelling speed of the left and right side tracks (Equation (1)), the instantaneous linear velocity of the center of mass v_0 in Equations (8) and (9) is replaced, and the kinematic model of the tracked tractor (Equation (3)) is combined to obtain the kinematic model of the orchard traction spraying robot, as shown in Equation (10).

$$\begin{pmatrix} \dot{x}_0 \\ \dot{y}_0 \\ \dot{\theta}_0 \\ \dot{x}_1 \\ \dot{y}_1 \\ \dot{\theta}_1 \end{pmatrix} = \begin{pmatrix} \frac{\cos(\theta_0)}{2} & \frac{\cos(\theta_0)}{2} \\ \frac{\sin(\theta_0)}{2} & \frac{\sin(\theta_0)}{2} \\ \frac{1}{B} & -\frac{1}{B} \\ \left(\frac{\cos(\theta_0-\theta_1)}{2} + \frac{L_0 \sin(\theta_0-\theta_1)}{B}\right)\cos(\theta_1) & \left(\frac{\cos(\theta_0-\theta_1)}{2} - \frac{L_0 \sin(\theta_0-\theta_1)}{B}\right)\cos(\theta_1) \\ \left(\frac{\cos(\theta_0-\theta_1)}{2} + \frac{L_0 \sin(\theta_0-\theta_1)}{B}\right)\sin(\theta_1) & \left(\frac{\cos(\theta_0-\theta_1)}{2} - \frac{L_0 \sin(\theta_0-\theta_1)}{B}\right)\sin(\theta_1) \\ \frac{\sin(\theta_0-\theta_1)}{2L_1} - \frac{L_0 \cos(\theta_0-\theta_1)}{BL_1} & \frac{\sin(\theta_0-\theta_1)}{2L_1} + \frac{L_0 \cos(\theta_0-\theta_1)}{BL_1} \end{pmatrix} \begin{pmatrix} v_R \\ v_L \end{pmatrix}, \tag{10}$$

Analysis of the kinematic model of the traction spraying robot shows that the relationship between the spraying trailer and the change in tractor heading is shown in Equation (11):

$$\frac{\dot{\theta}_1}{\dot{\theta}_0} = \frac{\sqrt{B^2 + {L_0}^2}}{L_1} \sin\left(\theta_0 - \theta_1 - \arctan\frac{L_0}{B}\right), \tag{11}$$

When $\theta_0 = \theta_1$, i.e., the spraying trailer heading angle is equal to the tractor heading angle, there is $\dot{\theta}_1/\dot{\theta}_0 = -L_0/L_1$, i.e., at this time, the heading change of the spraying trailer is opposite to that of the tractor, and the size is inversely proportional to the distance from the respective steering center to the hanging point.

2.3. *Double-DQN Model Development*

2.3.1. Double-DQN Network Architecture

We used the Double-DQN network structure, in which each action of the traction spraying robot had a separate output unit, i.e., turn left, turn right, and go straight. The state of the traction spraying robot was taken as the input of the neural network. The main advantage of this architecture is the ability to calculate the Q value of all actions in a given state with only one forward pass in the network [19].

In this study, the scanning result of a virtual radar on a virtual boundary was defined as a virtual radar map. The scanning of the virtual radar on the virtual path boundary could reflect the path deviation and path change of the tractor [9]. As shown in Figure 4, the starting point of the straight part of the reference path is $S(x_s, y_s)$, and the endpoint is $E(x_e, y_e)$. The width of the virtual path is d. The green graph in the figure shows the image detected by the virtual radar, which consists of four edges, of which the straight line segment is the detected virtual path boundary, and the circular arc segment is the maximum detection boundary of the radar. As shown in Figure 5, the virtual radar map positioning point is the midpoint of the connection between the tractor positioning point and the spraying trailer positioning point, for which θ is the course of the tractor. Using the virtual radar map as input to the Double-DQN network, the network structure and composition of the input and output layers are shown in Figure 6. To solve the path-tracking control problem, we made use of the optimal Q-value (action value) for each action. The Q-values for a series of given actions (the desired action in this study) were the output of the network, and a given state s was used as the network input. Each Q-value represented the expected future tracking performance of the corresponding steering action. Therefore, the action with the largest *Q*-value was selected.

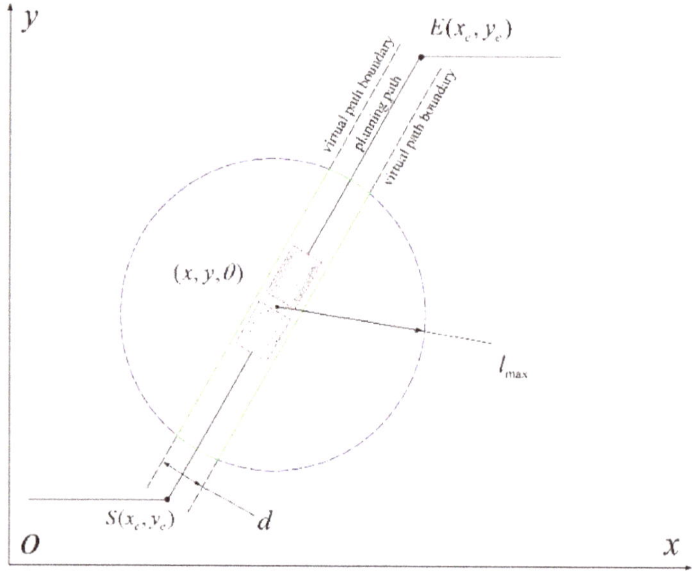

Figure 4. Schematic illustration of the virtual radar map generation.

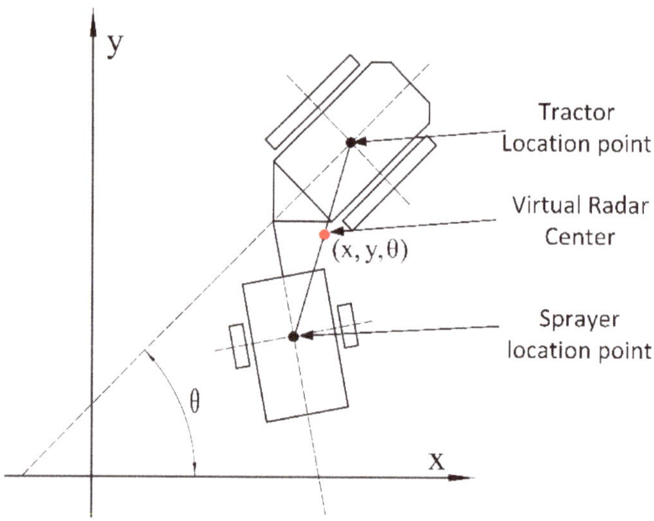

Figure 5. Schematic illustration of the main parameters of the virtual radar diagram of the orchard traction spraying robot.

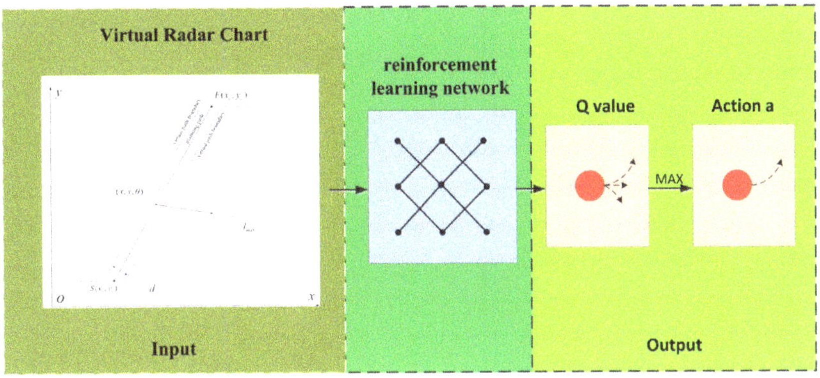

Figure 6. Overview of the network structure and the formation of its input and output layers.

The network architecture proposed in this study for training and control has six layers, including an input layer, four fully connected hidden layers, and an output layer, as shown in Figure 7. Based on Python 3.5, this study constructed a neural network model based on Windows10 using Google's open-source deep learning framework, TensorFlow2.0 (GPU). Specifically, the network was built using the Sequential model in Keras. There are 360 neurons in the input layer and 3 neurons in the output layer, corresponding to 3 predefined actions. The activation function is the sigmoid function [6]. Each hidden fully connected layer has 100 neurons, and the activation function is a rectification linear function. In this study, we determined the network structure by balancing the generalization ability and computational cost of the model. A simple network has fast calculation speed but poor generalization ability. Complex networks have strong generalization ability but need a long training time. When determining the model structure, we used the upper limit and rising speed of reward scores in network training as evaluation criteria. The values of the network hyperparameters are shown in Table 2.

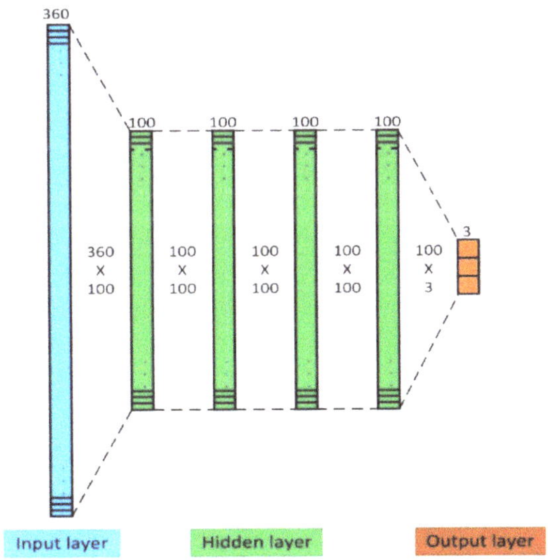

Figure 7. Schematic illustration of the Double-DQN network.

Table 2. List of hyperparameters and their values.

Hyperparameter	Value
Minibatch size	16
Replay memory size	20,000
update frequency	200
Discount factor γ	0.99
Learning rate	0.0001
Initial exploration ε	0.9

2.3.2. Double-DQN Training Algorithm

Double-DQN is an improved algorithm of the classical deep reinforcement learning algorithm DQN (Deep Q Network). Double-DQN solves the overestimation problem in DQN to a certain extent and improves the stability of the algorithm [20]. The flow of the Double-DQN agent training algorithm is shown in Figure 8.

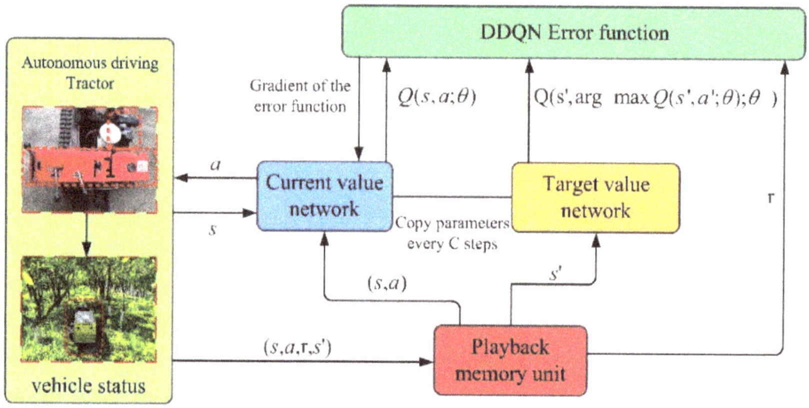

Figure 8. Agent training algorithm flowchart.

The agent (traction spraying robot) was trained for autonomous learning by first initializing the memory playback unit and the network parameters θ and θ^- of the current value network and the target value network. Then, s was initialized as the first state of the current state sequence; we subsequently selected action a and executed it using the ε-greedy algorithm to obtain the next state s', immediately return r, and store (s, a, r, s') in the memory playback unit. Each time a certain number of parameters were adjusted, a certain number of samples were randomly collected in the memory playback unit for gradient descent to solve the current value network parameters. Because the weight of the network is updated by gradient backpropagation, if the input data are sequential, it will lead to overestimation. Collecting a certain number of samples can improve training efficiency by disrupting the batch of input data. At regular intervals, the current value network parameters were assigned to the target value network so that the agent training was completed iteratively [21].

A reward is an incentive mechanism designed by people using prior knowledge to encourage reinforcement learning algorithms to learn in the direction of people's expectations. For path-following autonomous learning, we designed a sparse reward [22]. The rules for the incentive mechanism (12) were established as shown below. If R is a reward for the agent's action in the current state, when the agent made the correct action for the state, a reward of 1 was given; otherwise, a reward of -1 was given, and when the lateral deviation between the agent and the reference path was 0, a reward of 0 was given.

The network requires a large amount of data to be trained. In order to do this, each time the simulated line tracking process was run until the agent successfully reached the endpoint, the process was stored and then used for empirical replay [23]. Δd_1 denotes the lateral deviation value of the intelligent body in state s', $\Delta\theta_1$ denotes the heading deviation value of the agent in state s', and $\Delta\theta_0$ denotes the heading deviation value of the agent in state s. The initial state s and action a are randomly established in the set range. Δd_1 is in the range $[-700, 700]$, and $\Delta\theta_1$ and $\Delta\theta_0$ are in the range $(-\pi/2, \pi/2)$. When the lateral deviation and heading deviation took a positive value, it meant that it was on the left side of the planned path and vice versa for the right side. The value of action a is [0, 1, 2] (0 for straight, 1 for left, 2 for right). When the agent was outside the set range, the agent reverted to the starting point.

$$R = \begin{cases} +1 & \Delta d_1 \neq 0, \Delta\theta_1 \in (-\pi/2, \pi/2), \text{ Predefined action driving} \\ 0 & \Delta d_1 = 0, \Delta\theta_1 \in (-\pi/2, \pi/2) \\ -1 & \Delta d_1 \neq 0, \Delta\theta_1 \in (-\pi/2, \pi/2), \text{ Non-predefined action driving} \end{cases}, \quad (12)$$

According to the rules of the reward mechanism, the reward and punishment values of specific actions in different states of the agent could be obtained, as shown in Tables 3 and 4. Here, $\Delta\theta_1 > \Delta\theta_0$ for left, $\Delta\theta_1 = \Delta\theta_0$ for straight, and $\Delta\theta_1 < \Delta\theta_0$ for right.

Table 3. Reward and punishment values for the left side of the planning path.

Δd_1	$\Delta\theta_1$	R
(0, 70]	(−90, −10]	1 (left turn) −1 (other actions)
	(−10, 0)	1 (straight) −1 (other actions)
	[0, 90)	1 (right) −1 (other actions)
(70, 150]	(−90, −20]	1 (left) −1 (other actions)
	(−20, −12)	1 (straight) −1 (other actions)
	[−12, 90)	1 (right) −1 (other actions)
(150, 700]	(−90, −30]	1 (left) −1 (other actions)
	(−30, −22)	1 (straight) −1 (other actions)
	[−22, 90)	1 (right) −1 (other actions)

Table 4. Reward and punishment values for the right side of the planning path.

Δd_1	$\Delta\theta_1$	R
[−70, 0)	(−90, 0]	1 (left) −1 (other actions)
	(0, 10)	1 (straight) −1 (other actions)
	[10, 90)	1 (right) −1 (other actions)
[−150, −70)	(−90, 12]	1 (left) −1 (other actions)
	(12, 20)	1 (straight) −1 (other actions)
	[20, 90)	1 (right) −1 (other actions)
[−700, −150)	(−90, 22]	1 (left) −1 (other actions)
	(22, 30)	1 (straight) −1 (other actions)
	[30, 90)	1 (right) −1 (other actions)

3. Simulation Test Results

3.1. Simulation Environment Creation

The simulation environment of this study was based on Python 3.5, the PyQt5 framework and Google's open-source deep learning program framework, TensorFlow2.0 (GPU version). The running environment was a Windows 10 system; the simulation environment is shown in Figure 9. The agent model represents the traction spraying robot, the reference path fold line represents the operation path, and the solid dots represent the starting and ending points of the planned path. The reference path parameters are shown in Figure 10, with the reference path starting point being (0, 0) m and the reference path endpoint being (5.0, 8.0) m.

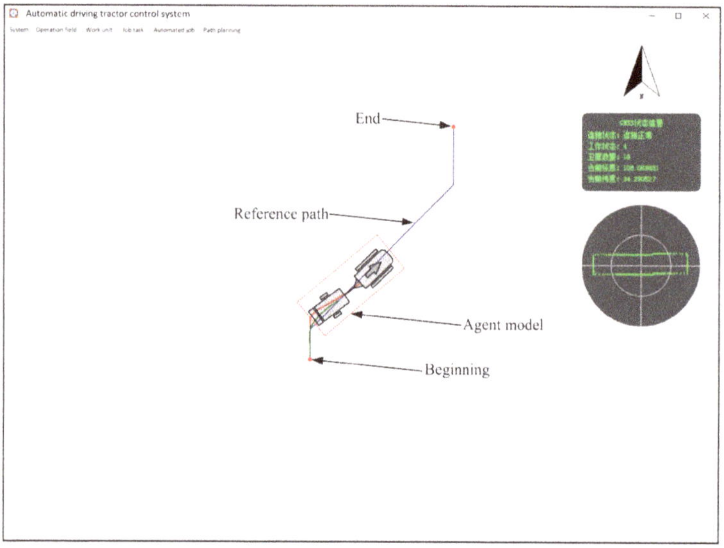

Figure 9. The agent simulation environment.

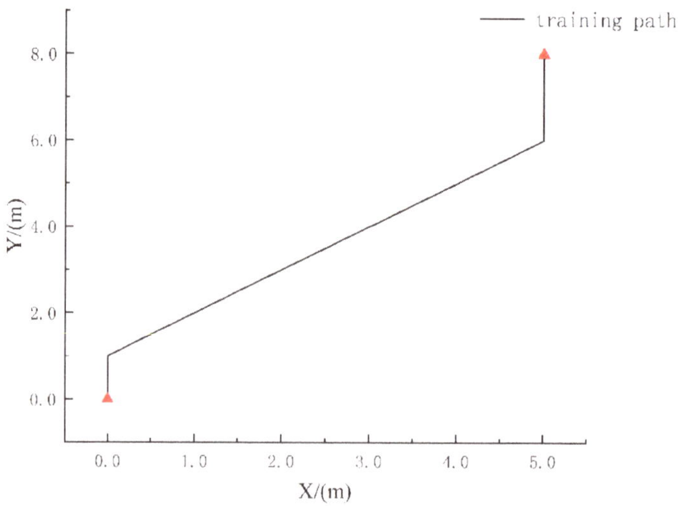

Figure 10. Schematic illustration of reference path.

The relevant network parameters and training hyperparameters were set according to the previous sections. The current value network and the target value network were structurally consistent. In the simulation experiment, the storage capacity of the memory playback unit was set to 90,000, and the network parameters of the target value network were copied to the current value network parameters once for each 200-parameter adjustment so as to reduce the correlation between the two networks and improve the network training efficiency. The ε value of the ε-greedy algorithm was set to 0.9 to improve the exploration ability of the intelligence, i.e., a 0.9 probability of selecting the action with the largest value of state behavior for execution; otherwise, a random action was selected for execution, and the decay factor of the ε-greedy algorithm was set to 0.0001. The training of the agent ended when the network converged to a certain extent.

The agent started from the starting point of the reference path. In order to make the agent's exploratory learning more comprehensive, we added random disturbance to the agent's kinematics model, and the random disturbance deviation range was [−20, 20] mm. If the lateral deviation or heading deviation of the agent from the reference path were greater than the set threshold, the agent ended the round and returned to the origin to start the next round of training again. The agent reaching the end of the reference path was regarded as the success of this round of learning.

3.2. Simulation Test Results

The single-step reward results of agent training are shown in Figure 11. The abscissa represents the number of training steps, and the ordinate represents the reward value of each step. The total number of agent training steps was 23,296 steps. With the increase in training steps, the reward value tended to be stable, and the score of each round increased gradually. The scoring results of each round of agent training are shown in Figure 12. At the beginning of the training, the score was low, and the score changed greatly. As the number of rounds increased, the score became larger, and the data tended to be stable. The success rate of agent training is shown in Figure 13. The agent reached the end of the reference path in the fifth round. As the training continued, the success rate of the agent continued to increase.

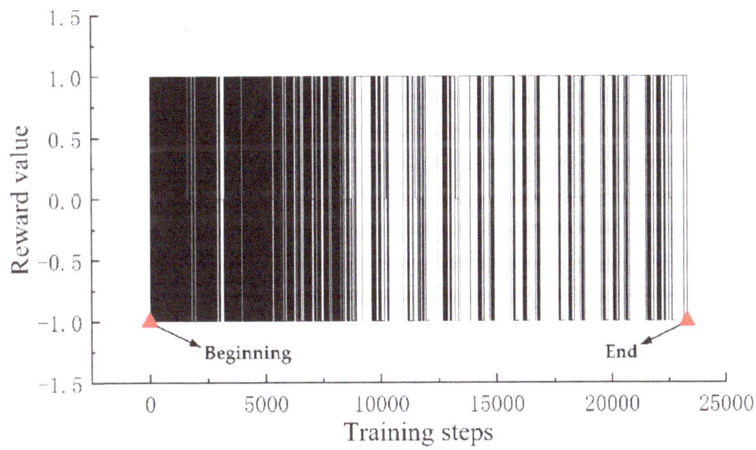

Figure 11. Agent training single-step rewards.

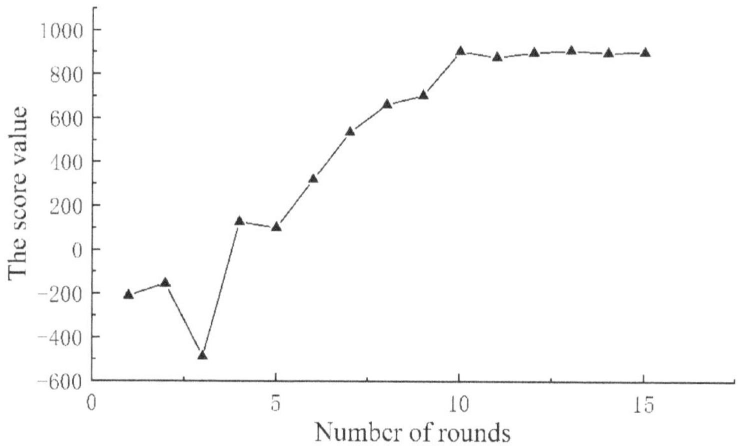

Figure 12. The agent trains to score points per turn.

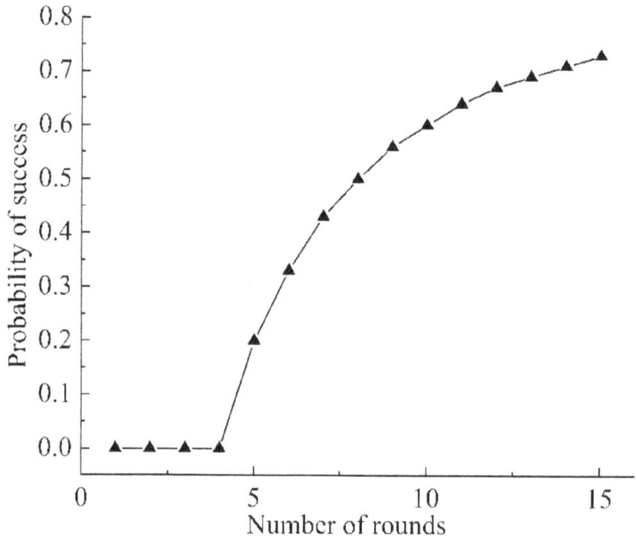

Figure 13. Agent training success rate.

The results of the training and testing of the agent showed the value and potential of this idea of autonomous learning using an agent. The advantage of reinforcement learning is that an agent without excessive a priori knowledge could learn some of the common experiences of artificial design through continuous interaction with the environment and may even discover some experiences that have not been learned by engineers [20,21].

3.3. Validation of Simulation Results

To verify the feasibility of the path-tracking algorithm proposed in the study, simulation experiments were carried out in a simulation environment. The path planned for the simulation is a 'U'-shaped folding path, 20 m long and 4 m wide (total path length is approximately 42.83 m), with a path starting point of (0, 0) m and a path end point of (0, 4.0) m. The initial lateral deviation of the kinematic model of the traction spraying robot was 0.5 m, and the heading angle was 180°. The simulation verification was carried out at two speeds of 0.36 m/s and 0.75 m/s. During the simulation process, the communication, display and data recording of the upper control system were normal, and the simulation test results are shown in Figure 14.

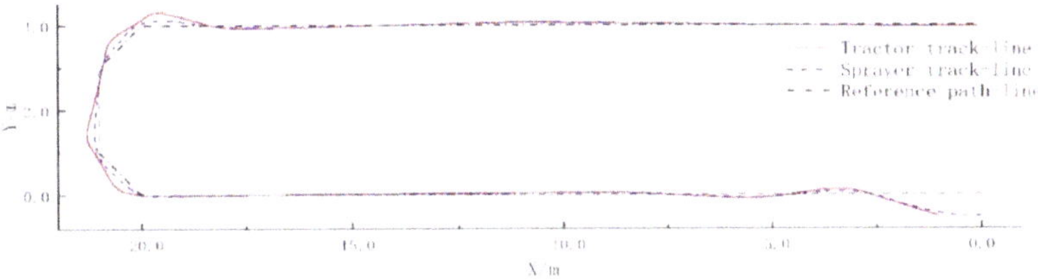

(**a**) The 0.36 m/s path-tracking track.

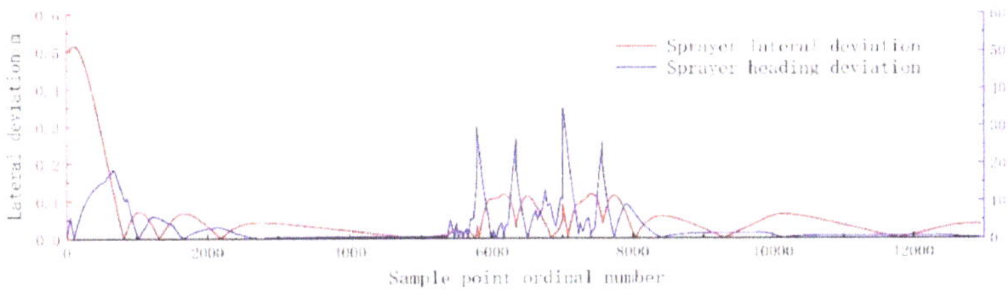

(**b**) The 0.36 m/s spraying trailer path-tracking lateral and heading deviations.

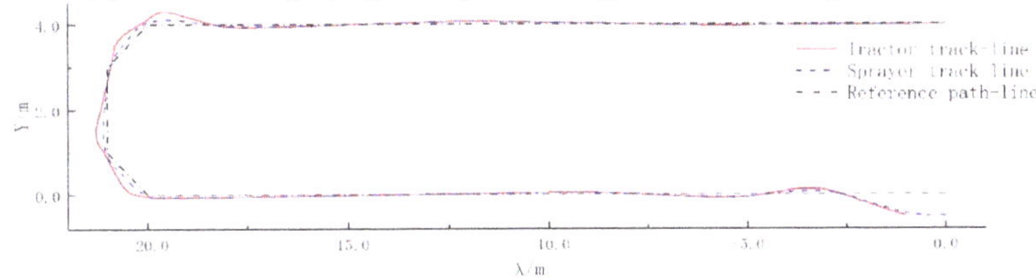

(**c**) The 0.75 m/s path-tracking track.

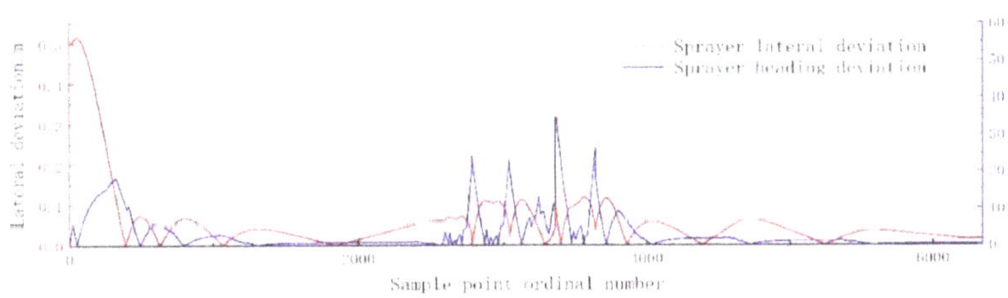

(**d**) The 0.75 m/s spraying trailer path-tracking lateral and heading deviations.

Figure 14. Simulation experimental results.

When the simulation speed was 0.36 m/s, the maximum lateral deviation of the spraying trailer path tracking was 0.117 m, the average lateral deviation was 0.038 m, and the standard deviation was 0.028 m. When the simulation speed was 0.75 m/s, the maximum lateral deviation of the spraying trailer path tracking was 0.119 m, the average lateral deviation was 0.040 m, and the standard deviation was 0.029 m. The simulation results showed that the Double-DQN-based orchard traction spraying robot navigation path-tracking control algorithm was feasible in principle for navigation path-tracking control, and the spraying trailer had high path-tracking accuracy to meet practical operational needs.

4. Field Test Results

Experimental Results of Robot Path-Tracking Control Algorithm

Our team used a portable differential positioning device (Figure 15) to verify the method of calculating the position of the spraying trailer by the established kinematic model of the traction spraying robot. For the verification test, our team carried out a portable differential positioning equipment accuracy test and kinematic model to calculate the spraying trailer pose information method validation test. The test results showed that the fixed differential signal of the portable differential positioning equipment was stable and met the requirements of this study. The position of the spraying trailer calculated by the kinematic model had the same trend as the position trajectory of the spraying trailer actually recorded by the portable differential positioning device. The experimental results showed that the kinematic model established in this study could better reflect the kinematic characteristics of the orchard traction spraying robot and had higher accuracy. To a certain extent, it could replace the sensor to calculate the position of the spraying trailer in the actual operation process of the orchard traction spraying robot, saving the cost of the equipment.

Figure 15. Portable differential positioning devices.

The experimental field of the path-tracking control algorithm for the traction spraying robot was located in the National Persimmon Germplasm Resources Garden of Northwest A&F University (108.066° E, 34.297° N). As shown in Figure 16, the row spacing of fruit trees is 2.4 m, and the plant spacing is about 2~3 m.

Figure 16. Orchard traction spraying robot test site.

The field test site is shown in Figure 17. Considering that the robot needed to follow a straight line and turn around to change rows during the orchard operation, the planned path was designed as a 20.0 × 2.4 m 'U'-shaped folding path, and the test speed was set at 0.36 m/s and 0.75 m/s. To ensure the validity of the test data, each set of tests was repeated three times (the lateral and heading deviations in the test results are absolute values). The results of the path-tracking tests are shown in Figure 18, the statistics of the repeated tests are shown in Table 5, and the three-dimensional diagrams of the path-tracking errors for the two control methods are shown in Figure 19.

Figure 17. Navigation path-tracking real vehicle experiment.

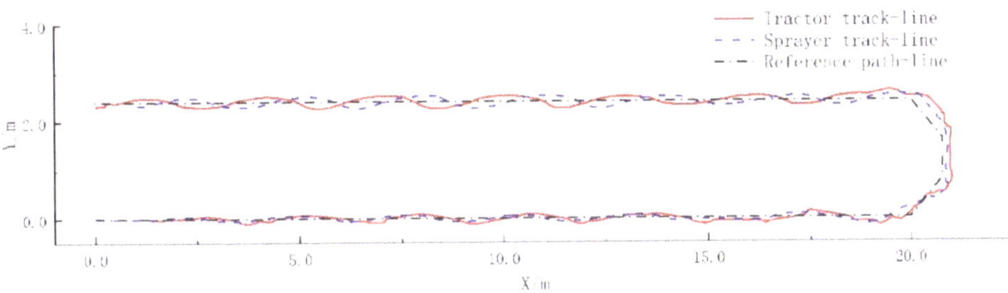

(**a**) The 0.36 m/s path-tracking track.

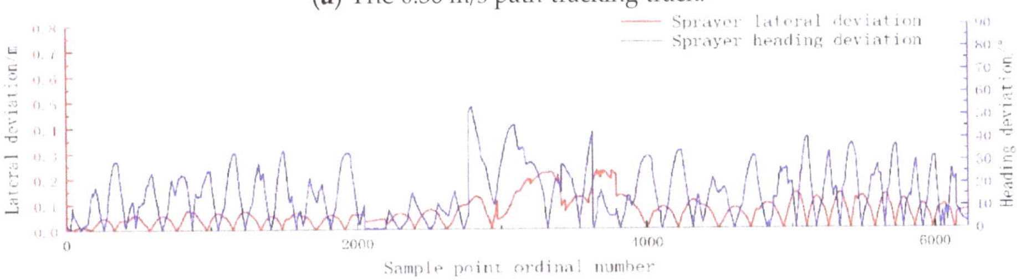

(**b**) The 0.36 m/s spraying trailer path-tracking lateral and heading deviations.

Figure 18. *Cont.*

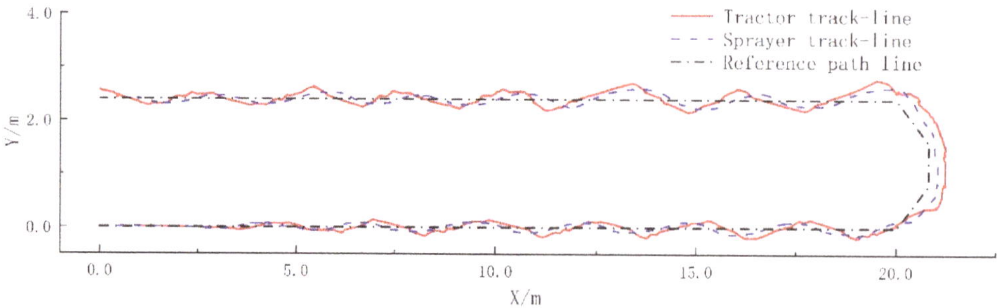

(**c**) The 0.75 m/s path-tracking track.

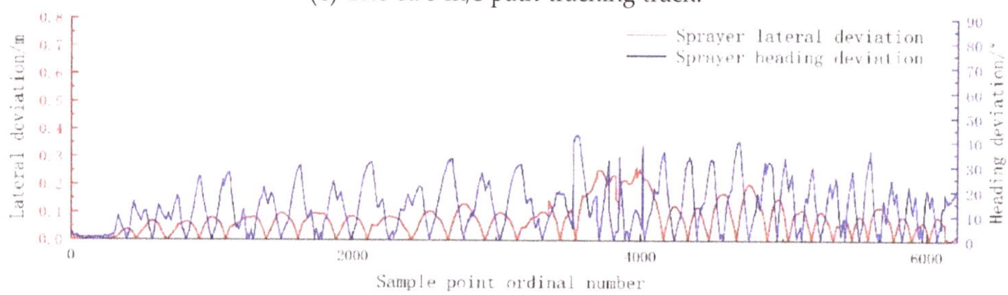

(**d**) The 0.75 m/s spraying trailer path-tracking lateral and heading deviations.

Figure 18. Results of the real vehicle experiment.

Table 5. Path-tracking experimental data.

v/(m/s)	Test No.	Sprayer Path-Tracking Deviation					
		Maximum Lateral Deviation /(m)		Average Lateral Deviation /(m)		Standard Deviation/(m)	
		Straight Line (0~20 m)	Full Range	Straight Line (0~20 m)	Full Range	Straight Line (0~20 m)	Full Range
0.36	r361	0.128	0.233	0.043	0.070	0.026	0.051
	r362	0.120	0.225	0.041	0.069	0.025	0.049
	r363	0.134	0.240	0.045	0.073	0.028	0.054
	$\overline{r36}$	0.127	0.233	0.043	0.071	0.026	0.051
	l361	0.126	0.536	0.051	0.084	0.021	0.059
	l362	0.146	0.558	0.061	0.080	0.023	0.069
	l363	0.113	0.509	0.056	0.066	0.026	0.063
	$\overline{l36}$	0.128	0.534	0.056	0.077	0.023	0.064
0.75	r751	0.145	0.263	0.055	0.076	0.033	0.056
	r752	0.155	0.275	0.057	0.079	0.034	0.058
	r753	0.140	0.259	0.054	0.074	0.032	0.056
	$\overline{r75}$	0.147	0.266	0.055	0.076	0.033	0.057
	l751	0.149	0.536	0.057	0.081	0.030	0.062
	l752	0.178	0.554	0.054	0.081	0.034	0.066
	l753	0.207	0.558	0.061	0.078	0.034	0.059
	$\overline{l75}$	0.178	0.549	0.057	0.080	0.033	0.062

Note: Test number r361 stands for the Double-DQN algorithm path-tracking test group 1 at 0.36 m/s speed, l361 stands for the virtual radar control algorithm path-tracking test group 1 at 0.36 m/s speed, $\overline{r36}$ stands for the average of the Double-DQN algorithm control at 0.36 m/s speed for 3 sets of test results, and $\overline{l36}$ stands for the average of the virtual radar model control at 0.36 m/s speed.

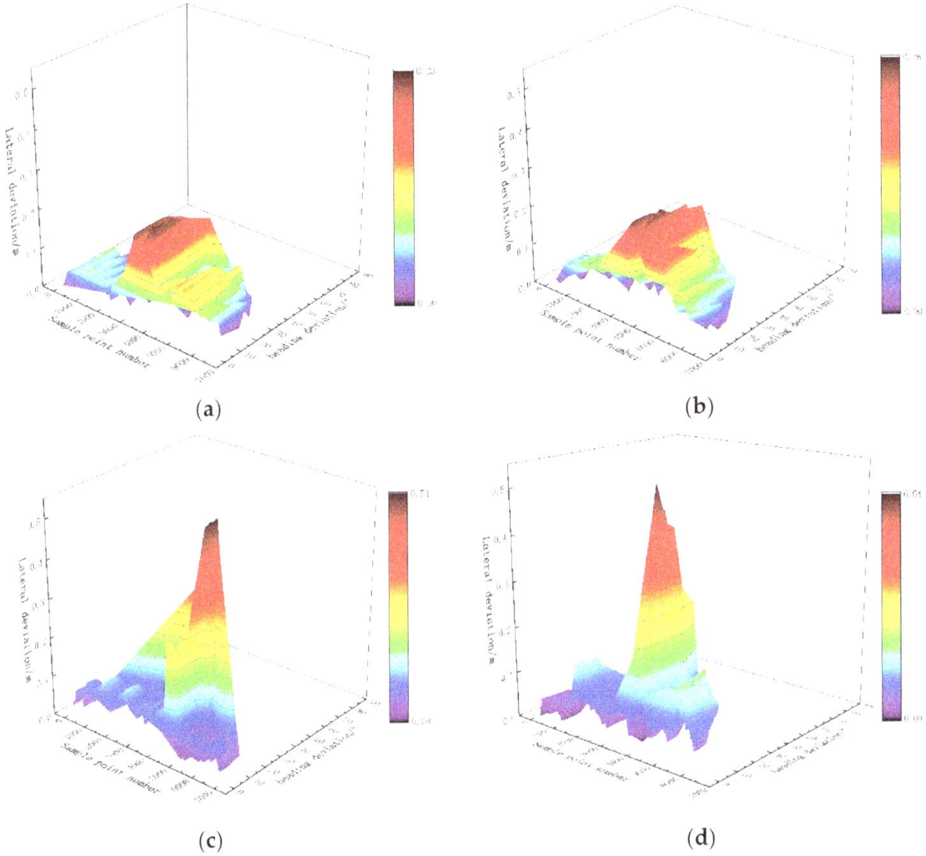

Figure 19. (**a**,**b**) The 3D maps of 0.36 m/s and 0.75 m/s path-tracking error for the Double-DQN method; (**c**,**d**) 3D maps of 0.36 m/s and 0.75 m/s path-tracking error for the virtual radar method.

From Figures 18 and 19, it can be seen that under the same speed, the path-tracking algorithm based on Double-DQN had a smaller overall lateral deviation in the whole path of the spraying trailer compared with the path-tracking algorithm based on the virtual radar model [9]. The spraying trailer had better path-tracking accuracy and driving stability in the turning process and the straight-line driving part after lane changing.

The results of comparing the tracking accuracy of the straight path portion ('U'-shaped fold path 0~20 m) of the spraying trailer for the two control algorithms are shown in Table 6. At the speed of 0.36 m/s, the virtual radar model algorithm was superior to the Double-DQN algorithm in terms of the driving stability (standard deviation) of the spraying trailer. However, the algorithm proposed in this study was superior to the virtual radar model algorithm in terms of the average lateral deviation and maximum lateral deviation. The average lateral deviation and maximum lateral deviation were reduced by 23.21% and 0.78%, respectively. At the speed of 0.75 m/s, the driving stability (standard deviation) of the spraying trailer proposed in this study was consistent, but it was superior to the virtual radar model algorithm in terms of average lateral deviation and maximum lateral deviation. The maximum lateral deviation and average lateral deviation were reduced by 17.41% and 3.51%, respectively.

Table 6. Comparison of straight-line path-tracking accuracy.

v/(m/s)	Test No.	Maximum Lateral Deviation /(m)	Average Lateral Deviation /(m)	Standard Deviation /(m)
0.36	$\overline{r36}$	0.127	0.043	0.026
	$\overline{l36}$	0.128	0.056	0.023
		−0.001 (0.78%)	−0.013 (23.21%)	0.003 (13.04%)
0.75	$\overline{r75}$	0.147	0.055	0.033
	$\overline{l75}$	0.178	0.057	0.033
		−0.031 (17.41%)	−0.002 (3.51%)	0 (0%)

Comparing the 'U'-shaped path-tracking accuracy of the spraying trailer under the two navigation path-tracking control algorithms, the results are shown in Table 7. All path-tracking indexes of the proposed algorithm were better than those of the virtual radar model algorithm. The maximum lateral deviation was reduced by 56.37%, the average lateral deviation was reduced by 7.8%, and the standard deviation was reduced by 20.31%. This showed that the overall path-tracking performance of the new algorithm in the 'U'-shaped path was significantly better than that of the tracking algorithm based on the virtual radar model.

Table 7. Comparison of 'U'-shaped path-tracking accuracy.

v/(m/s)	Test No.	Maximum Lateral Deviation /(m)	Average Lateral Deviation /(m)	Standard Deviation /(m)
0.36	$\overline{r36}$	0.233	0.071	0.051
	$\overline{l36}$	0.534	0.077	0.064
		−0.301 (56.37%)	−0.006 (7.8%)	0.013 (20.31%)
0.75	$\overline{r75}$	0.266	0.076	0.057
	$\overline{l75}$	0.549	0.080	0.062
		−0.283 (51.54%)	−0.004 (5.0%)	−0.005 (8.1%)

Comparing the tracking accuracy of the 'U'-shaped path and the straight path of the spraying trailer under the two algorithms, the results are shown in Table 8. The tracking accuracy indexes of the 'U'-shaped path and straight path based on the virtual radar model were quite different, especially the maximum lateral deviation. When the speed was 0.36 m/s and 0.75 m/s, the maximum lateral deviation of the 'U'-shaped path based on the virtual radar model was 317.19% and 208.43% higher than the maximum lateral deviation of the straight path, while the relevant indicators of the proposed algorithm were relatively similar. When the speed was 0.36 m/s and 0.75 m/s, the maximum lateral deviation of the 'U'-shaped path was 83.46% and 80.95% higher than the maximum lateral deviation of the straight path. It showed that the deviation control ability of the new algorithm was better than that of the virtual radar model control and had better steering path adaptability.

Table 8. Comparison of 'U'-shaped and straight line path-tracking accuracy.

v/(m/s)	Path Shape	Test No.	Maximum Lateral Deviation /(m)	Average Lateral Deviation /(m)	Standard Deviation /(m)
0.36	Straight	$\overline{r36}$	0.127	0.043	0.026
	'U'-shaped	$\overline{r36}$	0.233	0.071	0.051
			0.106 (83.46%)	0.028 (65.12%)	0.025 (96.15%)
	Straight	$\overline{l36}$	0.128	0.056	0.023
	'U'-shaped	$\overline{l36}$	0.534	0.077	0.064
			0.406 (317.19%)	0.021 (37.5%)	0.041 (178.26%)
0.75	Straight	$\overline{r75}$	0.147	0.055	0.033
	'U'-shaped	$\overline{r75}$	0.266	0.076	0.057
			0.119 (80.95%)	0.021 (38.18%)	0.024 (72.72%)
	Straight	$\overline{l75}$	0.178	0.057	0.033
	'U'-shaped	$\overline{l75}$	0.549	0.080	0.062
			0.371 (208.43%)	0.023 (40.35%)	0.029 (87.88%)

The statistical results of the path-tracking data of the spraying trailer and tractor were analyzed, as shown in Table 9. The path-tracking accuracy indexes of the spraying trailer and tractor based on the virtual radar model were quite different, especially the maximum lateral deviation. When the speed was 0.36 m/s and 0.75 m/s, the maximum lateral deviation of the spraying trailer path tracking was 256.0% and 187.43% higher than that of the tractor path tracking. The relevant indicators of the proposed algorithm were relatively similar. When the speed was 0.36 m/s and 0.75 m/s, the maximum lateral deviation of the spraying trailer path tracking was 39.0% and 45.5% lower than the maximum lateral deviation of the tractor path tracking. It showed that the new algorithm had better control ability than the virtual radar model control for spraying trailer path tracking.

Table 9. Comparison of spraying trailer and tractor path tracking.

v/(m/s)	Model	Test No.	Maximum Lateral Deviation /(m)	Average Lateral Deviation /(m)	Standard Deviation /(m)
0.36	Tractor	$\overline{r36}$	0.382	0.087	0.075
	Trailer	$\overline{r36}$	0.233	0.071	0.051
			−0.149 (39.0%)	−0.016 (18.39%)	−0.024 (32%)
	Tractor	$\overline{l36}$	0.150	0.031	0.025
	Trailer	$\overline{l36}$	0.534	0.077	0.064
			0.384 (256.0%)	0.046 (148.39%)	0.039 (156.0%)
0.75	Tractor	$\overline{r75}$	0.448	0.110	0.095
	Trailer	$\overline{r75}$	0.266	0.076	0.057
			−0.222 (45.5%)	−0.034 (30.9%)	−0.038 (40%)
	Tractor	$\overline{l75}$	0.191	0.051	0.036
	Trailer	$\overline{l75}$	0.549	0.080	0.062
			0.358 (187.43%)	0.029 (56.86%)	0.026 (72.22%)

5. Discussion

The automatic navigation of the traction spraying robot should not only focus on the position change of the tractor during its navigation and driving but also on whether the spraying trailer can effectively track the planned orchard operation path. In order to improve the path-tracking control accuracy and driving stability of the orchard traction spraying robot, this study proposed a navigation path-tracking control algorithm based on the Double Deep Q-Network (Double DQN). The algorithm was tested in both simulations and on a field to follow a typical 'U'-shaped path. The simulation results showed that the proposed algorithm was able to achieve accurate path-tracking control of the spraying trailer. The field tests showed that when the vehicle speed was 0.36 m/s and 0.75 m/s, the maximum lateral deviation of the algorithm was 0.233 m and 0.266 m, the average lateral deviation was 0.071 m and 0.076 m, and the standard deviation was 0.051 m and 0.057 m, respectively. Compared with the algorithm based on the virtual radar model [6], the maximum lateral deviation was reduced by 56.37% and 51.54%, the average lateral deviation was reduced by 7.8% and 5.0%, and the standard deviation was reduced by 20.31% and 8.1%, respectively.

Although the trajectory of the spraying trailer is controlled, compared with the simulation test, the indicators of the path tracking of the spraying trailer in the field test were quite different. The test comparison data are shown in Table 10. When the vehicle speed was 0.36 m/s and 0.75 m/s, the maximum lateral deviation of the spraying trailer path tracking in the field experiment increased by 99.147% and 123.53%, the average lateral deviation increased by 86.84% and 90%, and the standard deviation increased by 82.14% and 96.55%, respectively. The main possible reasons for this are as follows: (1) in order to save costs, we used a set of positioning systems to obtain the position information of the tractor and estimate the position information of the spraying trailer through the kinematic model. The angle difference between the heading angle of the tractor and the spraying trailer was not taken into account, which may lead to deviation errors. (2) The tractor control device was simple, and the tractor had a reaction time from receiving the signal to performing the corresponding action, which may lead to deviation errors. (3) With the increase in the speed of the traction spraying robot, the navigation path-tracking accuracy of the spraying trailer was reduced. Due to the simple control device of the tractor in this study, the influence of different speeds on the control accuracy was not considered, which may lead to deviation errors. (4) When the robot was working in the orchard, the GNSS signal drifted briefly, which may lead to deviation errors.

Table 10. Comparison of simulation and field experiment path-tracking accuracy.

v/(m/s)	Test Type	Maximum Lateral Deviation /(m)	Average Lateral Deviation /(m)	Standard Deviation /(m)
0.36	Simulation	0.117	0.038	0.028
	Field	0.233	0.071	0.051
		0.116 (99.14%)	0.033 (86.84%)	0.023 (82.14%)
0.75	Simulation	0.119	0.040	0.029
	Field	0.266	0.076	0.057
		0.147 (123.53%)	0.036 (90%)	0.028 (96.55%)

Future work will focus on: (1) in the stage of establishing the kinematics model of the traction spraying robot, the relevant parameters at the connection of the spraying trailer will be added, such as the angle difference between the yaw angle of the tractor and the spraying trailer, so as to include this interaction in the design stage, which may achieve a better control effect; (2) in the aspect of optimizing the control algorithm, we suggest studying the influence of virtual radar model parameters, the neural network structure and parameter changes on navigation effect; (3) we propose studying the influence of speed changes on navigation effectso that the control model can adapt to the effect of speed variation; and (4) we recommend using 3D radar sensors to sense the environment, as through multi-sensor data fusion, we can deal with the interference caused by trunk occlusion.

6. Conclusions

In this paper, in order to improve the precision and driving stability of a spraying trailer tracking a predefined trajectory in a traction spraying robot, this study proposed a navigation path-tracking control algorithm based on the Double-DQN method. The Double-DQN method is an optimal decision-making method based on state space. This method is different from the supervised learning method. It does not need to be trained according to the sample. The control standard (using the reward function) learned in the simulation environment can optimize the tracking performance. It does not depend on complex mathematical models and is easy to expand. In order to reduce the cost of use and facilitate practical application, we used a set of positioning systems to obtain the pose information of the tractor, combined the kinematics model of the traction spraying robot to calculate the pose information of the spraying trailer in real-time, and analyzed the path-tracking effect of the spraying trailer.

The algorithm was tested in both simulations and on a field to follow a typical 'U'-shaped path. The simulation results showed that when the speed was 0.36 m/s, the maximum lateral deviation of the spraying trailer path tracking was 0.117 m; when the speed was 0.75 m/s, the maximum lateral deviation of the spraying trailer path tracking was 0.119 m. This showed that the proposed algorithm could realize the navigation path-tracking control of the spraying trailer in principle and had a high control accuracy, which could meet the needs of the orchard's autonomous spraying operation. The field test results showed that when the vehicle speed was 0.36 m/s and 0.75 m/s, compared with the control algorithm of the virtual radar model, the maximum lateral deviation of the spraying trailer was reduced by 56.37% and 51.54%, respectively, the average lateral deviation was reduced by 7.8% and 5.0%, respectively, and the standard deviation was reduced by 20.31% and 8.1%, respectively. It showed that the overall performance of the algorithm was better than the control algorithm based on the virtual radar model, which effectively improved the tracking accuracy of the navigation path of the spraying trailer and the stability of the automatic operation and had better adaptability to the typical operation path of the orchard.

Author Contributions: Z.R., Z.L. and F.Y. started the work, completed the detailed investigations, and prepared the paper with the support of all the co-authors; H.L., M.Y., W.W. and J.Q. helped us with orchard trials. All authors have read and agreed to the published version of the manuscript.

Funding: This research was funded by the Major Science and Technology Project of Shaanxi Province of China (Program No. 2020zdzx03-04-01) and the National Key R&D Program of China "the 13th Five-Year Plan" (Program No. 2016YFD0700503).

Informed Consent Statement: Informed consent was obtained from all subjects involved in the study.

Data Availability Statement: Not applicable.

Acknowledgments: We are grateful to Fei Mu and Bing Yan for their help with the format of our writing.

Conflicts of Interest: The authors declare no conflict of interest.

References

1. Zhao, Y.; Xiao, H.; Mei, S.; Song, Z.; Ding, W.; Jin, Y.; Han, Y.; Xia, X.; Yang, G. Current status and development strategies of mechanized orchard production in China. *J. China Agric. Univ.* **2017**, *22*, 116–127.
2. Zheng, Y.J.; Jiang, W.; Chen, B.T.; Lu, H.; Wan, C.; Kang, F. Advances in mechanization technology and equipment for orchards in hilly mountainous areas. *J. Agric. Mach.* **2020**, *51*, 1–20.
3. He, X. Research status and development suggestions on precision application technology and equipment in China. *Smart Agric.* **2020**, *2*, 133–146.
4. Lu, C.; Nie, P.; Wang, L.; Wang, J.; Tao, J. Overview of orchard mechanization development in China Deciduous Fruit Trees. *J. Deciduous Fruits* **2018**, *50*, 30–31.
5. Jing, Y.; Liu, G.; Jin, Z. Navigation sideslip estimation and adaptive control method for farm grader. *J. Agric. Mach.* **2020**, *51*, 26–33.
6. Liu, Z.J.; Wang, S.L.; Ren, Z.G.; Mao, W.J.; Yang, F.Z. A virtual radar model-based navigation path tracking control algorithm for crawler tractors. *J. Agric. Mach.* **2021**, *52*, 376–385.
7. Wang, S.; Song, J.; Qi, P.; Yuan, C.; Wu, H.; Zhang, L.; Liu, W.; Liu, Y.; He, X. Design and development of orchard autonomous navigation spray system. *Front. Plant Sci.* **2022**, *13*, 960686. [CrossRef] [PubMed]
8. Murillo, M.; Sánchez, G.; Deniz, N.; Genzelis, L.; Giovanini, L. Improving path-tracking performance of an articulated tractor-trailer system using a non-linear kinematic model. *Comput. Electron. Agric.* **2022**, *196*, 106826. [CrossRef]

9. Backman, J.; Oksanen, T.; Visala, A. Navigation system for agricultural machines: Nonlinear Model Predictive path tracking. *Comput. Electron. Agric.* **2012**, *82*, 32–43. [CrossRef]
10. Rawlings, J.B.; Risbeck, M.J. Model predictive control with discrete actuators: Theory and application. *Automatica* **2017**, *78*, 258–265. [CrossRef]
11. Kayacan, E.; Kayacan, E.; Ramon, H.; Saeys, W. Distributed nonlinear model predictive control of an autonomous tractor–trailer system. *Mechatronics* **2014**, *24*, 926–933. [CrossRef]
12. Yue, M.; Wu, X.; Guo, L.; Gao, J. Quintic Polynomial-based Obstacle Avoidance Trajectory Planning and Tracking Control Framework for Tractor-trailer System. *Int. J. Control Autom. Syst.* **2019**, *17*, 2634–2646. [CrossRef]
13. Kayacan, E.; Ramon, H.; Saeys, W. Robust Trajectory Tracking Error Model-Based Predictive Control for Unmanned Ground Vehicles. *IEEE/ASME Trans. Mechatron.* **2015**, *21*, 806–814. [CrossRef]
14. Tang, L.; Yan, F.; Zou, B.; Wang, K.; Lv, C. An Improved Kinematic Model Predictive Control for High-Speed Path Tracking of Autonomous Vehicles. *IEEE Access* **2020**, *8*, 51400–51413. [CrossRef]
15. Mondal, K.; Rodriguez, A.A.; Manne, S.S.; Das, N.; Wallace, B. Comparison of Kinematic and Dynamic Model Based Linear Model Predictive Control of Non-Holonomic Robot for Trajectory Tracking: Critical Trade-offs Addressed. In Proceedings of the Control and Optimization of Renewable Energy Systems/860: Mechatronics and Control, Anaheim, CA, USA, 6–7 December 2019.
16. Kong, J.; Pfeiffer, M.; Schildbach, G.; Borrelli, F. Kinematic and dynamic vehicle models for autonomous driving control design. In Proceedings of the 2015 IEEE Intelligent Vehicles Symposium (IV), Seoul, Korea, 28 June–1 July 2015; pp. 1094–1099.
17. Werner, R.; Mueller, S.; Kormann, G. Path Tracking Control of Tractors and Steerable Towed Implements Based On Kinematic and Dynamic Modeling. In Proceedings of the 11th International Conference on Precision Agriculture, Indianapolis, IN, USA, 16 July 2012; pp. 15–18.
18. Ding, Y.; He, Z.; Xia, Z.; Peng, J.; Wu, T. Design of a navigation-immune PID controller for a small tracked rape planter. *J. Agric. Eng.* **2019**, *35*, 12–20.
19. Mnih, V.; Kavukcuoglu, K.; Silver, D.; Rusu, A.A.; Veness, J.; Bellemare, M.G.; Graves, A.; Riedmiller, M.; Fidjeland, A.K.; Ostrovski, G.; et al. Human-level control through deep reinforcement learning. *Nature* **2015**, *518*, 529–533. [CrossRef] [PubMed]
20. Huang, X.; Liu, J.R.; Luo, I. Parameter design of launch vehicle attitude controller based on DDQN Space Control. *Aerosp. Control* **2020**, *38*, 3–8. [CrossRef]
21. Guo, S.; Zhang, X.; Du, Y.; Zheng, Y.; Cao, Z. Path Planning of Coastal Ships Based on Optimized DQN Reward Function. *J. Mar. Sci. Eng.* **2021**, *9*, 210. [CrossRef]
22. Yang, Q.; Wang, S.; Sang, J.; Wang, C.; Huang, G.; Wu, C.; Song, S. Intelligent ship path planning and obstacle avoidance methods in complex open water. *Comput. Integr. Manuf. Syst.* **2022**, *28*, 2030–2040. [CrossRef]
23. Shan, Y.; Zheng, B.; Chen, L.; Chen, L.; Chen, D. A Reinforcement Learning-Based Adaptive Path Tracking Approach for Autonomous Driving. *IEEE Trans. Veh. Technol.* **2020**, *69*, 10581–10595. [CrossRef]

agronomy

MDPI

Article

Interactive Influence of Soil Erosion and Cropland Revegetation on Soil Enzyme Activities and Microbial Nutrient Limitations in the Loess Hilly-Gully Region of China

Fangwang Tang [1], Yufei Yao [1,*], Jinxi Song [1,2,*], Chengcheng Wang [1] and Yu Liu [1]

[1] Shaanxi Key Laboratory of Earth Surface System and Environmental Carrying Capacity, College of Urban and Environmental Sciences, Northwest University, Xi'an 710127, China
[2] State Key Laboratory of Soil Erosion and Dryland Farming on the Loess Plateau, Institute of Soil and Water Conservation, Chinese Academy of Sciences and Ministry of Water Resources, Xianyang 712100, China
* Correspondence: yufyao@nwu.edu.cn (Y.Y.); jinxisong@nwu.edu.cn (J.S.)

Abstract: Soil erosion is a major form of land degradation, especially in agroecosystems, which has been effectively controlled by vegetation restoration. However, the interactive role of erosion and cropland revegetation on soil enzyme activities and microbial nutrient limitations is less understood. To address this issue, we examined carbon (C), nitrogen (N), and phosphorus (P) in bulk soils and microbial biomass, enzyme activities, and microbial nutrient limitations in the 0–200 cm soils in eroded and deposited landscapes occupied by cropland, revegetated forest, and grassland. The results showed that the activities of C-, N-, and P-acquiring enzymes were larger in the deposited landscape than in the eroded landscape for 0–20 cm soils in forest and grassland but not in cropland. Microbial metabolism was co-limited by N and P, and the threshold element ratio (TER_L) indicated that P was the most limiting factor. Microbial N limitation was lower in the deposited than the eroded zone, especially in surface soils in revegetated forest and grassland. The TER_L value was larger at the deposited than at the eroded zone, and a greater difference was found in the surface soils of forest and grassland. Microbial nutrient limitations were mostly explained by C/P and N/P. Conclusively, the deposited areas were characterized by ameliorated enzyme activities, decreased microbial N limitation but relatively strengthened microbial P limitation compared to the eroded area, and such variations existed in the revegetated forest and grassland but not in the cropland, which thus contributes to a better understanding of C and nutrient cycling for agroecosystems and revegetation ecosystems in eroded environments.

Keywords: agroecosystem; revegetation; soil erosion; deposition; enzyme activity; microbial metabolic limitation

Citation: Tang, F.; Yao, Y.; Song, J.; Wang, C.; Liu, Y. Interactive Influence of Soil Erosion and Cropland Revegetation on Soil Enzyme Activities and Microbial Nutrient Limitations in the Loess Hilly-Gully Region of China. *Agronomy* **2022**, *12*, 2796. https://doi.org/10.3390/agronomy12112796

Academic Editors: Xiongkui He, Fuzeng Yang and Baijing Qiu

Received: 4 October 2022
Accepted: 6 November 2022
Published: 10 November 2022

1. Introduction

Soil erosion is the major driver of land surface processes and has caused environmental problems such as soil and nutrient losses, soil siltation, vegetation degradation, and water eutrophication [1–3]. Globally, approximately 84% of the Earth's land surface suffers from soil erosion, half of which occurs in agricultural ecosystems [4]. Cropland revegetation has been deemed the most effective and economical way to curb severe soil erosion and has achieved remarkable effects [5–7]. Understanding the coupled role of soil erosion and vegetation restoration in microbial metabolism is of vital importance to the biogeochemical cycling of essential elements in an erosive environment [8,9].

Soil erosion leads to the distribution of soil particles and nutrients and alters conditions, and thus has an important effect on soil microorganisms' metabolism [3,9–11]. Microorganisms secrete extracellular enzymes catalyzing the decomposition of complex organic compounds to get soil nutrients, thus playing a critical role in biogeochemical nutrient cycling [12–15]. Soil extracellular enzyme activities are affected by multiple factors, such as substrate concentration and availability, soil moisture, temperature, and ventilation conditions [15,16]. Research has shown that extracellular enzyme activities are lower in eroded than in deposited sites due to the redistribution of soil nutrients and changes in the microclimate during the eroded-deposited process [17,18]. The combination of ecoenzymatic stoichiometric theory with the metabolic theory of ecology and ecological stoichiometric theory was accustomed to assess microorganisms' metabolism limitations in soil ecosystems from enzyme activity patterns [19]. A common method used to evaluate and measure microbial metabolic limitations is the threshold element ratio (TER) model. This model characterizes the carbon (C), nitrogen (N), and phosphorous (P) ratios in which microbial metabolic management is effective in converting from energy to nutrients [20]. Recent studies have proven that microbial metabolic limitations differ based on land-use patterns, tillage management, and fertilization treatments, which are driven by soil moisture, soil physical properties (i.e., clay content and pH), and nutrient availability [21–24]. However, how soil microbial metabolic limitations vary in eroded and deposited landscapes remains less understood, despite the well-acknowledged variations in soil physiochemical properties.

Vegetation type is an important factor affecting soil extracellular enzyme activities and microbial nutrient limitations. Plant community characteristics and soil physicochemical properties (e.g., plant species and litter nutrient input, respectively) have a profound influence on soil extracellular enzyme activities and microbial nutrient limitations [12,14,25]. For example, Wobeng [21] recommended soybean cultivation in agroecosystems in South Africa since it can alleviate microbial C limitation compared with the other three leguminous species (*Phaseolus vulgaris* L., *Glycine max* L. *and Arachis hypogea* L.). Yan [26] demonstrated that reafforestation worsens soil microbial metabolic N and P limitation, especially at medium and high *Pinus tabuliformis* plantation densities on China's Loess Plateau. Cropland revegetation is an effective measure for ecosystem restoration and soil erosion control with the increasing plant coverage and less human disturbance, such as tillage activities [27,28]; however, the interactive role of soil erosion and cropland revegetation on soil microbial metabolism is less understood in eroding landscapes.

It has been estimated that 27–77% of organic C (OC) is stored below 20 cm, and any small change in deep soil C would also have an important impact on global C cycling [29,30]. With increasing soil depth, the input of plant organic matter decreases gradually, and the soil nutrient content, microbial biomass, and microbial activity also decrease correspondingly [31]. In addition, compared with topsoil, the environmental conditions, including moisture, porosity, and pH, are significantly different in the subsoil, which has a significant influence on soil microbial metabolism [32]. For example, soil enzyme activities have been reported to significantly decrease with depth and respond significantly to water-enrichment treatment in the 40–160 cm layers along the 0–340 cm soil profile [33]. While previous studies on microorganisms' nutrient limitations have principally targeted on surface soils [12,14,34], information about soil microbial metabolism in deep soils is currently limited, although the effects of soil erosion and vegetation restorage can influence deep soil carbon and nutrient cycling [9,35].

China's Loess Plateau is one of the "most liable to erosion" regions globally [7]. The "Grain-for-Green" project and the construction of check dams have been implemented to decelerate soil erosion on the Loess Plateau over the past few decades [6]; thus, the region provides an ideal platform for our study. The objective of this study was to elucidate the coupled role of erosion and vegetation restoration on soil enzyme activities and microbial nutrient limitations. To achieve this goal, cropland and revegetated forest and grassland were selected in both the eroded and the deposited landscapes, and 0–200 cm soil and sediment samples were collected to analyze the C-, N-, and P-acquiring enzyme activities and microbial nutrient limitation. We hypothesized that (H1) the deposited landscape would have higher enzyme activities and lower nutrient limitations than the eroded landscape due to the accumulation of C, nutrients, and moisture [20,36]; (H2) variations in these indicators would be greater in cropland since revegetation reduces the effect of erosion;

and (H3) variations in these indicators would be higher in surface soils since the influences of erosion and vegetation restoration are more obvious.

2. Materials and Methods

2.1. Study Site and Sampling

The study was conducted in the Jiuyuangou watershed (37°33′–37°38′ N, 110°16′–110°26′ E) in Suide County, Shaanxi Province, China, which belongs to the loess hilly area on the Loess Plateau. This area features a temperate, semi-arid, continental monsoon climate. The annual average temperature is 8 °C, the annual mean rainfall is 475 mm, and most of the rainy season is concentrated from July to September [37].

The upper and middle parts of the slopes were selected as the eroded zones, and the neighboring check dams were chosen as the deposited zones. Cropland, revegetated forest, and grassland were selected in both the eroded and the deposited zones. The main plant species of cropland, forest, and grassland were *Zea mays*, *Robinia pseudoacacia* L., and *Phragmites communis* in the deposited zone and *Solanum tuberosum*, *Prunus armeniaca* L., and *Cynanchum chinense* in the eroded zone, respectively. The chemical N and P fertilizers used were diammonium hydrogen phosphate and dipotassium hydrogen phosphate for cropland and no fertilizers were applied in the forest and grassland.

Site surveys and soil sample collection were conducted in July 2021. Three 20 × 20 m plots with similar elevations and slope aspects were selected for each land-use type in the eroded and deposited landscapes. In each plot, three subplots (3 × 3 m) were randomly established, and both surface soils (0–10 and 10–20 cm) and deep soils (60–80 and 180–200 cm) were sampled with a soil auger. The three samples for each subplot were mixed into one sample at each depth. Subsequently, visible plant organic matter was removed and each soil sample was divided into two parts. One part of the samples was immediately sieved to pass through a 2 mm sieve in the field, sent back to the laboratory, and stored at 4 °C to determine the biological indices. The other part of the bulk soil (without sieving) was put in sealed bags and sent back to the laboratory to measure the soil's physical and chemical properties. A subsample of these soils was oven-dried at 105 °C for 24 h until constant weight to measure soil moisture (SM) and the rest was air-dried.

2.2. Laboratory Analysis

The soil particle size distribution was measured with a laser particle size analyzer (Mastersizer 2000, Malvern Instruments Ltd., Malvern, UK). Soil pH was determined with a pH-meter (PHS-3C, Leici, Shanghai, China) at a soil:water ratio of 1:2.5. After extracting with 2 M KCl, soil nitrate nitrogen (NO_3^-) and ammonium nitrogen (NH_4^+) were analyzed with a continuous flow autoanalyzer (AutoAnalyzer-AA3, Seal Analytical, Norderstedt, Germany). The soil extractable P (EP) content was measured with the Olsen method. Soil organic C (OC) was measured with the Walkley Black method and total N (TN) with the Kjeldahl methods. Soil total P (TP) content was measured colorimetrically following digestion with $HClO_4$-H_2SO_4 [38].

The fresh soil samples were used to measure dissolved organic carbon (DOC), microbial biomass C (MBC), microbial biomass N (MBN), microbial biomass P (MBP), and extracellular enzyme activities. The DOC was measured through extraction with deionized water after shaking and then filtered through a Millipore 0.45-μm filter [39]. Soil MBC, MBN, and MBP were measured using the chloroform fumigation-extraction method [14]. The potential activities of C-acquiring enzymes (β-D-cellobiosidase (CBH), β-1,4-glucosidase (BG)), N-acquiring enzyme (β-1,4-N-acetylglucosaminidase (NAG), L-leucine aminopeptidase (LAP)), and P-acquiring enzyme (alkaline phosphatase (AP)) were determined with the microplate fluorescence method [40,41].

2.3. Calculation of Microbial Nutrient Limitation

Based on the stoichiometric microbic C:N:P values obtained from the enzyme information, the $TER_{C:N}$ and $TER_{C:P}$ values were obtained via the threshold element model according to the subsequent formulas projected by Sinsabaugh [42]:

$$TER_{C:X} = [A_X B_{C:X}]/CUE_{C:X} = B_{C:X}/S_{C:X} = L_{C:X} EEA_{C:X} \quad (1)$$

$$A_X = CUE_{C:X}/S_{C:X} \quad (2)$$

$$CUE_{C:X} = CUE_{max}[S_{C:X}/(S_{C:X} + K_X)] \quad (3)$$

$$S_{C:X} = (1/EEA_{C:X})(B_{C:X}/L_{C:X}) \quad (4)$$

where the CUE is the carbon use efficiency of microorganisms, EEA is the soil extracellular activity, A_X represents the apparent assimilation efficiency of N or P, the ratios of soil MBC:MBN and MBC:MBP were used as estimates of $B_{C:X}$, $S_{C:X}$ is a scalar indicating the extent to which the allocation of ecoenzymatic activities offsets the difference between the elemental composition and microbial biomass composition of the available resources, K_X is the half-saturation constant and has a value of 0.5, the molar ratios of soil OC:TN and soil OC:TP were used as estimates of $L_{C:N}$ and $L_{C:P}$, respectively, and $EEA_{C:N}$ and $EEA_{C:P}$ were expressed by (CBH + BG)/(NAG + LAP) and (CBH + BG)/AP, respectively [42].

In this study, we used the corresponding soil available nutrient ratios and subtracted the TER to calculate the microbial nutrient limitations, and the positive values indicated that microorganisms were subjected to nutrient restriction and vice versa. We also calculated the TER_L to identify the most limiting nutrient to microbial metabolism [20].

$$X_{limitation} = L_{C:X} - TER_{C:X} \quad (5)$$

$$TER_L = P_{limitation}/B_{C:P} - N_{limitation}/B_{C:N} \quad (6)$$

where X represents N or P.

2.4. Statistical Analysis

Multiway analysis of variance (three-way ANOVA) and LSD tests were used to examine the significant differences in landform position (eroded and deposited), land-use type (cropland, revegetated forest, and grassland), soil depth (0–10, 10–20, 60–80, 180–200 cm), and their interactions with soil physicochemical properties, microbial biomass, extracellular enzyme activities, and microbial nutrient limitations. Statistical analyses were performed using SPSS 20.0 (SPSS Company, Chicago, IL, USA). Redundancy analysis (RDA) was used to identify the contributions of soil physic-chemical properties to soil enzyme activities and microbial nutrient limitation in Canoco 5.0 (Microcomputer Power, Inc., Ithaca, NY, USA). The random forest (RF) model was used to identify the key factors influencing soil extracellular enzymatic activity and microbial nutrient limitations in R software v.4.1.3. Before the RF analysis, extracellular enzymatic activities were extracted from the principal components (Figure S1), and the first principal component was used to represent the microbial enzyme activity.

3. Results

3.1. Soil Physicochemical Properties

Soil moisture was significantly higher in the deposited zone (10.9%) than in the eroded zone (5.0%) ($p < 0.05$), with more dramatic variations at 60–80 and 180–200 cm than at 0–10 and 10–20 cm (Tables S1 and S2). The deposited sediments had a higher clay content than the eroded soils ($p < 0.05$), and such variations were higher in grassland (+70.4%) than in cropland (+37.5%) and forest (+28.0%). The deposited zone had a higher pH ($p < 0.1$), especially in cropland.

There were no significant variations in the OC and TN between the two landform positions ($p > 0.05$, Table S1). The DOC concentration was significantly higher in the deposited zone (55.88 mg kg^{-1}) than in the eroded zone (38.83 mg kg^{-1}) in cropland ($p < 0.05$) but not in grassland or forestland ($p > 0.05$). The soil NH_4^+ and NO_3^- concentrations did not differ significantly between the two landform positions for each land use ($p > 0.05$) (Tables S1 and S3). The soil TP concentration was significantly lower in the deposited zone than in the eroded zone (−9%) for each land use, while the EP concentration was significantly higher in the deposited zone than in the eroded zone for forest (+53%) and grassland (+143%) but was significantly lower in the deposited zone for cropland (−56%). The soil C/N did not significantly differ between the two landforms ($p > 0.05$). However, the C/P and N/P were larger at the deposited site than at the eroded site in forest ($p < 0.05$).

3.2. Soil Microbial C, N, P and Enzyme Activity

Generally, the differences in MBC and MBN were not significant between the deposited site and eroded site, and such variation was not dependent on land use and soil depth ($p > 0.05$, Figure 1a,b). The MBP content was considerably larger at the deposited site than at the eroded site in the surface but not in the deep soils in the forest and grassland, but not in cropland. Therefore, the deposited zone had a higher microbial P at the 0–10 cm depth in the forest and grassland (Figure 1c).

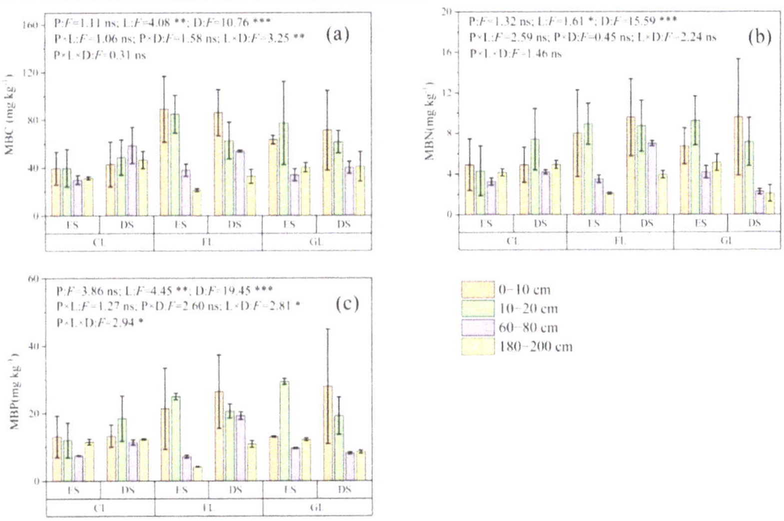

Figure 1. (**a**) Microbial biomass carbon (C), (**b**) nitrogen (N), and (**c**) phosphorous (P) from different soil sites. MBC: microbial biomass carbon; MBN: microbial biomass nitrogen; MBP: microbial biomass phosphorous. CL: cropland; FL: forestland; GL: grassland; ES: eroded site; DS: deposited site. Values are expressed as the mean ± standard error. P: landform position; L: land use type; D: soil depth. ***, **, * mean the significant differences at $p < 0.001$, $p < 0.01$, and $p < 0.05$, respectively, and ns means no significant differences ($p > 0.05$).

Landform position significantly influenced soil extracellular enzyme activities, and the effects depended on land-use type and soil depth ($p < 0.05$, Figure 2). For example, the C-acquiring enzyme CBH and the N-acquiring enzyme NAG were higher in the deposited zone (4.843 and 2.185 nmol g^{-1} h^{-1}, respectively) than in the eroded zone (2.831 and 1.254 nmol g^{-1} h^{-1}, respectively) (Figure 2a,c). The variations in the CBH and NAG were significant at the 0–20 cm depth in forest and grassland ($p < 0.05$) and not significant in cropland for either surface or deep soils ($p > 0.05$). The P-acquiring enzyme AP was also significantly higher in the deposited zone than in the eroded zone at the 0–20 cm depth in forest and grassland ($p < 0.05$) but similar in the two landform positions at the 0–2 m soil profile in cropland ($p > 0.05$, Figure 2e). Thus, the deposited area had higher C-, N-, and P-acquiring enzyme activities in surface soils in forest and grassland but were similar in cropland.

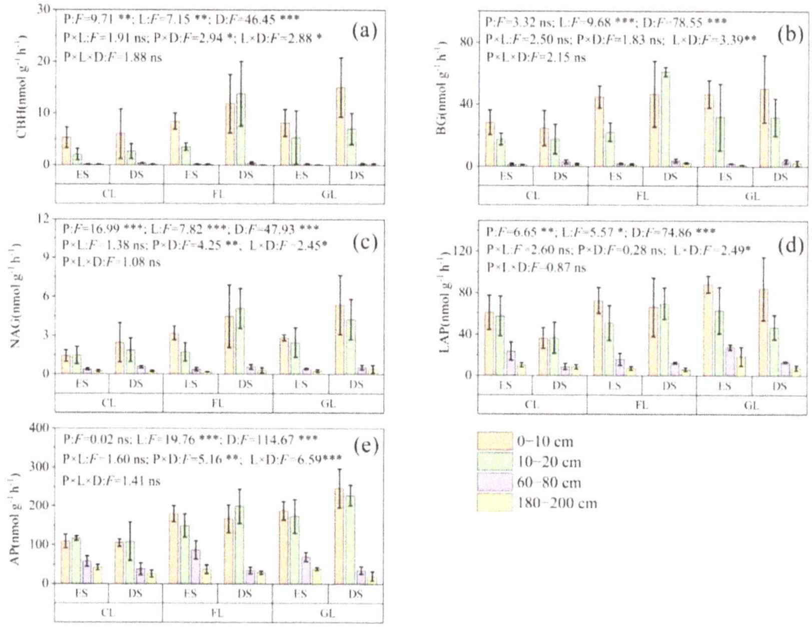

Figure 2. Soil microbial enzymatic activity of (**a**) CBH, (**b**) BG, (**c**) NAG, (**d**) LAP, and (**e**) AP from different soil sites. CBH: β-D-cellobiosidase; BG: β-1,4-glucosidase; NAG: β-1,4-N-acetylglucosaminidase; LAP: leucine aminopeptidase; AP: alkaline phosphatase. CL: cropland; FL: forestland; GL: grassland; ES: eroded site; DS: deposited site. Values are expressed as the mean ± standard error. P: landform position; L: land use type; D: soil depth. ***, **, * mean the significant differences at $p < 0.001$, $p < 0.01$ and $p < 0.05$, respectively, and ns means no significant differences ($p > 0.05$).

3.3. Soil Microbial Nutrient Limitation

The study area suffered from both N limitation and P limitation, with average values of 5.41 and 10.81, respectively (Figure 3a,b). The positive TER_L value (0.76) indicated that P was the most limiting factor for microbial metabolism (Figure 3c). The microbial N limitation in the eroded zone was 6.525, which was significantly higher than the value of 4.301 in the deposited zone ($p < 0.05$, Figure 3a). Although the variations in microbial N limitation between the two landform positions were independent of land use and soil depth, the response ratios of microbial N limitation were higher in surface soils (−45%) than in deep soils (−27%) and higher in forest (−39%) and grassland (−36%) than in cropland (−28%). Microbial P limitation did not differ significantly between the eroded and deposited zones, regardless of land use and soil depth ($p > 0.05$, Figure 3b), which indicated a similar P limitation between landform positions. However, the TER_L was significantly higher in the deposited zone than in the eroded zone at the 0–20 cm depth in forest and grassland ($p < 0.05$) but was not significantly different at the 0–2 m soil profile in cropland ($p > 0.05$, Figure 3c).

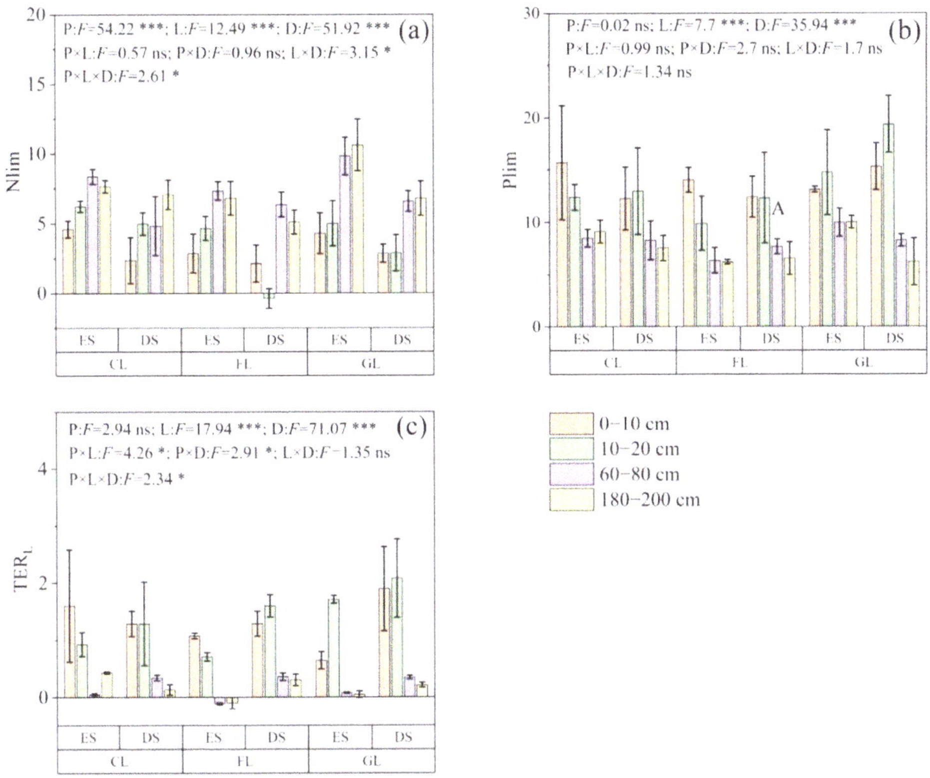

Figure 3. (**a**) Microbial N limitation, (**b**) microbial P limitation and (**c**) Threshold elemental ratio (TER_L) from different soil sites. CL: cropland; FL: forestland; GL: grassland; ES: eroded site; DS: deposited site. Different capital letters represent significant difference ($p < 0.05$) between different landform positions in the same land use type and soil depth. Values are expressed as the mean ± standard error. P: landform position; L: land use type; D: soil depth. ***, * mean the significant differences at $p < 0.001$ and $p < 0.05$, respectively, and ns means no significant differences ($p > 0.05$).

3.4. Factors Influencing Extracellular Enzyme Activities and Microbial Nutrient Limitations

The redundancy analysis (RDA) showed that the RDA 1 and 2 axes explained 85.47% and 85.17% of the soil extracellular enzyme activity and microbial nutrient limitation, respectively (Figure 4a,b). The forward selection and RF results indicated that the OC, MBP, SM, clay, NO_{3-}, and MBC were the most powerful factors affecting soil extracellular enzyme activities (Table 1 and Figure 5a). The RDA results showed that the OC, MBP, clay, NO_{3-}, and MBC were positively associated with soil extracellular enzyme activities, while the SM was negatively associated with extracellular enzyme activities (Figure 4a). Furthermore, the forward selection and RF results implied that the C/P, clay, C/N, OC, N/P, DOC, and MBP were the most influential factors affecting the microbial nutrient limitations (Table 2, Figure 5b,c). The RDA results showed that the C/P, clay, OC, N/P, DOC, MBP, and P limitation were negatively related to microbial N limitation (Figure 4b). The RF results also highlighted that soil nutrients and their stoichiometry, MBC, MBN, and MBP were affecting the TER_L (Figure 5d).

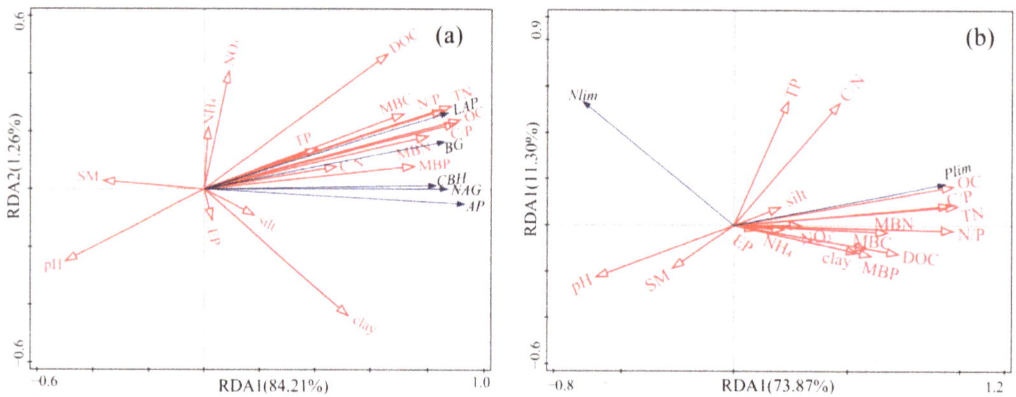

Figure 4. The redundancy analysis (RDA) was used to identify the (**a**) relationship between enzyme activities and soil physicochemical and microbial properties and (**b**) the relationship between microbial nutrient limitations and soil physicochemical and microbial properties. SM: soil moisture; OC: soil organic carbon; TN: total nitrogen; TP: total phosphorus; MBC: microbial carbon; MBN: microbial nitrogen; MBP: microbial phosphorus; DOC: dissolved organic carbon; EP: soil available phosphorus; NO_3^-: nitrate nitrogen; NH_4^+: ammonia nitrogen; Nlim: microbial N limitation; Plim: microbial P limitation.

Table 1. Simple term effects and forward selection of environmental variables on soil enzyme activities from the redundancy analysis (RDA).

Variable	Simple Term Effects		Forward Selection	
	Contributions %	*p*	Explains %	*p*
OC	81.8	0.002	70.2	0.002
C/P	79.5	0.002		
TN	76.2	0.002		
N/P	71.8	0.002		
MBN	62.9	0.002		
MBP	55.7	0.002	5.1	0.002
MBC	50.6	0.002	0.8	0.058
DOC	43.1	0.002		
clay	26.7	0.002	1.3	0.036
pH	24.5	0.002		
C/N	22.3	0.002	1	0.048
TP	16.4	0.002		
SM	13.1	0.002	2.5	0.012
sand	13.0	0.008		
silt	3.2	0.168		
NO_3^-	1.0	0.482	1.8	0.034
EP	0.1	0.88		
NH_4^+	<0.1	0.904		

Note: SM: soil moisture; OC: soil organic carbon; TN: total nitrogen; TP: total phosphorus; EP: extractable phosphorus; DOC: dissolved organic carbon; NH_4^+: ammonia nitrogen; NO_3^-: nitrate nitrogen; MBC: microbial carbon; MBN: microbial nitrogen; MBP: microbial phosphorus.

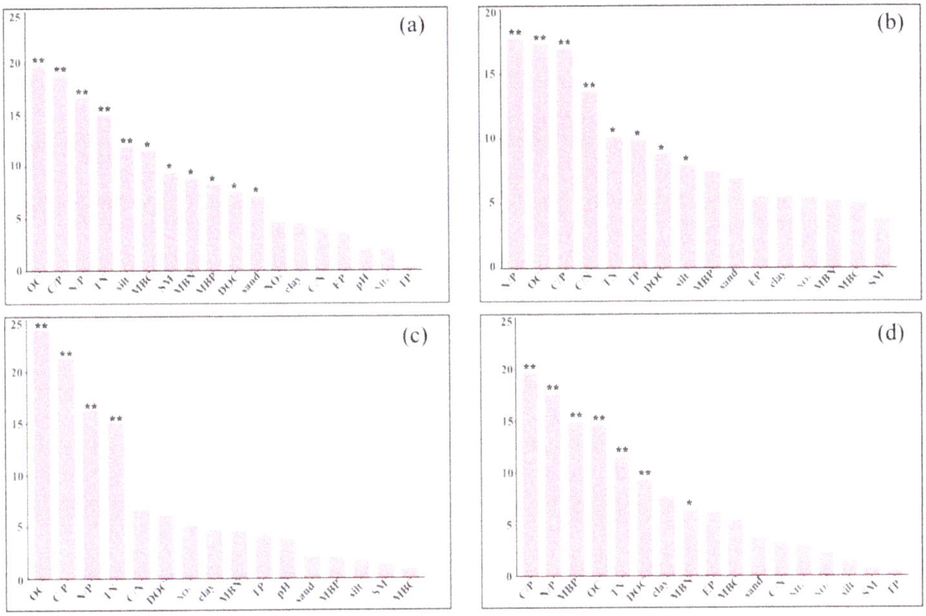

Figure 5. The Random Forest model (RF) was used to determine the key factors affecting (**a**) soil extracellular enzymatic activity, (**b**) soil microbial N limitation, (**c**) P limitation, and (**d**) Threshold elemental ratio (TER_L). SM: soil moisture; OC: soil organic carbon; TN: total nitrogen; TP: total phosphorus; DOC: dissolved organic carbon; EP: soil available phosphorus; NO_3^-: nitrate nitrogen; NH_4^+: ammonia nitrogen. **, * mean the significant differences at $p < 0.01$ and $p < 0.05$, respectively.

Table 2. Simple term effects and forward selection of environmental variables on microbial nutrient limitations from the redundancy analysis (RDA).

Variable	Simple Term Effect		Forward Selection	
	Contributions %	*p*	Explains %	*p*
C/P	84.9	0.002	51.3	0.002
OC	82.0	0.002	1.9	0.006
N/P	81.2	0.002	1.3	0.054
TN	79.4	0.002		
DOC	45.7	0.002	1	0.008
MBN	39.7	0.002		
pH	33.2	0.002		
MBP	31.9	0.002	0.9	0.08
MBC	29.9	0.002		
clay	26.0	0.002	22.1	0.002
C/N	22.4	0.002	3.6	0.006
sand	13.4	0.002		
TP	8.9	0.01		
NO_3^-	7.7	0.022		
SM	6.9	0.042		
NH_4^+	4.2	0.092		
silt	3.8	0.108		
EP	0.9	0.62		

Note: SM: soil moisture; OC: soil organic carbon; TN: total nitrogen; TP: total phosphorus; EP: extractable phosphorus; DOC: dissolved organic carbon; NH_4^+: ammonia nitrogen; NO_3^-: nitrate nitrogen; MBC: microbial carbon; MBN: microbial nitrogen; MBP: microbial phosphorus.

4. Discussion

4.1. Variations in Soil Enzyme Activities between the Eroded and Deposited Zones and Their Dependence on Land Use and Soil Depth

The activities of CBH, NAG, and AP were greater in the deposited area than in the eroded area, and such variations were significant at the 0–20 cm depth in revegetated forest and grassland but not in cropland in either surface or deep soils, which was consistent with Hypotheses H1 and H3, but not H2. Differences in the extracellular enzyme activities between the eroded and deposited areas may be due to the OC concentration; stoichiometry of C, N, and P; microbial biomass C, N, and P; clay content; and SM (Table 1, Figure 5a). Previous studies have shown that enzyme activities were considerably lower in the eroded zone than the deposited zone, mainly due to the depletion of OC and other nutrients by erosion [11]. Forward selection and RF results revealed that OC was the key factor influencing enzyme activities (Table 1, Figure 5a). However, the difference in OC between landform positions was not significant in our study, which may be explained by the positive effect of vegetation restoration in forest and grassland and fertilization in cropland on OC [22,43]. Previous studies also illustrated that soil enzyme activity tends to be closely associated with the microbial biomass [44,45] and lower MBP in the eroded site, which can have a greater inhibitory effect on extracellular enzyme activities by downregulating the quantity of enzyme released by unit of biomass [11]. The significantly higher MBP in the surface soils of forest and grassland in the deposited zone than in the eroded zone (Figure 1c) contributed to the larger discrepancy of soil enzyme activities between landform positions. In addition, the soil enzyme activity per unit of MBC can also assess microorganisms' metabolism [46]. In the present study, especially in the surface soils and forests and grasslands, we found that the CBH/MBC and NAG/MBC values were considerably higher at the deposited site than at the eroded site (Table S4); thus, soil microorganism biomass released more enzymes in the surface soils of forests and grasslands at the deposited site [47]. Previous studies found that changes in soil nutrient content altered the stoichiometric ratio of C, N, and P, and the relative shortage of certain elements promoted the secretion of corresponding soil enzymes [48,49]. In our study, the higher soil C/P and N/P at the deposited site than at the eroded site were only significant in surface soils in the forest and grassland (Table S3), which contributed to increasing the AP (Figure 2e).

The SM and clay content were two other important factors influencing soil extracellular enzyme activities (Table 2). The clay and silt contents within the eroded zone were considerably lower than those within the deposited zone, which may be because soil erosion destroyed the aggregate structure and selectively transported fine particles to the deposited site [1,9,22]. Reported studies illustrated that the removal of the clay and silt content caused by erosion will harm extracellular enzyme activities because these are easily absorbable on the surface of the fine particles [11,50]. In this study, the clay and silt contents were positively related to enzyme activities (Figure 4a); thus, the greater differences of clay and silt contents between two landform positions in the surface soils of forest and grassland (Table S2) may explain the greater variations of enzyme activities in these land uses and depths. A previous study indicated that a decrease in soil moisture could enhance the oxygen content, which stimulated extracellular enzyme activity [51]. In our study, we also found that SM was negatively correlated with CBH and NAG (Figure 4a). Compared with topsoils, the higher SM in the deposited site, as compared to the eroded site, was prominent in deep soils, which would attenuate the response of enzymes to erosion and deposition. Therefore, given the differences in soil enzyme activities between landform positions, its interactions with land use and soil depth may be explained by a combination of factors of the ratios of C, N, and P, MBP, clay content, and soil moisture.

4.2. Variations in Soil Microbial Nutrient Limitations between Eroded and Deposited Zones and Their Dependence on Land Use and Soil Depth

Our results implied that soil microorganisms' metabolism was restricted by N and P, which was in line with reported studies on the Loess Plateau [23,44]. Microbial N limitation was considerably larger in the eroded zone than in the deposited zone ($p < 0.05$, Figure 3a), and this change was higher in surface soils than in deep soils and higher in forest and grassland than in cropland, which was consistent with our Hypotheses H1 and H3 but did not support H2. Our forward selection and RF results indicated that nutrient and nutrient stoichiometry were the vital factors influencing soil microbial N limitation (Table 2 and Figure 5b). N and P are thought of as the dominant nutrient elements limiting vegetation growth and microorganism metabolism [14,52,53]. Studies have shown that soil microorganisms' nutrient limitation was vulnerable to nutrient ratio imbalance [12,54]. For example, Cui et al. (2019) [12] implied that the soil C/N, C/P, and N/P considerably affected microorganisms' C and P limitations in temperate grassland ecosystems. In surface soils, we found that the C/P and N/P were higher at the deposited site than at the eroded site, which had a positive impact on alleviating the microbial N limitation (Figure 4b). Another vital factor affecting N limitation was clay content (Table 2), which had a negative effect on N limitation (Figure 4b). A review of nutrient limitations in terrestrial ecosystems showed a shift from relative P to N limitation with lower soil clay content at global scales [52]. This was because high clay contents in soils are strongly weathered and tend to be deficient in P availability. Thus, the greater differences of clay content in grassland and forest between the eroded and deposited zones (Table S2) would result in more prominent variations of the N limitation. Furthermore, soil DOC concentration was negatively related to N limitation (Table 2, Figure 4), meaning that the higher the soil DOC concentration, the lower the microbial N limitation. Zhang et al. (2022) [55] implied that the soil microbial N limitation was strongly dependent on soil available C. The higher DOC concentration indicated that there were more labile C fractions and the labile C would provide energy for microbes to decompose organic matter, which would release available N and alleviate the microbial N limitation. Thus, in cropland, the higher DOC at the eroded site than at the deposited site was conducive to alleviating the N limitation at the eroded site (Table S3), which may be why the smallest difference was observed in cropland between landform positions relative to revegetated forest and grassland. Therefore, the change in N limitation affected by landform positions and its dependence on land-use types and soil depth may be attributed to the combined effect of soil nutrient stoichiometry, clay content, and DOC.

There was no significant difference in microorganism P limitation between the eroded and deposited sites. The TER_L indicated that the study area was primarily limited by P relative to N, which was in line with previous studies. Studies have reported that microorganism metabolism is widely restricted by soil P in several terrestrial ecosystems, such as subtropic forests and semiarid grasslands [14,23,56]. This phenomenon was mainly attributed to the soil nutrient ratio, especially in surface soils. Soil C/P and N/P values were higher at the deposited zone than at the eroded zone, which may enhance microbial P limitation (Figure 4b). That is, nutrient limitations rely not only on the supply of nutrients but also on the supply of other nutrients. The increase in soil C/P and N/P could also be attributed to the fact that soil P was principally gained through weathering. Higher vegetation aboveground biomass and root systems would assimilate more P after vegetation restoration, and soil eroded would accelerate the loss of P, which can hardly be replenished by plant inputs and resulted in a relatively stable P compared with soil OC and N [14,23,57]. The increase of C/P and N/P meant the relative dilution of P compared with C or N, which would potentially result in microbial P limitation and, in turn, stimulate P-acquisition enzyme activities [20,23]. Xiao et al. (2020) [23] and Jiao et al. (2013) [58] additionally reported that the ratio of N to P increased after farmland abandonment and implied that soil microorganisms were principally restricted by P in the Loess Plateau. Furthermore, the higher P limitation at the deposited site than at the eroded site was prominent in revegetated forests, which may be due to the larger difference in MBP between landform positions in forests as a positive relation between MBP and P limitation (Figures 4b and 5d). These results implied that soil nutrient stoichiometry has a powerfully regulatory impact on microorganism metabolism by influencing fundamental balance and the nutrient ratio would thus potentially be the important threshold resulting in microbic metabolic limitations.

5. Conclusions

Elucidating how soil enzyme activities and microbial nutrient limitations respond to soil erosion and their dependence on land use is essential to understanding the microbial metabolisms and nutrient cycling of agriculture and revegetated ecosystems in eroded and deposited landscapes. In our study, the deposited sites had higher enzyme activities and lower microbial N limitation than the eroded sites, and such variations were more pronounced in forest and grassland than in cropland and were greater in surface soils (0–10 and 10–20 cm) than in deep soils (60–80 and 180–200 cm). The terminal threshold ratio model illustrated that the study area was mostly restricted by P, and such restrictions were greater in the deposited zone than in the eroded zone for the surface soils in forest and grassland. Our study highlighted the variations in microbial nutrient limitations between two landform positions and their dependence on land uses in a hilly-gully area of the northern Loess Plateau. However, our study did not capture the changes in the metabolic metabolism characteristics in the eroded and deposited processes, and further studies are recommended to investigate the dynamic features with the aid of rainfall simulation experiments.

Supplementary Materials: The following supporting information can be downloaded at: https://www.mdpi.com/article/10.3390/agronomy12112796/s1, Table S1: ANOVA results for soil physicochemical properties as affected by landform position (P), land-use type (L), soil depth (D) and their interactions. Table S2: Soil physicochemical properties in erosion and deposition landform positions with different land-use types and soil depths. Table S3: Soil nutrients in erosion and deposition landform positions with different land-use types and soil depths. Table S4: The ratios of soil microbial C, N, P and ratios of soil extracellular enzymes to MBC in different landform positions within land use types and soil depths. Figure S1: Results of principle component analysis of soil extracellular enzyme activities. CBH: β-D-cellobiosidase; BG: β-1,4-glucosidase; NAG: β-1,4-N-acetylglucosaminidase; LAP: leucine aminopeptidase; AP: alkaline phosphatase.

Author Contributions: Conceptualization, F.T.; Data curation, F.T., C.W. and Y.L.; Formal analysis, F.T.; Supervision, Y.Y. and J.S.; Writing—original draft, F.T.; Writing—review & editing, F.T. and Y.Y. All authors have read and agreed to the published version of the manuscript.

Funding: This study was supported by the National Natural Science Foundation of China (42107344, 42041004), the China Postdoctoral Science Foundation (2021T140558, 2020M683699XB) and the State Key Laboratory of Soil Erosion and Dryland Farming on the Loess Plateau (A314021402-202106).

Data Availability Statement: The authors confirm that the data supporting the findings of this study are available within the article and its Supplementary Materials.

Acknowledgments: We would thankful for the support of the above funding programs.

Conflicts of Interest: The authors declare no conflict of interest.

References

1. Quinton, J.N.; Govers, G.; Van Oost, K.; Bardgett, R.D. The Impact of Agricultural Soil Erosion on Biogeochemical Cycling. *Nat. Geosci.* **2010**, *3*, 311–314. [CrossRef]
2. Petito, M.; Cantalamessa, S.; Pagnani, G.; Degiorgio, F.; Parisse, B.; Pisante, M. Impact of Conservation Agriculture on Soil Erosion in the Annual Cropland of the Apulia Region (Southern Italy) Based on the RUSLE-GIS-GEE Framework. *Agronomy* **2022**, *12*, 281. [CrossRef]
3. Lal, R. Accelerated Soil Erosion as a Source of Atmospheric CO_2. *Soil Tillage Res.* **2019**, *188*, 35–40. [CrossRef]
4. Borrelli, P.; Robinson, D.A.; Fleischer, L.R.; Lugato, E.; Ballabio, C.; Alewell, C.; Meusburger, K.; Modugno, S.; Schütt, B.; Ferro, V.; et al. An Assessment of the Global Impact of 21st Century Land Use Change on Soil Erosion. *Nat. Commun.* **2017**, *8*, 2013. [CrossRef] [PubMed]
5. Lasanta, T.; Nadal-Romero, E.; Arnáez, J. Managing Abandoned Farmland to Control the Impact of Re-Vegetation on the Environment. The State of the Art in Europe. *Environ. Sci. Policy* **2015**, *52*, 99–109. [CrossRef]

6. Wang, S.; Fu, B.; Piao, S.; Lü, Y.; Ciais, P.; Feng, X.; Wang, Y. Reduced Sediment Transport in the Yellow River Due to Anthropogenic Changes. *Nat. Geosci.* **2016**, *9*, 38–41. [CrossRef]
7. Zhu, G.; Shangguan, Z.; Hu, X.; Deng, L. Effects of Land Use Changes on Soil Organic Carbon, Nitrogen and Their Losses in a Typical Watershed of the Loess Plateau, China. *Ecol. Indic.* **2021**, *133*, 108443. [CrossRef]
8. Dungait, J.A.J.; Ghee, C.; Rowan, J.S.; McKenzie, B.M.; Hawes, C.; Dixon, E.R.; Paterson, E.; Hopkins, D.W. Microbial Responses to the Erosional Redistribution of Soil Organic Carbon in Arable Fields. *Soil Biol. Biochem.* **2013**, *60*, 195–201. [CrossRef]
9. Berhe, A.A.; Barnes, R.T.; Six, J.; Marín-Spiotta, E. Role of Soil Erosion in Biogeochemical Cycling of Essential Elements: Carbon, Nitrogen, and Phosphorus. *Annu. Rev. Earth Planet. Sci.* **2018**, *46*, 521–548. [CrossRef]
10. De Nijs, E.A.; Cammeraat, E.L.H. The Stability and Fate of Soil Organic Carbon during the Transport Phase of Soil Erosion. *Earth-Sci. Rev.* **2020**, *201*, 103067. [CrossRef]
11. Du, L.; Wang, R.; Hu, Y.; Li, X.; Gao, S.; Wu, X.; Gao, X.; Yao, L.; Guo, S. Contrasting Responses of Soil C-Acquiring Enzyme Activities to Soil Erosion and Deposition. *Catena* **2021**, *198*, 105047. [CrossRef]
12. Cui, Y.; Fang, L.; Guo, X.; Han, F.; Ju, W.; Ye, L.; Wang, X.; Tan, W.; Zhang, X. Natural Grassland as the Optimal Pattern of Vegetation Restoration in Arid and Semi-Arid Regions: Evidence from Nutrient Limitation of Soil Microbes. *Sci. Total Environ.* **2019**, *648*, 388–397. [CrossRef]
13. Guan, H.; Fan, J.; Zhang, H.; Harris, W. Comparison of Drivers of Soil Microbial Communities Developed in Karst Ecosystems with Shallow and Deep Soil Depths. *Agronomy* **2021**, *11*, 173. [CrossRef]
14. Xu, M.; Li, W.; Wang, J.; Zhu, Y.; Feng, Y.; Yang, G.; Zhang, W.; Han, X. Soil Ecoenzymatic Stoichiometry Reveals Microbial Phosphorus Limitation after Vegetation Restoration on the Loess Plateau, China. *Sci. Total Environ.* **2022**, *815*, 152918. [CrossRef]
15. Ahmed, I.U.; Mengistie, H.K.; Godbold, D.L.; Sandén, H. Soil Moisture Integrates the Influence of Land-Use and Season on Soil Microbial Community Composition in the Ethiopian Highlands. *Appl. Soil Ecol.* **2019**, *135*, 85–90. [CrossRef]
16. Hsiao, C.-J.; Sassenrath, G.F.; Zeglin, L.H.; Hettiarachchi, G.M.; Rice, C.W. Vertical Changes of Soil Microbial Properties in Claypan Soils. *Soil Biol. Biochem.* **2018**, *121*, 154–164. [CrossRef]
17. Park, J.-H.; Meusburger, K.; Jang, I.; Kang, H.; Alewell, C. Erosion-Induced Changes in Soil Biogeochemical and Microbiological Properties in Swiss Alpine Grasslands. *Soil Biol. Biochem.* **2014**, *69*, 382–392. [CrossRef]
18. Sarapatka, B.; Cap, L.; Bila, P. The Varying Effect of Water Erosion on Chemical and Biochemical Soil Properties in Different Parts of Chernozem Slopes. *Geoderma* **2018**, *314*, 20–26. [CrossRef]
19. Sinsabaugh, R.L.; Lauber, C.L.; Weintraub, M.N.; Ahmed, B.; Allison, S.D.; Crenshaw, C.; Contosta, A.R.; Cusack, D.; Frey, S.; Gallo, M.E.; et al. Stoichiometry of Soil Enzyme Activity at Global Scale. *Ecol. Lett.* **2008**, *11*, 1252–1264. [CrossRef]
20. Cui, Y.; Moorhead, D.L.; Guo, X.; Peng, S.; Wang, Y.; Zhang, X.; Fang, L. Stoichiometric Models of Microbial Metabolic Limitation in Soil Systems. *Glob. Ecol. Biogeogr.* **2021**, *30*, 2297–2311. [CrossRef]
21. Wobeng, N.B.M.; Banfield, C.C.; Megueni, C.; Mapongmetsem, P.M.; Dippold, M.A. Impact of Legumes on Soil Microbial Activity and C Cycle Functions in Two Contrasting Cameroonian Agro-Ecological Zones. *Pedobiologia* **2020**, *81–82*, 150662. [CrossRef]
22. Wang, X.; Cui, Y.; Wang, Y.; Duan, C.; Niu, Y.; Sun, R.; Shen, Y.; Guo, X.; Fang, L. Ecoenzymatic Stoichiometry Reveals Phosphorus Addition Alleviates Microbial Nutrient Limitation and Promotes Soil Carbon Sequestration in Agricultural Ecosystems. *J. Soils Sediments* **2022**, *22*, 536–546. [CrossRef]
23. Xiao, L.; Liu, G.; Li, P.; Li, Q.; Xue, S. Ecoenzymatic Stoichiometry and Microbial Nutrient Limitation during Secondary Succession of Natural Grassland on the Loess Plateau, China. *Soil Tillage Res.* **2020**, *200*, 104605. [CrossRef]
24. Bai, X.; Dippold, M.A.; An, S.; Wang, B.; Zhang, H.; Loeppmann, S. Extracellular Enzyme Activity and Stoichiometry: The Effect of Soil Microbial Element Limitation during Leaf Litter Decomposition. *Ecol. Indic.* **2021**, *121*, 107200. [CrossRef]
25. Kanté, M.; Riah-Anglet, W.; Cliquet, J.-B.; Trinsoutrot-Gattin, I. Soil Enzyme Activity and Stoichiometry: Linking Soil Microorganism Resource Requirement and Legume Carbon Rhizodeposition. *Agronomy* **2021**, *11*, 2131. [CrossRef]
26. Yan, B.; Duan, M.; Wang, R.; Li, J.; Wei, F.; Chen, J.; Wang, J.; Wu, Y.; Wang, G. Planted Forests Intensified Soil Microbial Metabolic Nitrogen and Phosphorus Limitation on the Loess Plateau, China. *Catena* **2022**, *211*, 105982. [CrossRef]
27. Liu, Y.; Fu, B.; Lü, Y.; Wang, Z.; Gao, G. Hydrological Responses and Soil Erosion Potential of Abandoned Cropland in the Loess Plateau, China. *Geomorphology* **2012**, *138*, 404–414. [CrossRef]
28. Kou, M.; Jiao, J.; Yin, Q.; Wang, N.; Wang, Z.; Li, Y.; Yu, W.; Wei, Y.; Yan, F.; Cao, B. Successional Trajectory Over 10 Years of Vegetation Restoration of Abandoned Slope Croplands in the Hill-Gully Region of the Loess Plateau. *Land Degrad. Dev.* **2016**, *27*, 919–932. [CrossRef]
29. Jobbágy, E.G.; Jackson, R.B. The Vertical Distribution of Soil Organic Carbon and Its Relation to Climate and Vegetation. *Ecol. Appl.* **2000**, *10*, 423–436. [CrossRef]
30. Koven, C.D.; Riley, W.J.; Subin, Z.M.; Tang, J.Y.; Torn, M.S.; Collins, W.D.; Bonan, G.B.; Lawrence, D.M.; Swenson, S.C. The Effect of Vertically Resolved Soil Biogeochemistry and Alternate Soil C and N Models on C Dynamics of CLM4. *Biogeosciences* **2013**, *10*, 7109–7131. [CrossRef]
31. Engelhardt, I.C.; Welty, A.; Blazewicz, S.J.; Bru, D.; Rouard, N.; Breuil, M.-C.; Gessler, A.; Galiano, L.; Miranda, J.C.; Spor, A.; et al. Depth Matters: Effects of Precipitation Regime on Soil Microbial Activity upon Rewetting of a Plant-Soil System. *ISME J.* **2018**, *12*, 1061–1071. [CrossRef]
32. Chen, X.; Feng, J.; Ding, Z.; Tang, M.; Zhu, B. Changes in Soil Total, Microbial and Enzymatic C-N-P Contents and Stoichiometry with Depth and Latitude in Forest Ecosystems. *Sci. Total Environ.* **2022**, *816*, 151583. [CrossRef]

33. Schütz, K.; Kandeler, E.; Nagel, P.; Scheu, S.; Ruess, L. Functional Microbial Community Response to Nutrient Pulses by Artificial Groundwater Recharge Practice in Surface Soils and Subsoils. *FEMS Microbiol. Ecol.* **2010**, *72*, 445–455. [CrossRef]
34. Mganga, K.Z.; Razavi, B.S.; Kuzyakov, Y. Land Use Affects Soil Biochemical Properties in Mt. Kilimanjaro Region. *Catena* **2016**, *141*, 22–29. [CrossRef]
35. Kong, W.; Wei, X.; Wu, Y.; Shao, M.; Zhang, Q.; Sadowsky, M.J.; Ishii, S.; Reich, P.B.; Wei, G.; Jiao, S.; et al. Afforestation Can Lower Microbial Diversity and Functionality in Deep Soil Layers in a Semiarid Region. *Glob. Chang. Biol.* **2022**, *28*, 6086–6101. [CrossRef]
36. Doetterl, S.; Berhe, A.A.; Nadeu, E.; Wang, Z.; Sommer, M.; Fiener, P. Erosion, Deposition and Soil Carbon: A Review of Process-Level Controls, Experimental Tools and Models to Address C Cycling in Dynamic Landscapes. *Earth-Sci. Rev.* **2016**, *154*, 102–122. [CrossRef]
37. Gao, H.; Li, Z.; Li, P.; Jia, L.; Zhang, X. Quantitative Study on Influences of Terraced Field Construction and Check-Dam Siltation on Soil Erosion. *J. Geogr. Sci.* **2012**, *22*, 946–960. [CrossRef]
38. Page, A.L. (Ed.) *Methods of Soil Analysis: Part 2 Chemical and Microbiological Properties*; Agronomy Monographs; American Society of Agronomy, Soil Science Society of America: Madison, WI, USA, 1983; ISBN 978-0-89118-977-0.
39. Jones, D.; Willett, V. Experimental Evaluation of Methods to Quantify Dissolved Organic Nitrogen (DON) and Dissolved Organic Carbon (DOC) in Soil. *Soil Biol. Biochem.* **2006**, *38*, 991–999. [CrossRef]
40. Saiya-Cork, K.R.; Sinsabaugh, R.L.; Zak, D.R. The Effects of Long Term Nitrogen Deposition on Extracellular Enzyme Activity in an Acer Saccharum Forest Soil. *Soil Biol. Biochem.* **2002**, *34*, 1309–1315. [CrossRef]
41. German, D.P.; Weintraub, M.N.; Grandy, A.S.; Lauber, C.L.; Rinkes, Z.L.; Allison, S.D. Optimization of Hydrolytic and Oxidative Enzyme Methods for Ecosystem Studies. *Soil Biol. Biochem.* **2011**, *43*, 1387–1397. [CrossRef]
42. Sinsabaugh, R.L.; Turner, B.L.; Talbot, J.M.; Waring, B.G.; Powers, J.S.; Kuske, C.R.; Moorhead, D.L.; Follstad Shah, J.J. Stoichiometry of Microbial Carbon Use Efficiency in Soils. *Ecol. Monogr.* **2016**, *86*, 172–189. [CrossRef]
43. Zeng, Y.; Fang, N.; Shi, Z. Effects of Human Activities on Soil Organic Carbon Redistribution at an Agricultural Watershed Scale on the Chinese Loess Plateau. *Agric. Ecosyst. Environ.* **2020**, *303*, 107112. [CrossRef]
44. Cui, Y.; Fang, L.; Guo, X.; Wang, X.; Zhang, Y.; Li, P.; Zhang, X. Ecoenzymatic Stoichiometry and Microbial Nutrient Limitation in Rhizosphere Soil in the Arid Area of the Northern Loess Plateau, China. *Soil Biol. Biochem.* **2018**, *116*, 11–21. [CrossRef]
45. Zhang, W.; Xu, Y.; Gao, D.; Wang, X.; Liu, W.; Deng, J.; Han, X.; Yang, G.; Feng, Y.; Ren, G. Ecoenzymatic Stoichiometry and Nutrient Dynamics along a Revegetation Chronosequence in the Soils of Abandoned Land and Robinia Pseudoacacia Plantation on the Loess Plateau, China. *Soil Biol. Biochem.* **2019**, *134*, 1–14. [CrossRef]
46. Lagomarsino, A.; Benedetti, A.; Marinari, S.; Pompili, L.; Moscatelli, M.C.; Roggero, P.P.; Lai, R.; Ledda, L.; Grego, S. Soil Organic C Variability and Microbial Functions in a Mediterranean Agro-Forest Ecosystem. *Biol. Fertil. Soils* **2011**, *47*, 283–291. [CrossRef]
47. Raiesi, F.; Beheshti, A. Soil Specific Enzyme Activity Shows More Clearly Soil Responses to Paddy Rice Cultivation than Absolute Enzyme Activity in Primary Forests of Northwest Iran. *Appl. Soil Ecol.* **2014**, *75*, 63–70. [CrossRef]
48. Kivlin, S.N.; Treseder, K.K. Soil Extracellular Enzyme Activities Correspond with Abiotic Factors More than Fungal Community Composition. *Biogeochemistry* **2014**, *117*, 23–37. [CrossRef]
49. Romanowicz, K.J.; Freedman, Z.B.; Upchurch, R.A.; Argiroff, W.A.; Zak, D.R. Active Microorganisms in Forest Soils Differ from the Total Community yet Are Shaped by the Same Environmental Factors: The Influence of PH and Soil Moisture. *FEMS Microbiol. Ecol.* **2016**, *92*, fiw149. [CrossRef]
50. Chen, H.; Luo, P.; Wen, L.; Yang, L.; Wang, K.; Li, D. Determinants of Soil Extracellular Enzyme Activity in a Karst Region, Southwest China. *Eur. J. Soil Biol.* **2017**, *80*, 69–76. [CrossRef]
51. Henry, H.A.L. Soil Extracellular Enzyme Dynamics in a Changing Climate. *Soil Biol. Biochem.* **2012**, *47*, 53–59. [CrossRef]
52. Du, E.; Terrer, C.; Pellegrini, A.F.A.; Ahlström, A.; van Lissa, C.J.; Zhao, X.; Xia, N.; Wu, X.; Jackson, R.B. Global Patterns of Terrestrial Nitrogen and Phosphorus Limitation. *Nat. Geosci.* **2020**, *13*, 221–226. [CrossRef]
53. Marklein, A.R.; Houlton, B.Z. Nitrogen Inputs Accelerate Phosphorus Cycling Rates across a Wide Variety of Terrestrial Ecosystems. *New Phytol.* **2012**, *193*, 696–704. [CrossRef]
54. Sinsabaugh, R.L.; Follstad Shah, J.J. Ecoenzymatic Stoichiometry and Ecological Theory. *Annu. Rev. Ecol. Evol. Syst.* **2012**, *43*, 313–343. [CrossRef]
55. Zhang, S.; Pan, Y.; Zhou, Z.; Deng, J.; Zhao, F.; Guo, Y.; Han, X.; Yang, G.; Feng, Y.; Ren, G.; et al. Resource Limitation and Modeled Microbial Metabolism along an Elevation Gradient. *Catena* **2022**, *209*, 105807. [CrossRef]
56. Vitousek, P.M.; Porder, S.; Houlton, B.Z.; Chadwick, O.A. Terrestrial Phosphorus Limitation: Mechanisms, Implications, and Nitrogen–Phosphorus Interactions. *Ecol. Appl.* **2010**, *20*, 5–15. [CrossRef]
57. Lemma, B.; Kebede, F.; Mesfin, S.; Fitiwy, I.; Abraha, Z.; Norgrove, L. Quantifying Annual Soil and Nutrient Lost by Rill Erosion in Continuously Used Semiarid Farmlands, North Ethiopia. *Environ. Earth Sci.* **2017**, *76*, 190. [CrossRef]
58. Jiao, F.; Wen, Z.-M.; An, S.-S.; Yuan, Z. Successional Changes in Soil Stoichiometry after Land Abandonment in Loess Plateau, China. *Ecol. Eng.* **2013**, *58*, 249–254. [CrossRef]

agronomy

MDPI

Article

Research on the Classification of Complex Wheat Fields Based on Multi-Scale Feature Fusion

Fei Mu [1,2,3,4], Hongli Chu [1,2,3,4], Shuaiqi Shi [1,2,3,4], Minxin Yuan [1,2,3,4], Qi Liu [1,2,3,4] and Fuzeng Yang [1,2,3,4,*]

1 College of Mechanical and Electronic Engineering, Northwest A&F University, Yangling 712100, China
2 Apple Mechanized Research Base, Yangling 712100, China
3 Shannxi Key Laboratory of Apple, Yangling 712100, China
4 State Key Laboratory of Soil Erosion and Dryland Farming on Loess Plateau, Yangling 712100, China
* Correspondence: yangfzkm@nwafu.edu.cn

Abstract: This study uses UAV multi-spectral remote sensing images to carry out ground object classification research in complex wheat field scenes with diverse varieties. Compared with satellite remote sensing, the high spatial resolution remote sensing images obtained by UAVs at low altitudes are rich in detailed information. In addition, different varieties of wheat have different traits, which makes it easy to misclassify categories in the process of semantic segmentation, which reduces the classification accuracy and affects the classification effect of ground object. In order to effectively improve the classification accuracy of ground object in complex wheat field scenes, two Multi-Scale U-Nets based on multi-scale feature fusion are proposed. Multi-Scale U-Net1 is a network model that adds a multi-scale feature fusion block in the copy process between U-Net encoding and decoding. Multi-Scale U-Net2 is a network model that adds a multi-scale feature fusion block before U-Net inputs an image. Firstly, the wheat field planting area of Institute of Water-saving Agriculture in Arid Areas of China (IWSA), Northwest A&F University was selected as the research area. The research area was planted with a variety of wheat with various types of traits, and some traits were quite different from one another. Then, multi-spectral remote sensing images of different high spatial resolutions in the study area were obtained by UAV and transformed into a data set for training, validation, and testing of network models. The research results showed that the overall accuracy (OA) of the two Multi-Scale U-Nets reached 94.97% and 95.26%, respectively. Compared with U-Net, they can complete the classification of ground object in complex wheat field scenes with higher accuracy. In addition, it was also found that within the effective range, with the reduction of the spatial resolution of remote sensing images, the classification of ground object is better.

Keywords: multi-scale feature fusion; U-Net; UAV; multi-spectral image; complex wheat fields; ground object classification

Citation: Mu, F.; Chu, H.; Shi, S.; Yuan, M.; Liu, Q.; Yang, F. Research on the Classification of Complex Wheat Fields Based on Multi-Scale Feature Fusion. *Agronomy* **2022**, *12*, 2658. https://doi.org/10.3390/agronomy12112658

Academic Editor: Francisco Manzano Agugliaro

Received: 24 September 2022
Accepted: 24 October 2022
Published: 27 October 2022

1. Introduction

The planting structure of wheat fields reflects the spatial distribution information of wheat fields in an area or production unit [1]. Obtaining the planting structure information of wheat fields efficiently and accurately is of great significance for wheat field yield estimation, agricultural condition monitoring and agricultural structure adjustment [2]. In the field of scientific research, due to the large differences in the research objectives of experts in different fields, the complex planting structure of wheat experimental fields, the large variety of wheat varieties, and the large differences in traits, it is difficult for the ground object classification method based on machine learning to meet the current needs. With the rapid development of UAV remote sensing technology and deep learning methods, the method of semantic segmentation of UAV remote sensing images based on fully convolutional neural network technology is increasingly applied to the research of farmland object classification [3–7]. The deep learning method can quickly and accurately realize the classification of ground object in complex wheat field planting areas and provide technical support for wheat field yield estimation, agricultural condition monitoring, and agricultural structure adjustment.

Remote sensing technologies include satellite remote sensing and UAV remote sensing. Satellite remote sensing can obtain large-scale remote sensing images of land parcels, and UAV remote sensing can obtain small-scale land remote sensing images with higher spatial resolution [8]. Based on farmland satellite remote sensing images, there have been many studies using traditional machine learning methods and deep learning-based semantic segmentation methods to realize the classification of farmland objects, which have achieved a good classification effect. For example, Al Awar B et al. [9] took a planting area in the Bekaa Valley of Lebanon, where the main crops are wheat and potato, as the research area, used Sentinel-1 and Sentinel-2 satellite remote sensing images as dataset, and adopted Support Vector Machine (SVM), Random Forest (RF), Classification and Regression Tree (CART) and Back Propagation Network (BPN) for crop classification research. The results found that the overall accuracy (OA) of these four classification models can reach more than 95%. Zheng Wenhui et al. [10] selected the Loess Plateau mulching dry agricultural planting area as the research area based on Google Earth Engine cloud platform data and Landsat-8 reflectivity data, using Random Forest (RF), Support Vector Machine (SVM), Decision Tree (DT), and Minimum Distance (MD) to classify farmland objects. The result showed that under the condition of artificial feature engineering, the overall accuracy (OA) of machine learning object classification can reach 95.5%. Song Tingqiang et al. [11] selected a crop planting area in Gaomi City, Shandong Province as the research area based on the GF-2 satellite high spatial resolution panchromatic images and multispectral images, and adopted an improved Multi-temporal Spatial Segmentation Network (MSSN) to realize the classification of farmland objects. The experimental result showed that the Pixel Accuracy (PA) of the model on the test set was 95%, the F1 score was 0.92, and the Intersection over Union (IoU) was 0.93. Xu Lu et al. [12] selected a farmland in Guangping County of Hebei Province and Luobei County of Heilongjiang Province as the research areas based on the high spatial resolution remote sensing image of Gaofen-2 (GF-2), firstly adopted the Depthwise Separable Convolution U-Net (DSCU-net) to realize the segmentation of the entire image, and then the extended Multi-channel Rich Convolutional Feature network (RCF) to further delineate the boundaries of cultivated land plots. The experimental result showed that the overall accuracy (OA) of classification in the two regions was around 90%. In summary, both the traditional machine learning algorithm and the semantic segmentation algorithm based on the fully convolutional neural network can show satisfactory results in the classification of agricultural satellite remote sensing images.

Compared with satellite remote sensing, UAV remote sensing has the characteristics of high flexibility, short period, less environmental impact, and easy access to small-scale agricultural remote sensing data [13]. In addition, satellite remote sensing can only obtain remote sensing image data with meter-level spatial resolution, while UAV remote sensing can obtain remote sensing image data with high spatial resolution at centimeter level. Therefore, UAV agricultural remote sensing image data has more detailed information, and it is easy to misclassify in the process of image classification, which brings great challenges to agricultural land classification to a certain extent. For such agricultural remote sensing images with more detailed information, due to the complexity and fragmentation of ground object and the surrounding environment at high spatial resolution, the accuracy of traditional classification methods has been unable to meet the standards of agricultural problems. The classification method based on convolutional neural network can effectively learn image features related to the target category [14]. Therefore, most of the current researches on the high spatial resolution UAV agricultural remote sensing images use the classification method based on the fully convolutional neural network to conduct ground object classification research. For example, Chen Yuqing et al. [15] used a remote sensing image of a farmland in Kaifeng City, Henan Province as a dataset. The dataset contained farmland and two other types of ground object. The improved Deeplab V3+ model was used for land classification. The experimental result showed that the Mean Pixel Accuracy (MPA) can reach 97.16%, which effectively improves the information extraction accuracy of farmland edges and small farmlands. Qinchen Yang et al. [16] used UAV remote sensing images of an agricultural planting area in the Hetao irrigation area of Inner Mongolia as a dataset. The dataset contained five types of land objects, including roads, green plants, cultivated land, wasteland and others. Two semantic segmentation algorithms, SegNet and FCN, and traditional Support Vector Machine (SVM) were used to conduct ground object classification research. The result showed that SegNet and FCN are significantly better than the traditional Support Vector Machine (SVM) in terms of accuracy and speed, and their Mean Pixel Accuracy (MPA) were 89.62% and 90.6%, respectively. Yang Shuqin et al. [17] used the UAV multispectral remote sensing images of Shahao Canal Irrigation Area in Hetao Irrigation District of Inner Mongolia Autonomous Region as a dataset. The dataset contained sunflower, zucchini, corn and other four types of ground object; the improved Deeplab V3+ and Support Vector Machine (SVM) to classify farmland objects were used. The result showed that the Mean Pixel Accuracy (MPA) of the improved Deeplab V3+ was 93.06%, which is 17.75% higher than the Support Vector Machine (SVM). To sum up, in terms of high spatial resolution UAV agricultural remote sensing images with more detailed information, the semantic segmentation model based on deep learning has satisfactory results in the classification of farmland objects.

The above-mentioned farmland object classification based on satellite remote sensing or UAV remote sensing was all carried out between different types of objects. So far, there have been few relevant reports on the classification of similar crops with different traits. At present, with the continuous advancement of breeding work, different varieties of crops of the same type may be planted in the same area, and some varieties possess large differences in traits. Especially in the classification of farmland objects on the high spatial resolution remote sensing images of UAVs, it is very easy to cause similar farmlands with large differences in traits to be incorrectly classified. Based on UAV remote sensing technology and the deep learning classification method of multi-scale feature fusion, this study carried out ground object classification for complex wheat fields planted with different varieties of wheat and with large differences in traits between some wheats.

2. Materials and Methods

As shown in Figure 1, the current research is the workflow of complex wheat field classification research based on multi-scale feature fusion. This study selects a suitable research area and conducts field surveys in the research area to understand the distribution information of ground object. The experimental plan is designed according to the results of the field investigation. First, the location information of Ground Control Points (GCPs) is collected in the study area for geometric correction of the later remote sensing image stitching process. Then, the UAV (DJ Inc., Shenzhen, China) flight parameters are set within the time range required by the experimental plan to collect multi-spectral remote sensing images in the study area. The collected remote sensing images and ground control points information are imported into Pix4Dmapper 4.5.6 (Pix4D Inc., Prilly, Switzerland) for image stitching and preprocessing, and finally the remote sensing images of the entire study area required by this study are generated. Since the remote sensing images collected by the UAV are multi-spectral data, in order to facilitate the later image labeling, it is necessary to fuse the five single-band images to generate a single image with five bands (that is, a five-channel remote sensing image). Next is labeling the image. Since it is difficult to directly import the entire remote sensing image into the deep learning network for model training, this study cropped the remote sensing images and their Labels to obtain regular small-sized images and Labels, and divided the cropped images and Labels into datasets, for training, validation, and testing of deep learning models. This research proposes two Multi-Scale U-Nets based on U-Net by adding a multi-scale feature fusion block, and selects Confusion Matrix (CM), Pixel Accuracy (PA), and Recall, F1-Score, and Intersection over Union (IoU) as the five indicators to evaluate the classification accuracy of the models in three categories (wheat, road and background). Three indicators (Overall Accuracy (OA), Mean Intersection over Union (MIoU), and Frequency Weighted Intersection over Union (FWIoU)) were used to comprehensively evaluate the models.

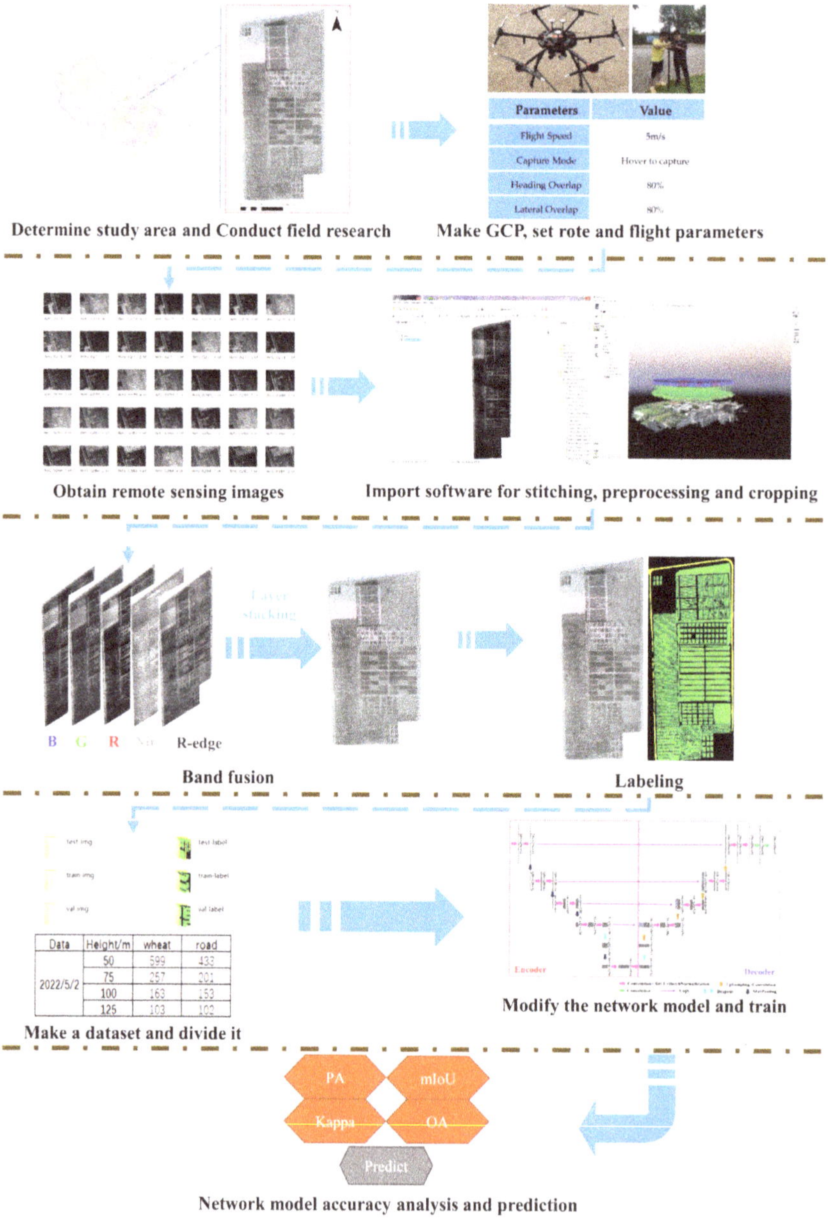

Figure 1. Workflow of complex wheat field classification research based on multi-scale feature fusion.

2.1. Study Area

The study area is located in the Institute of Water-saving Agriculture in Arid Areas of China (IWSA), Northwest A&F University, as shown in Figure 2. The study area is located in Yangling Demonstration Zone, Xianyang City, Shaanxi Province (34°17′51.11″ N–34°17′58.72″ N, 108°4′4.10″ E–108°4′7.64″ E). This area has fertile soil, relatively flat terrain, and an altitude between 560 and 790 m. It belongs to the semi-humid and semi-arid climate in the warm temperate zone. It has obvious continental monsoon climate characteristics such as warm and windy spring, hot and rainy summer, and cold and dry winter. The annual average temperature is about 12 °C, the frost-free period is 211 days, the average annual sunshine hours amount to about 2163 h, and the average annual precipitation is 635 mm. The area of the study area is about 17,553 m^2, including 14 varieties of wheat, some of which have great differences in wheat traits.

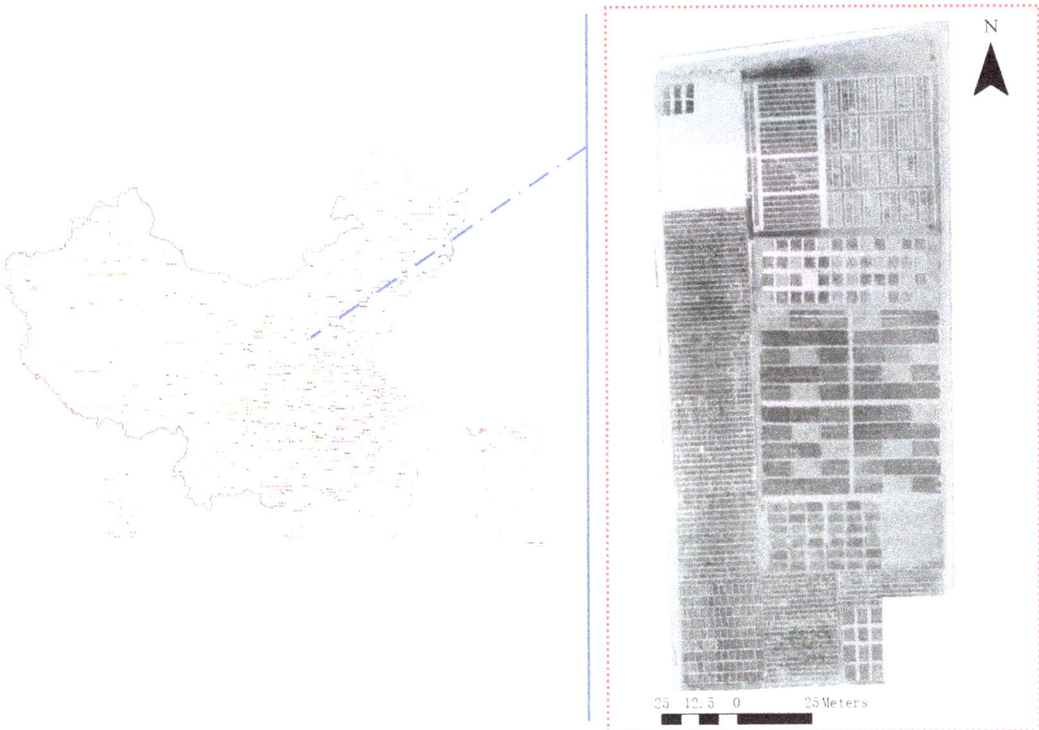

Figure 2. Multi-variety wheat planting area, Institute of Water-saving Agriculture in Arid Areas of China (IWSA), Northwest A&F University.

2.2. Data Acquisition and Preprocessing

2.2.1. UAV Acquires Multispectral Images with Different High Spatial Resolution

In this study, UAV remote sensing data collection was conducted on 2 May 2022 at the Institute of Water-saving Agriculture in Arid Areas of China (IWSA), Northwest A&F University. As shown in Figure 3, it is a UAV remote sensing image data acquisition system, which consists of a DJI M600 PRO UAV (DJ Inc., Shenzhen, China), battery packs (DJ Inc., Shenzhen, China), a RedEdge-MX Five-channel multispectral camera (MicaSense Inc., Seattle, WC, USA), a FLIR Duo Pro Thermal Infrared Imager (FLIR Systems Inc., Washington, DC, USA), and a Zenmuse X3 visible camera (DJ Inc., Shenzhen, China). The system customizes a gimbal that can carry the above three cameras at the same time to meet the requirements of the three cameras working independently. The system is equipped with four battery packs to meet the long-term work requirements. This study mainly uses the RedEdge-MX Five-channel multispectral camera to obtain remote sensing image data.

Some parameters of the RedEdge-MX Five-channel multispectral camera are shown in Table 1.

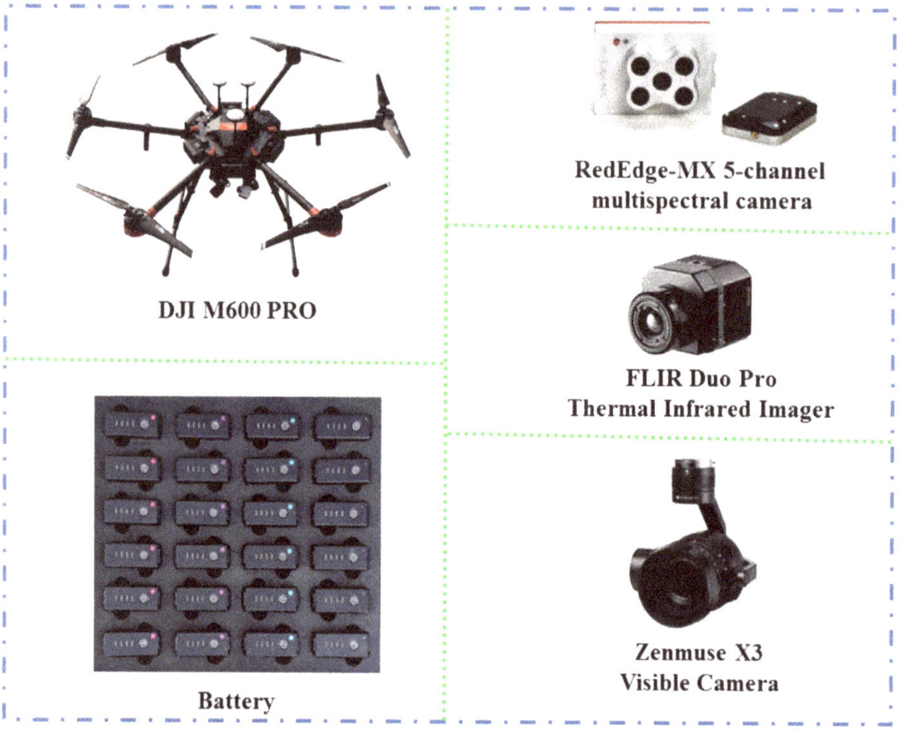

Figure 3. UAV remote sensing image data acquisition system.

Table 1. Part of the parameters of the multispectral camera.

Parameter Name	Parameter Value		
	Band Name	**Central Wavelength (nm)**	**Bandwidth (nm)**
Spectral band	Blue	475	32
	Green	560	27
	Red	668	16
	Red edge	717	12
	Near infrared	842	57
Spatial Resolution	8 cm @ 120 m flight height		

Ground Control Points (GCPs) are important data source for geometric correction and geolocation of UAV remote sensing images [18]. As shown in Figure 4, this study uses the Trimble GNSS system to obtain the coordinates of the GCPs. The Trimble GNSS system includes a Trimble R8 GNSS receiver (Trimble Inc., Westminster, CO, USA), a Trimble R2 GNSS receiver (Trimble Inc., Westminster, CO, USA), a radio, an antenna (Trimble Inc., Westminster, CO, USA), a lead-acid battery (Trimble Inc., Westminster, CO, USA), a mobile terminal (Trimble Inc., Westminster, CO, USA), and a connection line (Trimble Inc., Westminster, CO, USA). The Trimble R8 GNSS receiver, radio, antenna, lead-acid battery, and connecting line form a base station for receiving satellite signals and also for sending signals to the Trimble R2 GNSS receiver. The Trimble R2 GNSS receiver is also called a mobile station, which can receive the signal of the base station. The mobile station sends the coordinate information of the base station and the relative position information between itself and the base station to the mobile terminal, and the mobile terminal obtains the coordinate information of the mobile station after calculation and processing. The base station is set up in the research area in advance, and the coordinate information of each GCP is obtained by using the mobile station and mobile terminal. Therefore, the coordinate accuracy of the GCPs completely depends on the accuracy of the base station and mobile station. The accuracy of the GCPs obtained in this research is controlled at the millimeter level, which fully meets the accuracy requirements of this experiment.

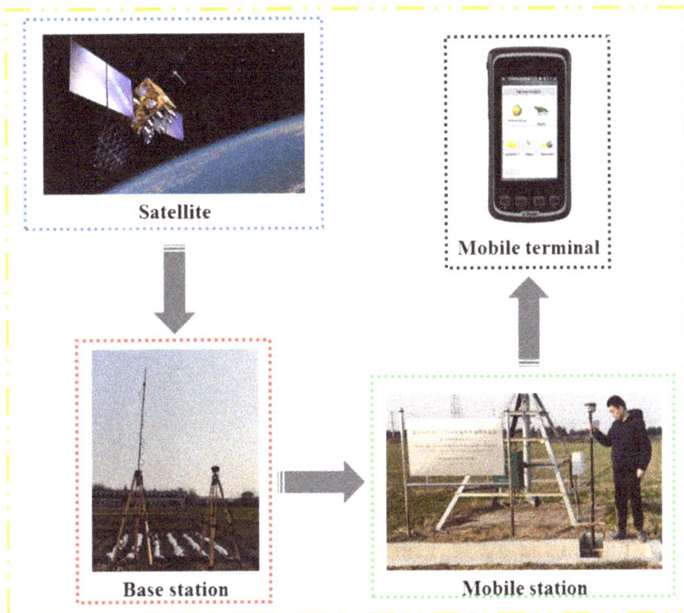

Figure 4. Use Trimble GNSS System to Obtain the Coordinates of Ground Control Points (GCPs).

Before the UAV collecting of remote sensing image data, it is necessary to perform radiometric calibration of the multispectral camera. As shown in Figure 5, this experiment conducted radiometric calibration in an open field in the study area at around 11:40 noon on 2 May 2022. After the radiometric calibration, the remote sensing image data was collected according to the expected experimental plan. The spatial resolution of remote sensing images refers to the size of the ground range from the perspective of pixels, that is, a pixel in a remote sensing image represents the area of the ground [19]. As shown in Formula (1), it is the calculation formula of image spatial resolution, in which the flying height of the UAV is inversely proportional to the spatial resolution of the image, and the remote sensing images with different spatial resolutions can be obtained by changing the different flying heights of the UAV. As shown in Table 2, it is the flight parameters of the UAV. According to these parameters, the remote sensing images of four different spatial resolutions required in this study are obtained. Figure 6 is the local detail map of remote sensing images with different spatial resolutions.

$$S = \frac{D \cdot f}{H}, \quad (1)$$

where D is the Ground Sampling Distance (GSD), S is the Image Spatial Resolution, H is the flying height, f is the focal length.

Figure 5. Multispectral camera radiometric calibration.

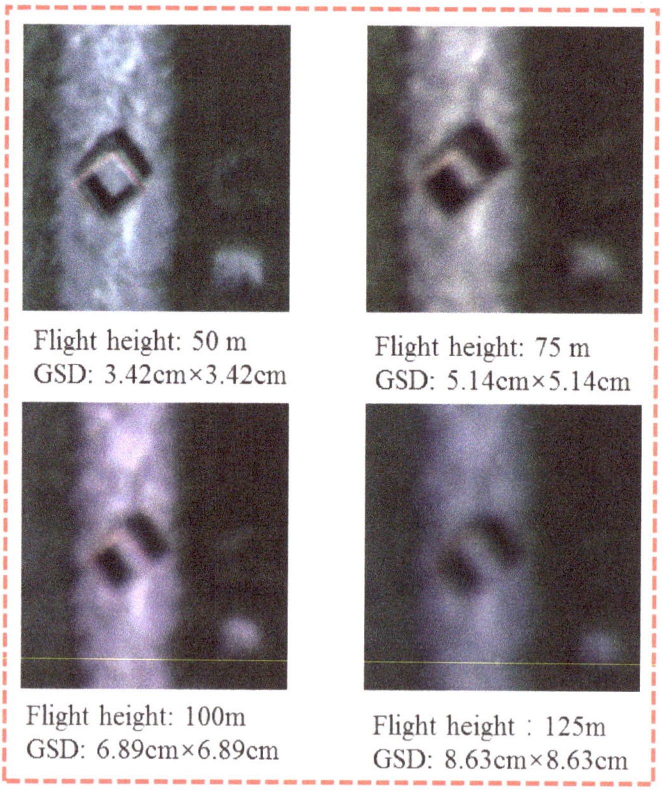

Figure 6. Local remote sensing images with different spatial resolutions (GSD: Ground Sampling Distance).

Table 2. UAV flight parameters.

Parameter		Value
Flight Speed		5 m/s
Capture Mode		Hover to capture
Heading Overlap		80%
Lateral Overlap		80%
Flying Height	The First Experiment	50 m
	The Second Experiment	75 m
	The Third Experiment	100 m
	The Fourth Experiment	125 m

2.2.2. Multispectral Image Data Processing and Dataset Making

The remote sensing images obtained by UAV are a large number of small-sized pictures taken according to the overlap rate. In order to facilitate the later labeling work, the remote sensing images obtained by the UAV are subjected to stitching, geometric correction, and other related preprocessing operations. The remote sensing image preprocessing software used includes Pix4Dmapper 4.5.6 (Pix4D Inc., Prilly, Switzerland) and ENVI 5.3 (EVIS Inc., Los Angeles, CA, USA). As shown in Figure 7, the usage of ENVI 5.3 performs band fusion on the remote sensing image and crops out the study area.

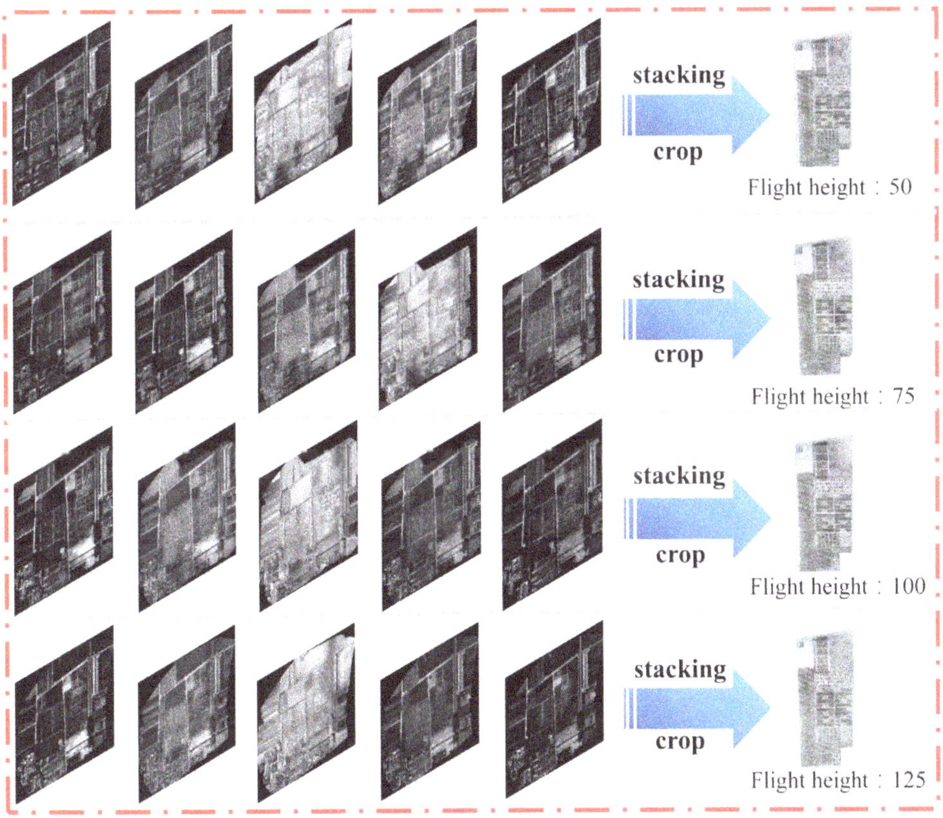

Figure 7. Remote Sensing Image Band stacking and Cropping.

After band fusion and cropping, the remote sensing image of the study area is obtained. Compared with the RGB image, the image contains 5 channels, namely Blue band, Green band, Red band, Near infrared band, and Red-edge band. Traditional RGB image labeling methods often use Labelme 4.5.7 (MIT, Cambridge, MA, USA) and other software for labeling, but the remote sensing images with 5 channels in this study cannot be labelled by Labelme. Therefore, the remote sensing images in this study were labelled with ENVI Classic 5.3 (EVIS Inc., Los Angeles, CA, USA). Figure 8 are Labels obtained after labeling. If the above images and their Labels are directly input into the deep learning network, it will cause memory overflow, so these images and labels need to be cropped into small-size images and input into the network. As shown in Figure 9, this study adopts the sliding window cropping method for image cropping, and uses Python language to write a program for cropping multispectral remote sensing images and their labels. The multispectral images and their Labels are cropped by presetting the size of the slider window and the overlap ratio. In this study, a slider window of 256 × 256 and an overlap rate of 20% were used to crop images of the study area and their labels to obtain the dataset of this study.

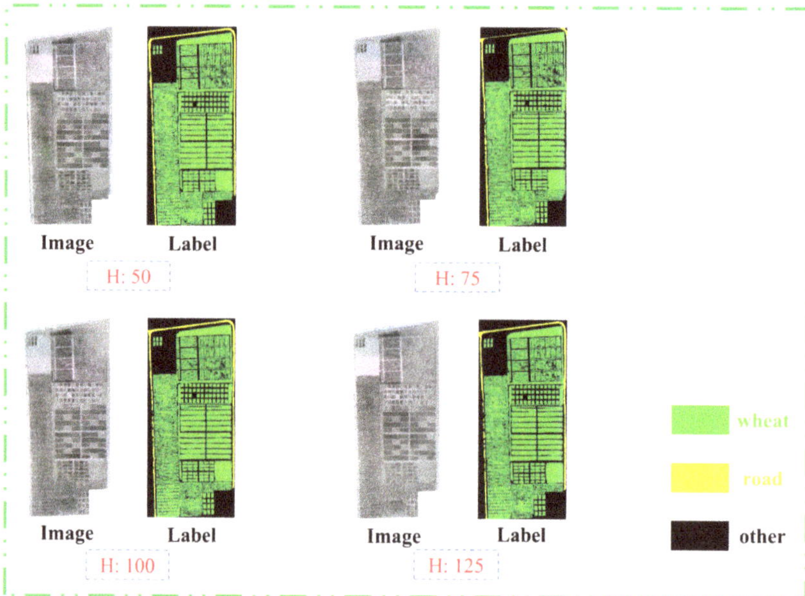

Figure 8. Multispectral remote sensing images and their labels.

The basic configuration of the dataset used in this study is shown in Figure 10 and Table 3. From Figure 10a, it can be observed that the number of wheat and road categories is unbalanced after data cropping. Unbalanced categories will cause the model learning effect to be biased towards the party with the larger category, which seriously affects the generalization ability of the model [20]. Therefore, this study balances the dataset categories by extracting images with road categories under different slider overlap ratios. As shown in Figure 10b, the number of dataset categories after data balance is relatively balanced. After the dataset is balanced, the dataset is integrated into the training set, validation set and test set according to the ratio of 60%, 20% and 20%. On the one hand, data augmentation can make the model have sufficient dataset for training, and on the other hand, it can improve the generalization ability of the model [21]. This study performs data augmentation on the training and validation set by horizontally flipping, vertically flipping, and diagonally mirroring the multispectral images and their labels. The dataset division before and after data enhancement is shown in Table 3, with 4828 training images and labels, 1608 validation images and labels, and 402 test images and labels.

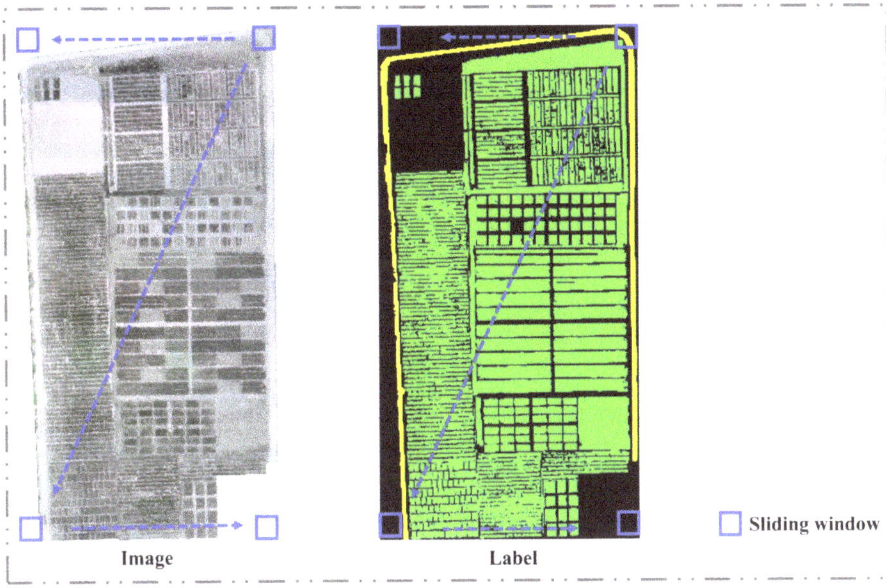

Figure 9. Cropping multispectral images and their labels.

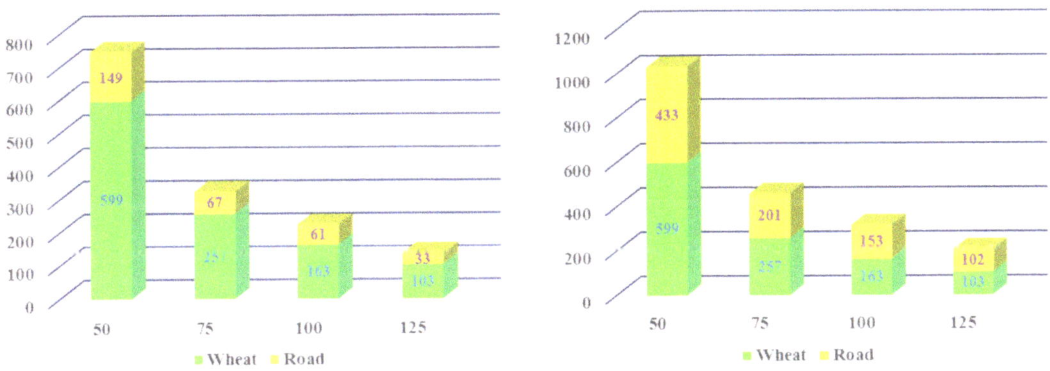

(**a**) (**b**)

Figure 10. Statistics chart of dataset categories, (**a**) Statistics chart of categories at different spatial resolutions before data balance, (**b**) Statistics chart of categories at different spatial resolutions after data balance.

Table 3. Overview after dividing the dataset.

Dataset	Number	The Number of Data Augmentation
Training set	1207	4828
Validation set	402	1608
Test set	402	402

USA). The Multi-Scale U-Net1 model and the Multi-Scale U-Net2 model were tested using the test set, and the classification accuracy of the two models in the complex wheat field scene of UAV multispectral images was analyzed and compared.

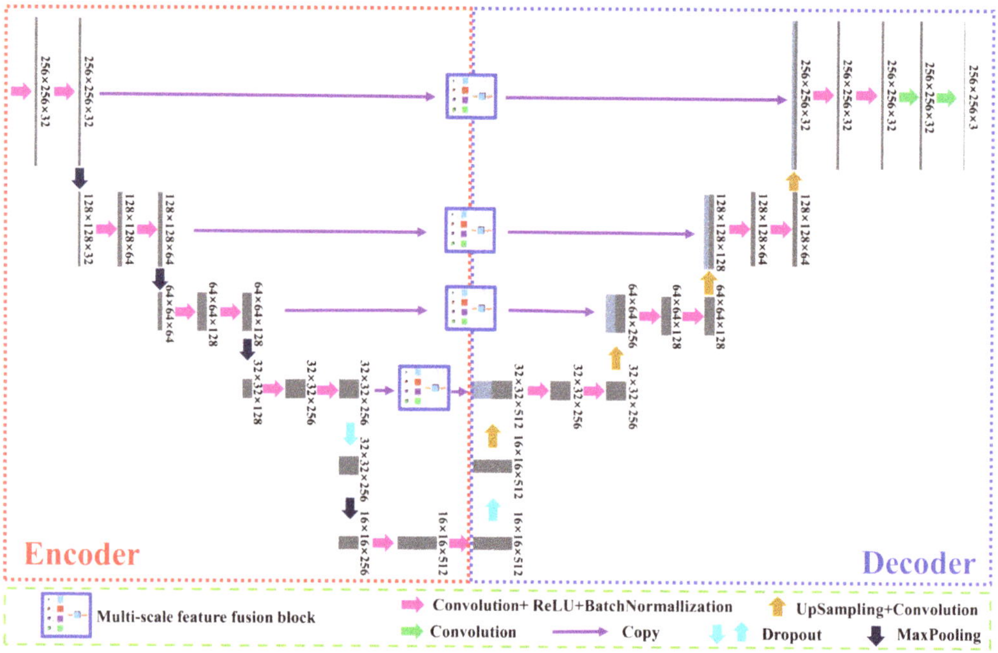

Figure 13. Multi-Scale U-Net1.

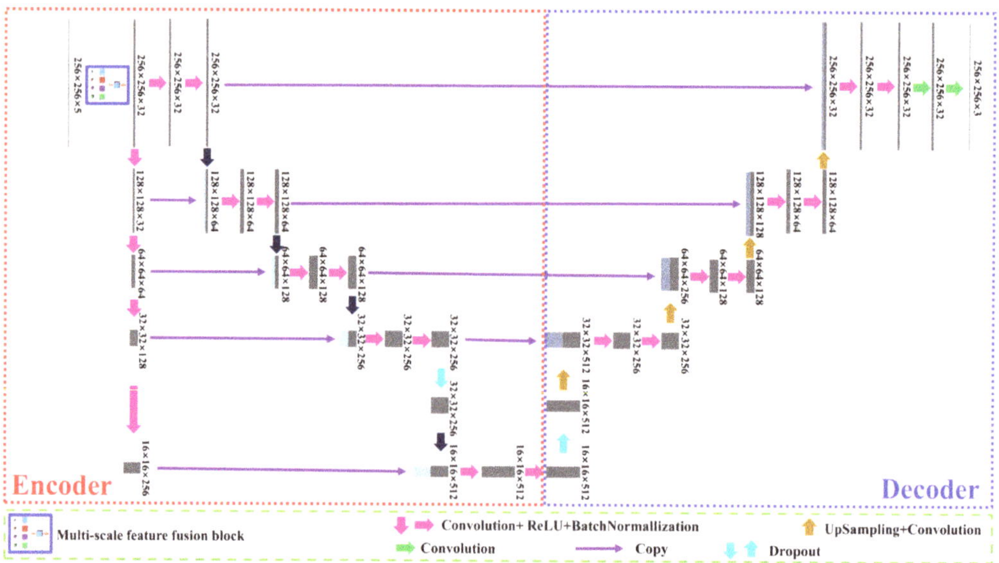

Figure 14. Multi-Scale U-Net2.

3. Results

3.1. Model Training

This study uses the training set and validation set to train the U-Net, Multi-Scale U-Net1, and Multi-Scale U-Net2, respectively. According to the aforementioned Wheat and Road accounting for 56% and 44%, respectively, the weights are balanced before training the model, and the training weights of 0.56 and 0.44 are set for the Wheat and Road categories, respectively. The training parameters set for the three models in this study are shown in Table 4.

Table 4. Model Training Parameters.

Model Training Parameters	Value	Model Training Parameters	Value
Batch size	8	Optimizer	Adam
Epochs	50	Loss	Categorical cross-entropy
Learning rate	1×10^{-5}	Programming Language Framework	Tensorflow keras

In this study, 4828 training set samples and 1608 validation set samples were input into three models for training. Figure 15 shows the accuracy curve and Loss function curve of U-Net, Multi-Scale U-Net1, and Multi-Scale U-Net2 model training, respectively. U-Net, Multi-Scale U-Net1, and Multi-Scale U-Net2 finish training at epoch 28, 29, and 44, respectively. As can be seen from Figure 15, Multi-Scale U-Net2 converges first and displays higher accuracy than the other two models, while the U-Net model is the slowest to converge, and the accuracy is not as good as the other two models.

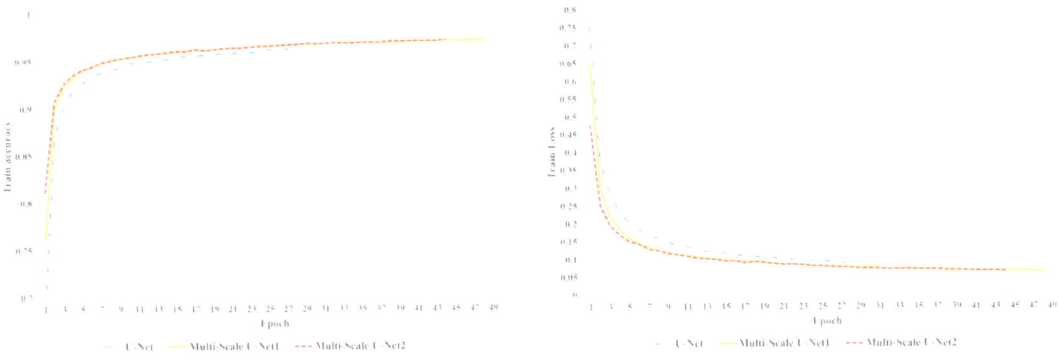

(a) (b)

Figure 15. Train accuracy and train loss. (**a**) Train accuracy, (**b**) Train Loss.

3.2. Model Testing and Prediction Results

In this study, 402 test set samples were used for the accuracy evaluation of the three models. Five indicators, Confusion Matrix (CM), Pixel Accuracy (PA), Recall (Recall), F1-Score, and Intersection over Union (IoU), were selected to evaluate the model in three categories. Three indicators, Overall Accuracy (OA), Mean Intersection over Union (MIoU), and Frequency Weighted Intersection over Union (FWIoU), were used to comprehensively evaluate models. The Confusion Matrix (CM) is an analysis table that summarizes the prediction results of the classification model in machine learning. The records in the dataset are summarized in the form of a matrix according to the two criteria of the real category and the category predicted by the classification model. The Pixel Accuracy (PA) is "the number of correctly predicted data in those data that are predicted to be positive examples". The

Recall is "the number of correctly predicted data in the real positive examples". F1-Score is the harmonic mean of precision and recall. Over Accuracy (OA) is the ratio between what the model predicted correctly on all test sets and the overall number. Mean Intersection over Union (MIoU) is the average of the ratio of the intersections and unions of all categories.

Tables 5–7 are the Confusion Matrix of U-Net, Multi-Scale U-Net1, and Multi-Scale U-Net2, respectively. It can be seen that the three models have relatively few misclassifications in the two categories of 'Wheat' and 'Road', but the three models all misclassify 'Wheat' and 'Road' into 'Other'; considering that the purpose of this study is to separate 'Wheat' and 'Road', the overall classification of the three models is good. Compared with U-Net, Multi-Scale U-Net1, and Multi-Scale U-Net2 with multi-scale feature fusion modules show better performance on 'Wheat' and 'Road' classification, and especially the classification accuracy between 'Wheat' and 'Road' categories is significantly improved.

Table 5. U-Net Confusion Matrix.

U-Net		**Predict**		
		Other	**Wheat**	**Road**
Real	Other	5,759,250	1,012,158	47,563
	Wheat	149,715	10,400,195	48
	Road	55,981	194	1,056,048

Table 6. Multi-Scale U-Net1 Confusion Matrix.

Multi-Scale U-Net1		**Predict**		
		Other	**Wheat**	**Road**
Real	Other	6,102,568	693,462	22,941
	Wheat	184,810	10,365,148	0
	Road	27,949	96	1,084,178

Table 7. Multi-Scale U-Net2 Confusion Matrix.

Multi-Scale U-Net2		**Predict**		
		Other	**Wheat**	**Road**
Real	Other	6137,736	602,737	78,498
	Wheat	183,612	10,366,342	4
	Road	11,132	78	1,101,013

This study evaluates U-Net, Multi-Scale U-Net1, and Multi-Scale U-Net2 on 'Other', 'Wheat', 'Road', and overall classification accuracy, respectively. Table 8 is the Accuracy evaluation of different categories under three network models. Figure 16 is the accuracy evaluation curve. According to the precision evaluation index, Multi-Scale U-Net1 is superior to the other two network models, and Multi-Scale U-Net2 is inferior to the other two models in the classification accuracy of the 'road' category, but the accuracy of Multi-Scale U-Net2 in the 'road' category is 93.34%, and the result is still high, which can complete the correct classification of the 'road' category with higher accuracy. According to the Recall evaluation index, the Recall of Multi-Scale U-Net1 and Multi-Scale U-Net2 in the 'road' and 'other' categories were significantly higher than that of U-Net. The Recall for the three network models in the 'wheat' category are almost close. According to the Intersection over Union (IoU) and Figure 16d, Multi-Scale U-Net1 and Multi-Scale U-Net2 are better than U-Net. It can be seen that the classification accuracy of the Multi-Scale U-Net1 and Multi-Scale U-Net2 is better than that of the U-Net. The Multi-Scale U-Net1 outperforms the Multi-Scale U-Net2 in both 'Wheat' and 'Road' classification. To sum up, a module with a multi-scale feature fusion block can effectively improve the classification accuracy.

Table 8. Accuracy evaluation of Different categories under three Network Models.

		Precision	Recall	F1-Score	IoU
U-Net	Other	96.55%	84.46%	90.10%	81.99%
	Wheat	91.13%	98.58%	94.71%	89.95%
	Road	95.69%	94.95%	95.32%	91.05%
Multi-Scale U-Net1	Other	96.63%	89.49%	92.93%	86.79%
	Wheat	93.73%	98.25%	95.94%	92.19%
	Road	97.93%	97.48%	97.70%	95.51%
Multi-Scale U-Net2	Other	96.92%	90.00%	93.34%	87.51%
	Wheat	94.50%	98.26%	96.35%	92.95%
	Road	93.34%	98.99%	96.09%	92.47%

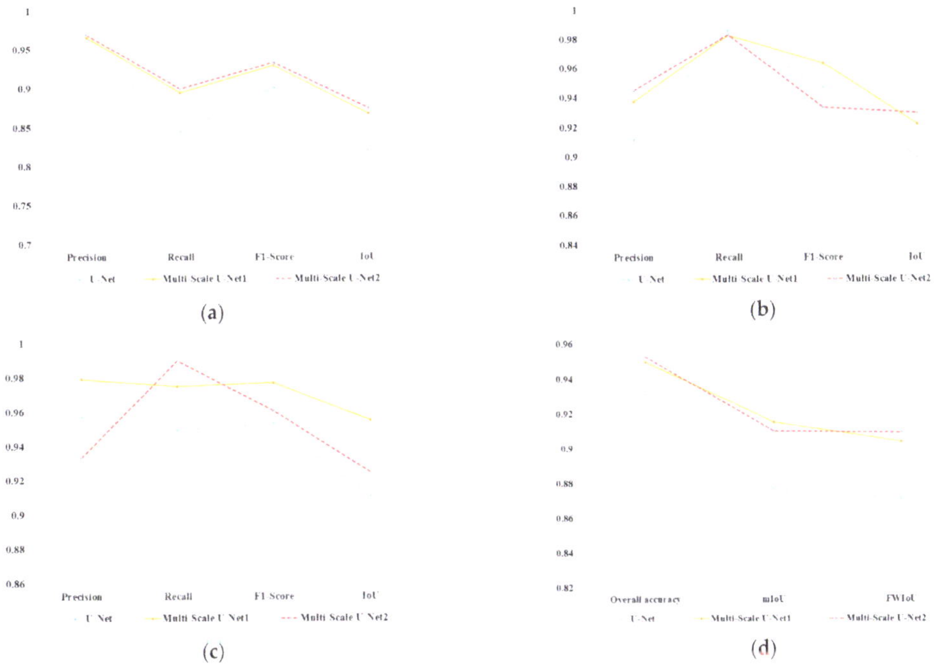

Figure 16. Accuracy evaluation curve. (**a**) Other, (**b**) Wheat, (**c**) Road, (**d**) Overall evaluation.

Figures 17–20 are the prediction pictures of the three models in different spatial resolution remote sensing images. The difference between the prediction pictures and the Labels are not big. It can be seen that the three models have a good classification effect on high spatial resolution remote sensing images, and as the spatial resolution increases, the classification effect of the model improves.

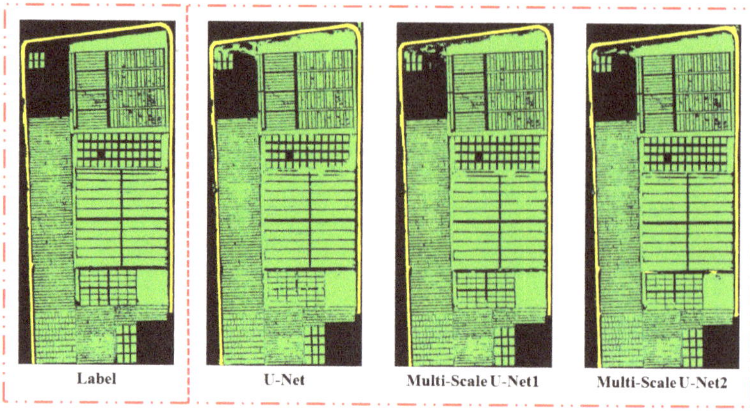

Figure 17. Three network models predict the results (GSD: 3.42 cm × 3.42 cm).

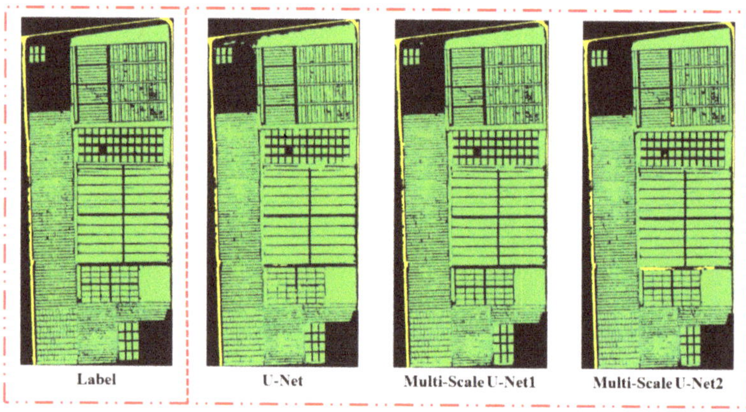

Figure 18. Three network models predict the results (GSD: 5.14 cm × 5.14 cm).

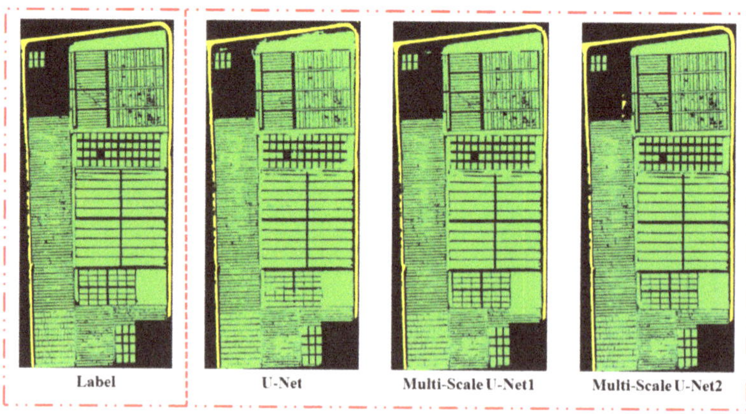

Figure 19. Three network models predict the results (GSD: 6.89 cm × 6.89 cm).

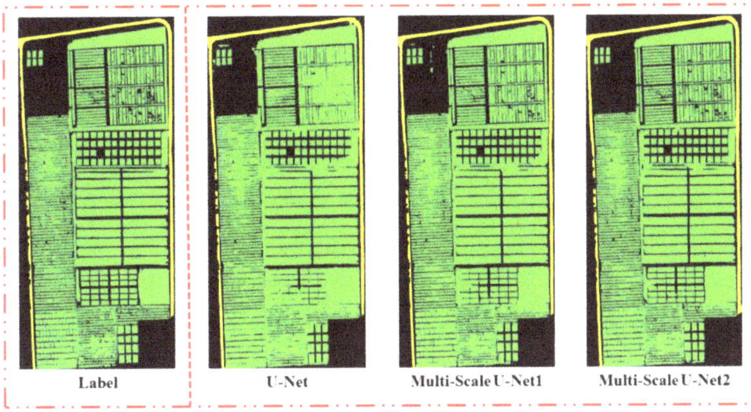

Figure 20. Three network models predict the results (GSD: 8.63 cm × 8.63 cm).

4. Discussion

UAV remote sensing images have the characteristics of high spatial resolution and a lot of detailed information, which brings great challenges to image semantic segmentation. Based on the remote sensing images of UAV, this study carried out ground object classification for the complex wheat fields with diverse varieties, which once again increased the difficulty of semantic segmentation of images. Based on U-Net, multi-scale feature fusion block is added to different model structures, and Multi-Scale U-Net1 and Multi-Scale U-Net2 are proposed, and the planting structure of complex wheat fields is predicted with high accuracy. The following conclusions can be drawn from the performance of the U-Net, Multi-Scale U-Net1, and Multi-Scale U-Net2 on the test set in this study:

(1) Adding a multi-scale feature fusion block between the Encoder and the Decoder can effectively extract multi-scale features of different depths, and to a certain extent, the model can learn multi-scale features of different depths, thereby effectively improving the semantic segmentation accuracy of the model.

(2) Adding a multi-scale feature fusion block before the image input model can extract the shallowest features of the original image, and input them to the convolutional layers of different depths of the Encoder after a series of feature mining. Although the classification accuracy of the model can be improved, the features mined by the multi-scale feature fusion module are relatively single, so the performance of Multi-Scale U-Net2 is not as good as Multi-Scale U-Net1 in the classification of 'Wheat' and 'Road'.

(3) From the perspective of remote sensing images with different spatial resolutions, within a certain range, reducing the spatial resolution of images can effectively improve the classification performance of the network. The reason is that with the reduction of the spatial resolution of the remote sensing image, the detailed information of the image is greatly reduced, the phenomenon of model misclassification is reduced, and the land classification effect is significantly improved.

5. Conclusions

In this study, the multi-scale feature fusion method based on U-Net has effectively improved the classification accuracy of high spatial resolution UAV remote sensing images and achieved satisfactory classification results in complex wheat fields with diverse varieties. The overall classification accuracy of Multi-Scale U-Net1 and Multi-Scale U-Net2 in complex wheat fields can reach 94.97% and 95.26%, respectively.

This study uses UAV to obtain remote sensing images of complex wheat fields with different high spatial resolutions, and uses Multi-Scale U-Net1 and Multi-Scale U-Net2 with multi-scale feature fusion block to test on the obtained remote sensing images. The results show that the classification accuracy of the two models is higher than 94%, and the classification accuracy is improved compared with the U-Net model. In this study, it is found that adding a multi-scale feature fusion block based on U-Net can achieve high-precision classification of UAV remote sensing images in complex wheat field scenes. Based on the prediction results of different high spatial resolution remote sensing images, this study found that with the increase of spatial resolution, the classification accuracy of the model will be reduced. Therefore, it is difficult to conduct ground object classification research based on UAV remote sensing images in complex wheat field scenes with diverse varieties. The study itself has certain limitations, but the multi-scale feature fusion method proposed in this study can effectively improve the classification accuracy of the model and accurately complete the classification of UAV remote sensing images in complex wheat field scenes.

Although the multi-scale feature fusion method proposed in this study can effectively improve the ground object classification accuracy of UAV remote sensing images in complex wheat field scenes, the model itself has a large number of parameters, and the study only focuses on one growth cycle of wheat. Therefore, in later research, the study of the classification of ground object can be carried out for the complex wheat field scenes with lightweight models and multiple growth cycles.

Author Contributions: F.M. and F.Y. started the work, completed the detailed investigations, and prepared the paper with support of all the co-authors; H.C. and S.S. helped us with remote sensing data collection; M.Y. helped us with orchard ground data measurement; Q.L. helped us with ground image control point measurements. All authors have read and agreed to the published version of the manuscript.

Funding: This research was funded by the Major Science and Technology Project of Shaanxi Province of China (Program No. 2020zdzx03-04-01), the National Key R&D Program of China "the 13th Five-Year Plan" (Program No. 2016YFD0700503).

Informed Consent Statement: Informed consent was obtained from all subjects involved in the study.

Acknowledgments: We are grateful to Bin Yan and Pan Fan for their help with my writing.

Conflicts of Interest: The authors declare no conflict of interest.

References

1. Hu, Q.; Wu, W.B.; Song, Q.; Yu, Q.Y.; Yang, P.; Tang, H.J. Recent progresses in research of crop patterns mapping by using remote sensing. *Sci. Agric. Sin.* **2015**, *5*, 14. [CrossRef]
2. Zhang, P.; Hu, S.G. Fine crop classification by remote sensing in complex planting areas based on field parcel. *Trans. Chin. Soc. Agric. Eng.* **2019**, *10*, 23. [CrossRef]
3. Hamer, A.M.; Simms, D.M.; Waine, T.W. Replacing human interpretation of agricultural land in Afghanistan with a deep convolutional neural network. *Int. J. Remote Sens.* **2021**, *42*, 3017–3038. [CrossRef]
4. Meyarian, A.; Yuan, X.; Liang, L.; Wang, W.; Gu, L. Gradient convolutional neural network for classification of agricultural fields with contour levee. *Int. J. Remote Sens.* **2022**, *43*, 75–94. [CrossRef]
5. Gao, L.; Luo, J.; Xia, L.; Wu, T.; Sun, Y.; Liu, H. Topographic constrained land cover classification in mountain areas using fully convolutional network. *Int. J. Remote Sens.* **2019**, *40*, 7127–7152. [CrossRef]
6. Zhang, D.; Pan, Y.; Zhang, J.; Hu, T.; Zhao, J.; Li, N.; Chen, Q. A generalized approach based on convolutional neural networks for large area cropland mapping at very high resolution. *Remote Sens. Environ.* **2020**, *247*, 111912. [CrossRef]
7. Lazin, R.; Shen, X.; Anagnostou, E. Estimation of flood-damaged cropland area using a convolutional neural network. *Environ. Res. Lett.* **2021**, *16*, 054011. [CrossRef]
8. Aliabad, F.A.; Reza, H.; Malamiri, G.; Shojaei, S.; Sarsangi, A.; Sofia, C.; Ferreira, S. Sentinel. Investigating the ability to identify new constructions in urban areas using images from unmanned aerial vehicles, Google earth, and sentinel-2. *Remote Sens.* **2022**, *14*, 3227. [CrossRef]
9. Al-Awar, B.; Awad, M.M.; Jarlan, L.; Courault, D. Evaluation of nonparametric machine-learning algorithms for an optimal crop classification using big data reduction strategy. *Remote Sens. Earth Syst. Sci.* **2022**, *5*, 141–153. [CrossRef]
10. Zheng, W.H.; Wang, R.H.; Cao, Y.X.; Jin, N.; Feng, H.; He, J.Q. Remote sensing recognition of plastic-film-mulched farmlands on Loess Plateau based on Google Earth engine. *Trans. Chin. Soc. Agric. Mach.* **2021**, *9*, 3. [CrossRef]
11. Song, T.Q.; Zhang, X.Y.; Li, J.X.; Fan, H.S.; Sun, Y.Y.; Zong, D.; Liu, T.X. Research on application of deep learning in multi-temporal greenhouse extraction. *Comput. Eng. Appl.* **2020**, *5*, 12. [CrossRef]

12. Xu, L.; Ming, D.; Du, T.; Chen, Y.; Dong, D.; Zhou, C. Delineation of cultivated land parcels based on deep convolutional networks and geographical thematic scene division of remotely sensed images. *Comput. Electron. Agric.* **2022**, *192*, 106611. [CrossRef]
13. Li, Z.M.; Zhao, J.; Lan, Y.B.; Cui, X.; Yang, H.B. Crop classification based on UAV visible image. *J. Northwest A F Univ. (Nat. Sci. Ed.).* **2019**, *11*, 27. [CrossRef]
14. Yao, C.; Zhang, Y.; Liu, H. Application of convolutional neural network in classification of high resolution agricultural remote sensing images. In Proceedings of the ISPRS—International Archives of the Photogrammetry, Remote Sensing and Spatial Information Sciences, Wuhan, China, 18–22 September 2017; pp. 989–992.
15. Chen, Y.Q.; Wang, X.X. Improved DeepLabv3+ model UAV image farmland information extraction. *Comput. Eng. Appl.* **2022**, *7*, 7. [CrossRef]
16. Yang, Q.; Liu, M.; Zhang, Z.; Yang, S.; Ning, J.; Han, W. Mapping plastic mulched farmland for high resolution images of unmanned aerial vehicle using deep semantic segmentation. *Remote Sens.* **2019**, *11*, 2008. [CrossRef]
17. Yang, S.Q.; Song, Z.S.; Yin, H.P.; Zhang, Z.T.; Ning, J.F. Crop classification method of UVA multispectral remote sensing based on deep semantic segmentation. *Trans. Chin. Soc. Agric. Mach.* **2020**, *12*, 21. [CrossRef]
18. Zhang, K.; Okazawa, H.; Hayashi, K.; Hayashi, T.; Fiwa, L.; Maskey, S. Optimization of ground control point distribution for unmanned aerial vehicle photogrammetry for inaccessible fields. *Sustainability* **2022**, *14*, 9505. [CrossRef]
19. Zhang, T.B.; Tang, J.X.; Liu, D.Z. Feasibility of satellite remote sensing image about spatial resolution. *J. Earth Sci. Environ.* **2006**, *28*, 79–82.
20. Temraz, M.; Keane, M.T. Solving the class imbalance problem using a counterfactual method for data augmentation. *Mach. Learn. Appl.* **2022**, *9*, 100375. [CrossRef]
21. Wagle, S.; Ramachandran, H.; Sampe, J.; Faseehuddin, M.; Ali, S. Effect of data augmentation in the classification and validation of tomato plant disease with deep learning methods. *Traitement Du Signal* **2021**, *38*, 1657–1670. [CrossRef]
22. Chen, Y.; Gao, W.; Widyaningrum, E.; Zheng, M.; Zhou, K. Building classification of VHR airborne stereo images using fully convolutional networks and free training samples. In Proceedings of the ISPRS Technical Comission II, Delft, The Netherlands, 18–22 September 2017; pp. 87–92.
23. Song, R.Z.; Zheng, H.Y.; Wang, D.C.; Shang, Z.; Wang, X.J.; Zhang, C.Y.; Li, J. Classification of features in open-pit mining areas based on deep learning and high-resolution remote sensing images. *China Min. Mag.* **2022**, *6*, 15.
24. Chu, B.C.; Gao, F.; Shuai, T.; Wang, S.C.; Chen, J.; Chen, J.Y.; Yu, W.D. Remote sensing image object classification by deep learning based on feature map set. *Radio Eng.* **2022**, *1*, 13.
25. Cerón, J.C.Á.; Ruiz, G.O.; Chang, L.; Ali, S. Real-time instance segmentation of surgical instruments using attention and multi-scale feature fusion. *Med. Image Anal.* **2022**, *81*, 102569. [CrossRef] [PubMed]
26. Zheng, S.; Zhang, X.L.; Deng, H.; Ren, H.W. 3D liver image segmentation method based on multi-scale feature fusion and grid attention mechanism. *J. Comput. Appl.* **2022**, *8*, 4. [CrossRef]
27. Wen, P.; Cheng, Y.L.; Wang, P.; Zhao, M.J.; Zhang, B.X. Ground object classification based on height-aware multi-scale graph convolution network. *J. Beijing Univ. Aeronaut. Astronaut.* **2021**, *11*, 22. [CrossRef]
28. Ronneberger, O.; Fischer, P.; Brox, T. U-Net: Convolutional Networks for Biomedical Image Segmentation. *arXiv* **2015**, arXiv:1505.04597. [CrossRef]

agronomy

Article

Canopy Segmentation Method for Determining the Spray Deposition Rate in Orchards

Shilin Wang [1,2], Wei Wang [1], Xiaohui Lei [1], Shuangshuang Wang [3], Xue Li [1,*] and Tomas Norton [4]

1 Institute of Agricultural Facilities and Equipment, Jiangsu Academy of Agricultural Sciences, Nanjing 210014, China; shilin@jaas.ac.cn (S.W.); jaaswangwei@163.com (W.W.); leixiaohui.2008@163.com (X.L.)
2 Key Laboratory of Agricultural Engineering in the Middle and Lower Reaches of Yangtze River, Nanjing 210014, China
3 College of Engineering, Huazhong Agricultural University, Wuhan 430070, China; wangshsh_123@mail.hzau.edu.cn
4 M3-BIORES—Measure, Model & Manage Bioresponses, KU Leuven, B-3001 Leuven, Belgium; tomas.norton@kuleuven.be
* Correspondence: lixue@jaas.ac.cn

Abstract: The effective quantification of deposition rate is of vital importance in optimizing the application performance and the utilization of pesticides; meanwhile, the canopies of fruit tree orchards are large, with dense branches and leaves shading each other, making it difficult to quantify spraying efficiency. Therefore, it is imperative to develop a facile methodology for assessing the performance of different spraying techniques in terms of distribution and utilization rate in orchards. To evaluate spraying efficacy in orchards, a canopy segmentation method was developed in to be able to determine the spray deposition rate. The distribution and deposition rate of spray liquid applied using three kinds of orchard sprayer were measured in a pear orchard and a peach orchard. The test results showed that the trailer sprayer had the highest deposition rates, with values of 31.54% and 56.92% on peach and pear trees, respectively. The deposition rates of the mounted sprayer in the peach and pear canopies were 21.75% and 40.61%, and the rates of the hand-held sprayer were 25.19% and 29.97%, respectively. The spray gun had the best droplet distribution uniformity, with CVs of the spray in the peach and pear canopies of 20.54% and 25.06%, respectively. The CVs in the peach and pear canopies were 35.98% and 26.54% for the trailer sprayer, and the CVs of the mounted sprayer were 92.52% and 94.90%, respectively. The canopy segmentation method could effectively be used to calculate the deposition rate and drioplet distribution in orchard application, while a great deal of time was consumed by counting the number of leaves in the different areas of the fruit tree canopies. Therefore, research on the density of branches and leaves in fruit tree canopies should be carried out in order to improve the efficiency of fruit tree canopy information extraction.

Keywords: orchard; sprayer; pesticide application; canopy; deposition rate

Citation: Wang, S.; Wang, W.; Lei, X.; Wang, S.; Li, X.; Norton, T. Canopy Segmentation Method for Determining the Spray Deposition Rate in Orchards. *Agronomy* **2022**, *12*, 1195. https://doi.org/10.3390/agronomy12051195

Academic Editor: José Casanova Gascón

Received: 6 April 2022
Accepted: 11 May 2022
Published: 16 May 2022

1. Introduction

China is a major producer and consumer of fruits, and the planting area of orchards has ranked first in the world for a number of consecutive years. According to official data from the National Bureau of Statistics, China's fruit output totaled 256.88 million tons, and the orchard planting area reached 11.87 million hectares [1]. The stable development of the rural orchard planting industry has made outstanding contributions to increasing farmers' income, optimizing the industrial structure, and accelerating the construction of new rural areas [2]. Meanwhile, due to the diversity of planting patterns and complicated ecological conditions, pests and diseases are infecting and spreading seriously in orchards, leading to large numbers of insects and diseases in orchards.

In current orchard cultivation systems, pests and diseases are primarily controlled using pesticides that are applied to the canopy with different types of sprayers [3,4]. Off-target losses of sprayed pesticide is a common phenomenon in tree-fruit orchards [5]. During pesticide application in orchards, only about 30% of the pesticide liquid is effectively deposited on the canopy, and a great deal of spray liquid runs into the soil and the environment [6], with such losses causing a serious waste of pesticide, which is harmful to the environment and to residents [7–9].

With the increasing emphasis on the negative impacts of pesticides, the Chinese government has issued several successive policies, making agricultural non-point source pollution a priority for the improvement of the agricultural ecological environment [10]. It is urgently necessary to improve the pesticide utilization rate and reduce pesticide usage in pesticide application. In 2015, the Chinese Ministry of Agriculture implemented a policy that sought to achieve zero growth in the use of pesticide by 2020. After five years of implementation, the utilization rate of pesticides in China's three major grain crops (rice, wheat, and maize) reached 40.6% in 2020, achieving the expected goal of reducing the amount of pesticide used, and increasing their efficiency.

The utilization rate (also called the deposition rate) of pesticides is a basic parameter for measuring the quality of pesticide application, which is expressed as the ratio of pesticide deposited on the crop after spraying to the total amount applied [11]. Therefore, the effective quantification of the deposition rate is of vital importance for optimizing the application performance and utilization of pesticides. The method for measuring the deposition rate of cereal crops and vegetables is relatively simple. The deposition rate can be calculated on the basis of the deposition volume of a single plant, the application volume per unit area, and the plant density [12,13]. Gao et al. developed an ultra-performance liquid chromatography (UPLC) analytical method for determining the deposition amount of Allura Red on rice plants. Laboratory testing and field trials confirmed that the developed method for assessing spray deposition rate was feasible and practicable [14]. Wang et al. evaluated the deposition rate for pesticide application on tomato crop using knapsack sprayers, mist sprayers, thermal foggers, and handheld spray guns [15].

Compared with cereal crops and vegetables, the canopy of tree-fruit orchards is large, with dense branches and leaves shading each other, making it difficult to quantify the spraying efficiency. Vercruysse et al. [16] measured the ground deposits in the orchard and downwind from the target area of pesticides applied with an air-blast sprayer on semi-dwarf pears and apples, and the total off-target deposits were used to evaluate application efficiency. However, this method does not calculate the specific utilization rate of spray liquid in orchard application. Zhang et al. [17] developed a new method for measuring the pesticide deposition rate on fruit trees using canopy characteristics and deposition. The new method quantified the canopy structure of orchard into "leaf area index", "canopy shadow area", and other indicators; however, it did not take the differences in leaf density of canopy regions into account.

Therefore, it is imperative to develop a facile methodology for assessing the performance of different spraying techniques in terms of distribution and utilization rate in orchards [18–20]. To evaluate spraying efficacy in orchards, a canopy segmentation method was developed to determine the spray deposition rate in orchards. The distribution and deposition rate of the sprayed liquid applied using three kinds of orchard sprayers were measured in a pear orchard and a peach orchard, respectively. This study can be used as a reference for the quantification of deposition rate in orchards, and can also provide guidance for the application of pesticides in orchards.

2. Materials and Methods

2.1. Sprayers and Orchards

In recent years, China's orchard sprayers have gradually changed from manual knapsack sprayers to motorized sprayers, with the main development trends for primary spraying equipment being motorized hand-held spray guns, mounted air-assisted orchard sprayers and trailer air-assisted orchard sprayers. Trailer air-assisted orchard sprayers have a large liquid tank and high operating efficiency, but the unit is relatively long as a whole and has a wide turning radius, making it suitable for large-scale orchards on flat ground [21]. The mounted orchard air-assisted sprayer is generally integrated with the tractor at three points. It is characterized by light weight and flexible maneuverability, and it can operate on small plots, but its tank is small and its working efficiency is lower than that of trailer sprayers [22]. Orchard hand-held spray guns have a simple structure and flexible operation, and they are currently the most widely used spraying equipment in orchards in developing countries. They can be transported by manual lifting or using a motorized tricycle, and their movement capacity is basically not limited by terrain or orchard conditions. At the same time, they can additionally be equipped with dozens of meters of spray hose. A handheld rod attached to the end can also expand the spraying range. In this research, a 3WQF-1000 trailer air-assisted sprayer (Figure 1a), a 3WG-500 mounted air-assisted sprayer (Figure 1b) and an orchard motorized handheld spray gun (Figure 1c) were selected for orchard spraying.

(a) (b) (c)

Figure 1. Orchard sprayers used for pesticide application. (**a**) 3WQF-1000 trailer air-assisted sprayer, (**b**) 3WG-500 mounted air-assisted sprayer, (**c**) handheld spray gun.

The spray deposition rates of the three sprayers were measured in a peach orchard and a pear orchard. The peach orchard test was carried out from 16 to 17 September 2019, and the test site was located at the Experimental Station of the Jiangsu Academy of Agricultural Sciences. The row spacing of the peach orchard was 4 m, and the spacing in the row was 2 m. The average plant height of the peach trees was about 3.0 m, the heading height was 0.6 m, the canopy width of the peach tree was about 2.0 m, the leaf area index of the peach orchard was 1.18, and the transmission coefficient was 0.22. The pear orchard test was carried out from 23 to 24 September 2020, and the test site was located in the modern agricultural industrial park of Xinjie Town, Taixing City, Jiangsu Province. The row spacing of the pear orchard was 5 m, and the spacing in the row was 2 m. The average plant height of the pear tree was about 3.8 m, the heading height was 0.6 m, the canopy width of the pear tree was about 2.5 m, the leaf area index of the pear orchard was 1.43, and the transmission coefficient was 0.29.

2.2. *Experimental Design*

2.2.1. Operating Parameters of Spraying Equipment

Before field application, the flow rate and travel speed of the three kinds of orchard sprayer were measured, and the preliminary field application parameters of each sprayer were determined. In each treatment, the fruit tree rows were sprayed from both sides, and the application distance was 100 m. All treatments were performed by well-trained pesticide applicators. Figure 2 shows pictures of the sprayers in the peach orchard.

(a) (b) (c)

Figure 2. Photos of the sprayers in the peach orchard during the experiment. (**a**) Trailer air-assisted orchard sprayer, (**b**) mounted air-assisted orchard sprayer, (**c**) orchard spray gun.

During the test, the operating distance, duration and application volume of each sprayer were accurately recorded, and the actual operating parameters of each sprayer were calculated. The specific operating parameters of the sprayers in the field test are shown in Table 1.

Table 1. Operating parameters of the sprayers in the field test.

Spraying Parameters of Sprayer	Peach Orchard			Pear Orchard		
	Trailer Sprayer	Mounted Sprayer	Spray Gun	Trailer Sprayer	Mounted Sprayer	Spray Gun
Forward velocity/(m·s^{-1})	1.14	1.20	-	1.04	1.07	-
Flow rate/(L·min^{-1})	4.96	6.30	6.25	11.28	5.49	6.83
Application volume/(L·ha^{-1})	180	220	3220	360	165	1920

2.2.2. Canopy Segmentation

Before the test, three typical peach trees and pear trees were selected in order to perform canopy segmentation and obtain leaf density statistics. Considering the external features of the peach and pear canopies, each canopy was divided into the top, middle upper, middle lower, and bottom layers (Figure 3a,b). Each layer was divided into external and internal sides, where the external side contained four areas A, B, C and D, and the internal side was identified as area E (Figure 3c). Thus, the canopy of each fruit tree was divided into 20 areas. According to the schematic diagram of canopy division in Figure 3, the canopy of the fruit trees was segmented using metallic wire to build a frame, and the number of leaves in each frame-selected area was counted. The average number of leaves of the three fruit trees in each frame-selected area was taken as the number of leaves in the corresponding area.

2.2.3. Spray Liquid and Sample Collection

To quantify the deposition of spray liquid on peach and pear leaves, BSF fluorescence with a mass fraction of 0.1% was added to tap water as a tracer. Photo-degradation tests of BSF verified that the fluorescence degradation after 120 min of solar exposure was less than 3.1% [23]. After spraying, the peach and pear leaves in each canopy area were randomly collected and individually placed in labeled plastic zip-lock bags (140 × 200 × 0.05 mm^3). To evaluate the droplet distribution and deposition of each sprayer, 5 fruit trees were selected from the application row as repetitions. Ten pieces of leaves were picked for each area of the segmented canopy and placed in a zip-lock bag. In this study, the deposition rate of the spray liquid was calculated based on the number of leaves, regardless of the area of a single leaf. Therefore, leaves of different sizes were randomly selected in each area to eliminate the influence of leaf area on the detection results. To reduce the photolysis of BSF, all of the leaf samples were collected within 1 h after application and then stored in the dark at −20 °C until analysis.

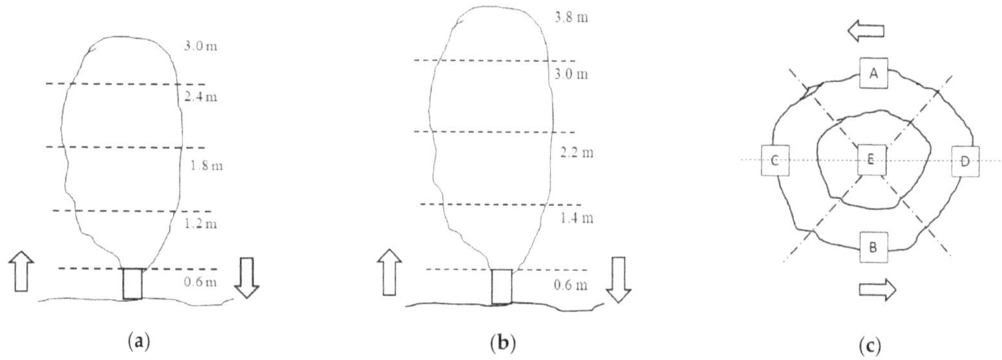

Figure 3. Schematic diagram of fruit tree canopy segmentation. (**a**) Height division of the peach canopy, (**b**) height division of the pear canopy, (**c**) horizontal division of the canopy.

2.2.4. Meteorological Conditions

A portable meteorological station (WatchDog, Spectrum Technologies, Inc., Aurora, CO, USA) was used to measure the weather conditions during the field experiments, and the station was located upwind of the peach and pear orchards at a height of 2 m. The temperature, relative humidity, wind speed and wind direction were collected at 1 Hz and averaged across the period of each treatment. The wind direction was expressed as the angle between the wind and the forward direction of the sprayers. The meteorological conditions of the treatments are shown in Table 2.

Table 2. Meteorological conditions during spraying in the field test.

Meteorological Parameter	Peach Orchard			Pear Orchard		
	Trailer Sprayer	Mounted Sprayer	Spray Gun	Trailer Sprayer	Mounted Sprayer	Spray Gun
Relative humidity (%)	47.5	43.7	42.6	73.8	65.9	54.8
Temperature (°C)	29.7	30.9	31.6	27.1	27.3	30.9
Mean wind velocity (m·s^{-1})	1.23	1.16	1.37	0.81	0.84	0.93
Wind direction (°)	110.3	112.3	111.7	18.75	11.56	3.83

2.3. *Sample Processing and Calculation*

2.3.1. Deposition on the Samples

A quantitative volume of deionized water was used to elute the collected leaves, and a fluorescence spectrometer (LS55, PerkinElmer Instruments, Waltham, MA, USA) was used to determine the amount of droplet deposition. In addition, the fluorescence value of each original spray liquid diluted 1000 times was also measured to calculate the volume of liquid on the samples. The deposition of the spray liquid was calculated using the following equation:

$$V_l = \frac{V_w \times FL_S}{n \times N_d \times FL_a} \times 10^3 \quad (1)$$

where V_l : deposition on a single leaf, μL; V_w: volume of eluent, mL; *FLs*: fluorescence value of eluent; *FLa*: fluorescence value of diluted liquid; *n*: number of leaves in sampling area; N_d: dilution times of spray liquid.

2.3.2. Deposition Rates in Orchard

The deposition rate of the spray liquid of each sprayer was calculated according to the number of leaves in the sampling area, the applied volume in the treatment area, and the fruit tree density. The deposition volume for a single plant was calculated using Equation (2), and the effective deposition rate of the spray liquid was calculated using Equation (3).

$$V_t = \sum_{i=1}^{20} V_{li} \times n_i \tag{2}$$

where V_t is the deposition of the spray liquid on a single tree, mL; V_{li} is the deposition of the liquid in the i-th sampling area; and n_i is the number of leaves in the i-th sampling area.

$$R = \frac{V_t \times S \times 10^{-3}}{V_0 \times l \times d} \times 100\% \tag{3}$$

where R is the deposition rate of the spray liquid, %; S is the unit area of the orchard, ha, m2; V_0 is the total spraying volume of the unit area, L; l is the row spacing, m; and d is the spacing in the row, m.

3. Results

3.1. Leaf Density of Fruit Trees

Table 3 shows the number of leaves in the different areas of the peach tree canopy. Compared to other areas of the canopy, the top layer of the canopy has a smaller volume and the least number of leaves. The number of leaves in the external area of the top layer was about 500, which was slightly lower than the number of leaves in the upper middle, lower middle, and lower layers. The number of leaves in the internal area of the top canopy was only 223, which was significantly lower than the number of leaves in the upper middle, lower middle, and lower layers. The number of leaves in the internal region of the middle and upper layers was 705, which was significantly higher than the values of 459 and 400 found for the internal region of the middle lower and bottom layers, respectively. The numbers of leaves in the middle upper, middle lower and external regions of the bottom layer were not much different, but the numbers of leaves in the A and B regions of the same layer were higher than those in the C and D regions. This may be due to the fact that areas A and B were closer to the application row, and the orchard has a row spacing that was larger than its plant spacing in the row, so areas A and B have better light, ventilation and more room for growth.

Table 3. The numbers of leaves in different areas of the peach canopy.

Layer of Canopy	A	B	C	D	E
Top	512	465	485	442	223
Middle upper	655	711	440	434	705
Middle lower	754	682	561	442	459
Bottom	702	668	455	406	400

Table 4 shows the numbers of leaves in the different areas of the pear canopy. The results were similar to the leaf distribution in the canopy of the peach trees. The top layer of the pear canopy was smaller in volume and had the lowest number of leaves, and the numbers of leaves in each area of the top layer were all less than 100. The numbers of leaves in the middle upper and the middle lower layers of the pear canopy were larger than those in the top and bottom layers. The number of leaves in the middle lower layer was the highest, with the four external regions (A, B, C, and D) of the canopy having about 300 leaves, while the number of leaves in the internal region reached up to 463. Compared with the middle upper and middle lower layers, the number of leaves in the bottom layer of the canopy was lower, being in the range of 133 to 201. In contrast to the distribution of leaves in the peach trees, the density of the pear branches was lower, thus having less negative effect on ventilation and light transmission; therefore, the numbers of leaves in areas A and B of the pear canopy were similar to those in areas C and D.

Table 4. The numbers of leaves in the different areas of the pear canopy.

Layer of Canopy	A	B	C	D	E
Top	65	74	70	61	82
Middle upper	179	195	165	150	239
Middle lower	280	326	301	264	463
Bottom	181	133	152	201	164

3.2. Liquid Distribution in the Canopy

Table 5 shows the deposition amount and average deposition of spray liquid in the peach canopy. The difference in the deposition amount of the liquid sprayed by the trailer sprayer in the four extenal areas of the top layer, the middle upper layer and the middle lower layer of the peach canopy is small. However, the deposition amounts of spray liquid in the internal area of the top layer and the middle lower layers were 2.09 μL and 2.02 μL, respectively, which were values significantly lower than that found for the external area of the corresponding layer. This might be caused by the occlusion of the peripheral blades. In the bottom layer of the peach trees, the deposition of the spray liquid in areas A and B was 6.07 μL and 5.11 μL, respectively, which was much larger than the 3.14 μL and 3.31 μL found for areas C and D. This is because the height of the spraying unit was similar to the height of the bottom layer of the tree canopy, and areas A and B of the bottom layer were closer to the operation row, increasing the adhesion probability of the spray liquid. In general, there was little difference between the deposition in each area of peach canopy and the average deposition in the whole canopy, and the deposition of the spray liquid applied by the trailer sprayer in the peach canopy was relatively uniform.

Table 5. Deposition amount (μL) and average deposition (μL) of spray liquids in the peach canopy.

Area of Canopy		Trailer Sprayer	Mounted Sprayer	Spray Gun
Top	A	3.86 ± 1.03	3.17 ± 2.38	49.10 ± 2.54
	B	3.74 ± 0.13	2.67 ± 2.09	73.60 ± 3.69
	C	3.14 ± 0.59	0.34 ± 0.22	59.05 ± 13.15
	D	2.50 ± 0.82	1.27 ± 0.61	66.73 ± 0.91
	E	2.09 ± 0.27	0.45 ± 0.38	53.49 ± 16.26
Middle upper	A	5.98 ± 1.60	6.15 ± 1.69	55.58 ± 5.57
	B	4.73 ± 1.61	5.63 ± 2.78	68.26 ± 16.18
	C	5.56 ± 2.17	0.88 ± 0.58	62.63 ± 10.82
	D	5.15 ± 1.75	1.91 ± 1.36	68.71 ± 5.65
	E	4.56 ± 1.31	1.14 ± 0.92	59.00 ± 11.67
Middle lower	A	4.65 ± 0.73	7.49 ± 5.26	65.67 ± 24.01
	B	4.53 ± 1.10	4.59 ± 3.09	67.34 ± 11.94
	C	3.92 ± 0.81	1.45 ± 1.02	65.01 ± 19.08
	D	4.24 ± 0.44	4.20 ± 2.05	70.65 ± 3.42
	E	2.02 ± 0.76	0.36 ± 0.24	58.63 ± 12.80
Bottom	A	6.07 ± 1.18	8.38 ± 2.21	49.98 ± 3.64
	B	5.11 ± 1.13	4.21 ± 2.77	55.40 ± 19.05
	C	3.14 ± 0.88	4.14 ± 2.49	48.00 ± 13.07
	D	3.31 ± 0.31	4.20 ± 2.73	68.83 ± 4.12
	E	3.46 ± 0.90	1.56 ± 1.18	59.12 ± 13.32
Average deposition		4.09	3.21	61.24

The deposition amount of the liquid spray when using the mounted sprayer showed a downward trend with increasing canopy height. Taking area A as an example, the deposition amounts of the spray liquid in the bottom, middle lower, middle upper and top layers of the peach canopy were 8.38, 7.49, 6.15 and 3.17 µL, respectively. Due to the lower air speed of the mounted orchard air-assisted sprayer, the deposition amounts of the spray solution in areas A and B were greater than those in areas C and D in the same layer of the peach canopy. Compared with the deposition amounts on the external areas, the deposition amounts of the liquid in the internal areas of the the canopy were lower, and the deposition amounts in the top, middle upper, middle lower and bottom layers were 0.45, 1.14, 0.36 and 1.56 µL, respectively, all of which were much lower than the average deposition of 3.21 µL.

Since the application volume of the spray gun is much greater than those of the trailer sprayer and the mounted sprayer, the average deposition of the spray gun throughout the whole peach canopy was 61.24 µL, which was also much higher than those of the other two sprayers. Meanwhile, the adhesive capacity of droplets applied using the high-volume spray gun exceeded the maximum retention of the leaves, causing a large amount of spray liquid runoff from the leaves of the fruit trees. Therefore, the deposition amount of the spray gun sprayed in different areas of the peach canopy was relatively uniform, and the deposition amounts of the liquid were greater than 50 µL in almost all areas. The differences in the deposition amounts in each area of the peach canopy when using spray gun application were mainly determined on the basis of leaf inclination and density in the area.

Table 6 shows the deposition amounts and the average deposition of spray liquids in the pear canopy. Since the leaf area of pear trees is larger than that of peach trees, the deposition amounts of the three sprayers on single pear leaves were greater than those on peach leaves. The deposition and distribution trends of liquids applied using the three sprayers in the pear canopy were similar to those in the peach canopy. Relatively speaking, the deposition amount of spray liquid applied by the trailer sprayer was lower in the top layer of the canopy, but the differences in the deposition amount of liquid between each of the layers were small. The deposition amounts of the liquid sprayed by the trailer sprayer in the internal areas of the pear canopy were the lowest, and the deposition amounts in the top, middle upper, middle lower and bottom layers of the pear canopy were 29.07, 28.91, 43.02 and 43.22 µL, respectively. All of these values were lower than the average deposition of 52.65 µL found throughout the whole canopy.

Table 6. Deposition amount (µL) and average deposition (µL) of spray liquids in pear canopy.

Area of Canopy		Trailer Sprayer	Mounted Sprayer	Spray Gun
Top	A	59.03 ± 4.55	12.24 ± 4.41	176.75 ± 28.66
	B	44.41 ± 15.32	1.20 ± 0.58	141.77 ± 22.80
	C	49.60 ± 11.60	0.68 ± 0.24	144.82 ± 76.52
	D	47.14 ± 0.92	2.08 ± 1.77	200.17 ± 29.48
	E	29.07 ± 2.02	0.60 ± 0.28	161.21 ± 13.99
Middle upper	A	81.52 ± 15.12	39.23 ± 12.74	170.14 ± 30.32
	B	77.06 ± 18.29	3.47 ± 3.35	209.66 ± 13.21
	C	55.79 ± 8.28	5.50 ± 5.22	147.32 ± 71.69
	D	61.28 ± 1.36	10.94 ± 11.67	251.93 ± 53.22
	E	28.91 ± 7.40	3.11 ± 3.36	129.31 ± 23.51
Middle lower	A	71.29 ± 15.13	25.40 ± 8.08	126.10 ± 48.10
	B	57.21 ± 13.97	5.09 ± 3.68	168.17± 73.50
	C	47.38 ± 8.79	11.47 ± 6.25	122.91 ± 64.15
	D	50.18 ± 3.74	18.69 ± 11.76	183.85 ± 8.33
	E	43.02 ± 7.33	10.57 ± 7.04	128.74 ± 54.17

Table 6. *Cont.*

Area of Canopy		Trailer Sprayer	Mounted Sprayer	Spray Gun
Bottom	A	70.35 ± 24.00	41.00 ± 20.93	140.94 ± 29.82
	B	60.76 ± 10.64	41.57 ± 23.29	162.38 ± 26.48
	C	57.22 ± 3.32	28.29 ± 4.87	76.53 ± 13.70
	D	57.20 ± 11.16	50.90 ± 10.74	189.40 ± 29.39
	E	43.22 ± 11.75	32.80 ± 13.26	130.54 ± 31.39
Average deposition		52.65	17.32	152.42

Similar to the deposition results on peach trees, the deposition amount of liquid applied using the mounted sprayer in the pear canopy also decreased with increasing canopy height. The highest deposition of liquid was found in the bottom layer, and the lowest deposition of liquid was found in the top layer. The liquid applied using mounted sprayer was relatively uniformly deposited in the bottom areas of the pear canopy, while the liquid in the other three layers was mainly deposited in the external areas. The deposition amounts of spray liquid in the internal regions of the top, middle upper, and middle lower layers were 0.60, 3.11, and 10.57 μL, respectively, which were significantly lower values than the average canopy deposition volume of 17.32 μL. The spray gun had the highest deposition amount of liquid in the pear canopy, and the deposition amount of the spray gun sprayed in different areas of the pear canopy was relatively uniform.

3.3. *Deposition Rate and Variable Coefficient*

Table 7 shows the deposition rates and coefficients of variation of spray liquid in the peach and pear canopies. The results showed that the deposition rate of the spray liquid applied using the trailer sprayer was the highest, with the deposition rate of liquid in peach and pear canopies being 31.54% and 56.92%, respectively. The deposition rates of liquid applied using the mounted sprayer in the peach and pear canopies were 21.75% and 40.61%, respectively. The deposition rates of liquid applied using the hand-held spray gun in the peach and pear canopies were relatively close, with the deposition rates of the pesticide solution being 25.19% and 29.97%, respectively.

Table 7. Deposition rates (%) and coefficients of variation (%) of spray liquid in the peach and pear canopies.

Sprayer	Peach Orchard		Pear Orchard	
	CV of Deposition	Deposition Rate	CV of Deposition	Deposition Rate
Trailer sprayer	35.98	31.54	26.54	56.92
Mounted sprayer	92.52	21.75	94.90	40.61
Spray gun	20.54	25.19	25.06	29.97

The uniformity of spray droplet distribution on the targets is commonly described by the coefficient of variation (CV) [24]. The lower the CV is, the better the uniformity of droplet distribution [25]. The spray gun with high volume spray had the best droplet distribution uniformity, and the CVs of the spray on peach and pear trees were 20.54% and 25.06%, respectively. The uniformity of the liquid distribution of the trailer sprayer was second, where the CVs of the spray liquid deposited on the peach and pear trees were 35.98% and 26.54%, respectively. The mounted sprayer had the worst performance in terms of the deposition and distribution of the liquid, with the CVs of the liquid deposited on the peach and pear trees being 92.52% and 94.90%, respectively.

4. Discussion

Pesticide application aims at depositing the highest possible amount of the active ingredient onto the target surface [26]. The performance of the sprayer, the meteorological conditions, the structure of the plant canopy, and the surface properties of the leaves all have an impact on this during the deposition of the pesticide solution from the nozzle to the surface of the leaves [27–30]. The orchards have the characteristics of continuous planting, large canopies, and dense foliage. To improve the penetration of the spray liquid, air-assisted sprayers are used to apply pesticides in orchards [16].

In this study, the assisted air speed of the trailer sprayer was higher than that of the mounted sprayer and the non-air-assisted spray gun, and its delivery performance was better for droplets, ensuring that more droplets were deposited on the target canopy and penetrated into the interior of the canopy. However, the assisted air speed of the mounted orchard sprayer was relatively lower, and its ability to transport droplets to the target canopy was weak, resulting in a large number of droplets being deposited under the canopy or in the operation row during the spraying process. Therefore, the deposition rate of spray liquid obtained when using the trailer sprayer was the highest, with a deposition rate on peach trees that was 1.45 times and 1.40 times those of the mounted sprayer and the hand-held spray gun, respectively. Similarly, the deposition rate of the trailer sprayer on pear trees was 1.40 times and 1.90 times those obtained using the mounted sprayer and the hand-held spray gun, respectively.

The leaf area index and canopy volume of fruit trees also have a significant effect on the deposition rate of spraying solution [16,31]. The peach trees selected in this experiment had a maximum width of 2.0 m, a plant height of 3.0 m, and a leaf area index of 1.18. The maximum width of the pear trees was 2.5 m, the plant height was 3.8 m, and the leaf area index was 1.43, which were all larger than the values obtained of the peach trees. Therefore, compared to the deposition rate of spraying solution on peach trees, the deposition rate of liquid sprayed using the trailer sprayer and the mounted sprayer on pear trees increased by about 20%. However, the deposition rates of the high volume applied using the hand-held spray gun on the two fruit trees only increased by about 5%.

Hand-held spray guns are used for spraying without air assistance. Its operation efficiency is low, the amount of runoff liquid is large, and the risk of the operator's exposure to pesticide is greatly increased [21]. Michael et al. [31] assessed the deposition on leaves and the losses to the ground for four different spraying treatments in trellis and goblet viticulture. The high-volume sprayer exhibited in the highest normalized deposit on the ground, suggesting that runoff is excessive compared to the other types of sprayers. Compared with hand-held spray gun, the air-assisted sprayers were highly efficient, showing the potential for replacing high-volume pesticide sprayings with low-volume applications. However, motorized air-assisted sprayers are not able to meet the requirements of mountainous orchards and non-standardized orchards. Therefore, hand-held spray guns are still the optimal choice for pesticide application in these areas.

In addition to the deposition rate of the pesticide solution, the uniformity of the liquid distribution in the canopy is another important indicator for evaluating the performance of spraying, and is of great significance for optimizing the spray parameters of equipment and improving the utilization rate of the chemical solution [32]. The airflow generated by air-assisted orchard sprayers is able to toss the branches and leaves of the fruit tree canopy, so that the droplets entrained by the airflow penetrate into the target, thereby greatly increasing the penetration of the droplets [33]. Compared with the trailer air-assisted sprayer, the weaker air speed of the mounted air-assisted sprayer resulted in a greater concentration of the spray liquid on the external and lower areas of the fruit tree canopy. Therefore, the CVs of the deposition amount of the spray liquid in the peach and pear canopy were 92.52% and 94.90%, respectively. The trailer air-assisted sprayer was able to blow the droplets to various areas of the fruit tree canopy, so the CVs of pesticide solution deposition in the peach and pear canopies were 35.98% and 26.54%, respectively, values which were much lower than the CVs of the mounted sprayer. It is worth mentioning that when the airflow of the air-assisted orchard sprayer is too weak, this will lead to deficient deposition in the canopies of fruit trees, while excessive airflow will result in the droplets being blown through the canopy, resulting in pesticide runoff and environmental pollution [3,30]. Therefore, in order to ensure the spraying performance of the air-assisted orchard sprayer, research should be carried out on airflow demand in the context of fruit tree canopies, and the air supply rate of sprayers should be regulated accordingly.

Admittedly, a great deal of time was consumed by counting the number of leaves in different areas of the fruit tree canopies. To study the distribution of the droplet penetration rate inside the tree canopy during air-assisted spraying, Sun et al. [34] chose cube-shaped frames to divide the canopy area of the fruit trees, and counted the number of leaves in each frame area. This method for the determination of the statistics of leaves in the fruit tree canopy also consumes a lot of time. At present, the most widely used target detection methods for fruit tree canopies are mainly aimed at calculating the outline and volume of the fruit tree [35–38]. Therefore, research on the density of branches and leaves in fruit tree canopies should be carried out to improve the efficiency of fruit tree canopy information extraction.

The quantitative indicators of the density of branches and leaves in the canopy of fruit trees include the leaf area index, leaf area density, point quadrat analysis and biomass density. Zhang et al. [39] developed a calculation method for leaf area density based on a three-dimensional laser point cloud, which provided a function relationship between the leaf number and the laser point. Sanz-Cortiella et al. [40] used the SICK LMS200 two-dimensional LiDAR sensor to scan the fruit tree canopy bilaterally, and obtained a three-dimensional structure map of the canopy. In addition, the linear relationship equation between the number of laser point clouds falling on the canopy branches and the leaf area was determined using statistical methods. Sanz et al. [41] found a linear relationship between the canopy leaf area density and the logarithm of the canopy volume based on the statistical results of the orchard test data and established a mathematical equation between the leaf area density and the canopy volume. The above studies could provide reference for the determination of fruit tree canopy information when determining the utilization rate of new orchard spray liquids.

5. Conclusions

To evaluate spraying efficacy in orchards, a canopy segmentation method was developed to determine the deposition rate and droplet distribution in orchard application. The distribution and deposition rate of spray liquid applied using three kinds of orchard sprayer were measured in a pear orchard and a peach orchard. The test results showed that the trailer sprayer had the highest deposition rates, with values of 31.54% and 56.92% on peach and pear trees, respectively. The spray gun had the best droplet distribution uniformity, and the CVs of the spray in the peach and pear canopies were 20.54% and 25.06%, respectively. Nevertheless, a great deal of time was consumed in counting the number of leaves in the different areas of the fruit tree canopies. Therefore, research on the density of branches and leaves in fruit tree canopies should be carried out to improve the efficiency of fruit tree canopy information extraction.

Author Contributions: S.W. (Shilin Wang), X.L. (Xue Li) and S.W. (Shuangshuang Wang) conceived the research idea and designed the experiments; S.W. (Shilin Wang), W.W. and X.L. (Xiaohui Lei) performed the experiments and analyzed the data; S.W. (Shilin Wang) and X.L. (Xue Li) wrote the paper; X.L. (Xue Li) and T.N. reviewed and revised the paper. All authors have read and agreed to the published version of the manuscript.

Funding: This research was funded by the Natural Science Foundation of Jiangsu Province (BK20200280) and the China Postdoctoral Science Foundation (2020M671390).

Institutional Review Board Statement: Not applicable.

Informed Consent Statement: Not applicable.

Data Availability Statement: The data presented in this study are available on request from the corresponding author.

Acknowledgments: The authors would like to give special thanks to the staff of Yejia pear orchard Co., Ltd. (Taizhou, China) and department of experiment station of JAAS for providing trial fields and operating sprayers.

Conflicts of Interest: The authors declare no conflict of interest.

References

1. Yearbook, C.S. *National Bureau of Statistics of China*; China Statistical Yearbook: Beijing, China, 2018.
2. Li, L. Research on Design and Application of Automatic Profiling Precision Variable-Rate Orchard Sprayer Based on LIDAR Detection. Doctor's Thesis, China Agricultural University, Beijing, China, 2018.
3. Jensen, P.K.; Olesen, M.H. Spray mass balance in pesticide application: A review. *Crop Prot.* **2014**, *61*, 23–31. [CrossRef]
4. He, X.; Yan, K.; Chu, J.; Wang, J.; Zeng, A.; Liu, Y. Design and Testing of the Automatic Target Detecting, Electrostatic, Air assisted, Orchard sprayer. *Trans. Chin. Soc. Agric. Eng.* **2003**, *19*, 78–80.
5. Sinha, R.; Ranjan, R.; Khot, L.R.; Hoheisel, G.-A.; Grieshop, M.J. Comparison of within canopy deposition for a solid set canopy delivery system (SSCDS) and an axial–fan airblast sprayer in a vineyard. *Crop. Prot.* **2020**, *132*, 105124. [CrossRef]
6. Ministry of Agriculture and Rural Affaies. China's Pesticide Fertilizer Utilization Rate of about 30% to Be Improved [EB/OL]. Available online: http://politics.people.com.cn/n/2015/0414/c1001-26843054.html (accessed on 14 April 2015).
7. Pivato, A.; Barausse, A.; Zecchinato, F.; Palmeri, L.; Raga, R.; Lavagnolo, M.C.; Cossu, R. An integrated model-based approach to the risk assessment of pesticide drift from vineyards. *Atmos. Environ.* **2015**, *111*, 136–150. [CrossRef]
8. Lammoglia, S.-K.; Kennedy, M.C.; Barriuso, E.; Alletto, L.; Justes, E.; Munier-Jolain, N.; Mamy, L. Assessing human health risks from pesticide use in conventional and innovative cropping systems with the BROWSE model. *Environ. Int.* **2017**, *105*, 66–78. [CrossRef] [PubMed]
9. Yang, L.Z.; Feng, Y.F.; Shi, W.M.; Xue, L.H.; Wang, S.Q.; Song, X.F.; Chang, Z.Z. Review of the advances and development trends in agricultural non-point source pollution control in China. *Chin. J. Eco-Agric.* **2013**, *21*, 96–101. [CrossRef]
10. Cui, Y.Z.; Gao, Y.; Zhao, G.S. Progress of differentiated ecological compensation policies related to agricultural non-point source pollution. *J. Agro-Environ. Sci.* **2017**, *36*, 1232–1241. [CrossRef]
11. Yuan, H.; Yang, D.; Yan, X. Pesticide efficiency and the way to optimize the spray application. *Plant. Prot.* **2011**, *37*, 14–20. [CrossRef]
12. Gu, Z.; Xu, G.; Xu, D. Analysis of pesticide efficiency of insecticides against brown planthopper, Nilaparvata lugens Stal. *Chin. J. Pestic. Sci.* **2018**, *20*, 704–714. [CrossRef]
13. Ebert, T.A.; Downer, R.A. A different look at experiments on pesticide distribution. *Crop. Prot.* **2006**, *25*, 299–309. [CrossRef]
14. Gao, S.; Wang, G.; Zhou, Y.Y.; Wang, M.; Yang, D.; Yuan, H.; Yan, X. Water-soluble food dye of Allura Red as a tracer to determine the spray deposition of pesticide on target crops. *Pest Manag. Sci.* **2019**, *75*, 2592–2597. [CrossRef] [PubMed]
15. Wang, S.; Li, X.; Liu, Y.; Lv, X.; Zheng, W. Comparison of a new knapsack mist sprayer and three traditional sprayers for pesticide application in plastic tunnel greenhouse. *Phytoparasitica* **2021**, *50*, 177–190. [CrossRef]
16. Vercruysse, F.; Steurbaut, W.; Drieghe, S.; Dejonckheere, W. Off target ground deposits from spraying a semi-dwarf orchard. *Crop. Prot.* **1999**, *18*, 565–570. [CrossRef]
17. Zhang, P.; Gao, Y.; Liu, Z.; Zhao, J.; Yang, J.; Fan, J.; Fan, R. Method for calculation of pesticide deposition rate using spraying machinery in the orchard. *Chin. J. Pestic. Sci.* **2020**, *22*, 277–284. [CrossRef]
18. Hoffmann, W.C.; Hewitt, A.J. Comparison of three imaging systems for water-sensitive papers. *Appl. Eng. Agric.* **2005**, *21*, 961–964. [CrossRef]
19. Hong, S.-W.; Zhao, L.; Zhu, H. CFD simulation of pesticide spray from air-assisted sprayers in an apple orchard: Tree deposition and off-target losses. *Atmos. Environ.* **2018**, *175*, 109–119. [CrossRef]
20. Wang, P.; Yu, W.; Ou, M.; Gong, C.; Jia, W. Monitoring of the Pesticide Droplet Deposition with a Novel Capacitance Sensor. *Sensors* **2019**, *19*, 537. [CrossRef]
21. Wang, S.; Fan, D.; Li, X.; Zhou, H.; Zhang, M.; Yan, T.; Lv, X. Effect of Orchard Sprayers on Dermal Exposure of Operators. *J. Ecol. R. Environ.* **2020**, *36*, 1612–1618. [CrossRef]
22. He, X. *Yaoxie Yu Shiyao Jishu*, 1st ed.; China Agricultural University Press: Bejing, China, 2013; pp. 167–172.
23. You, K.; Zhu, H.; Abbott, J.P. Assessment of Fluorescent Dye Brilliant Sulfaflavine Deposition on Stainless Steel Screens as Spray Droplet Collectors. *Trans. ASABE* **2019**, *62*, 495–503. [CrossRef]
24. Qin, W.-C.; Qiu, B.-J.; Xue, X.-Y.; Chen, C.; Xu, Z.-F.; Zhou, Q.-Q. Droplet deposition and control effect of insecticides sprayed with an unmanned aerial vehicle against plant hoppers. *Crop. Prot.* **2016**, *85*, 79–88. [CrossRef]
25. Smith, D.B. Uniformity and Recovery of Broadcast Sprays Using Fan Nozzles. *Trans. ASAE* **1992**, *35*, 39–44. [CrossRef]
26. Matthews, G.; Bateman, R.; Miller, P. *Pesticide Application Methods*, 4th ed.; Wiley & Blackwell: Hoboken, NJ, USA, 2014.
27. Delele, M.A.; Nuyttens, D.; Duga, A.T.; Ambaw, A.; Lebeau, F.; Nicolai, B.M.; Verboven, P. Predicting the dynamic impact behaviour of spray droplets on flat plant surfaces. *Soft Matter* **2016**, *12*, 7195–7211. [CrossRef] [PubMed]
28. Li, L.; He, X.; Song, J.; Liu, Y.; Zeng, A.; Liu, Y.; Liu, C.; Liu, Z. Design and experiment of variable rate orchard sprayer based on laser scanning sensor. *Int. J. Agric. Biol. Eng.* **2018**, *11*, 101–108. [CrossRef]

29. Li, H.; Zhai, C.; Weckler, P.; Wang, N.; Yang, S.; Zhang, B. A Canopy Density Model for Planar Orchard Target Detection Based on Ultrasonic Sensors. *Sensors* **2016**, *17*, 31. [CrossRef] [PubMed]
30. Zhai, C.; Zhu, R.; Sui, S.; Xue, S.; Shangguan, Z. Design and experiment of control system of variable pesticide application machine hauled by tractor. *Trans. Chin. Soc. Agric. Eng.* **2009**, *25*, 105–109. [CrossRef]
31. Michael, C.; Gil, E.; Gallart, M.; Stavrinides, M. Influence of Spray Technology and Application Rate on Leaf Deposit and Ground Losses in Mountain Viticulture. *Agriculture* **2020**, *10*, 615. [CrossRef]
32. Sun, C.; Qiu, W.; Ding, W.; Gu, J. Parameter optimization and experiment of air-assisted spraying on pear trees. *Trans. Chin. Soc. Agric. Eng.* **2015**, *31*, 30–38. [CrossRef]
33. Wiles, T.L.; Sharp, D.G.; Dobson, H. *Guidelines on Minimum Requirements for Agricultural Pesticide Application Equipment*; Food & Agriculture Org.: Rome, Italy, 2004; Volume 4.
34. Sun, C.; Liu, C. Construction and application of droplet canopy penetration model for air-assisted spraying pattern. *Trans. Chin. Soc. Agric. Eng.* **2019**, *35*, 25–32. [CrossRef]
35. Miranda-Fuentes, A.; Llorens, J.; Gamarra-Diezma, J.L.; Gil-Ribes, J.A.; Gil Moya, E. Towards an Optimized Method of Olive Tree Crown Volume Measurement. *Sensors* **2015**, *15*, 3671–3687. [CrossRef]
36. Liu, H.; Zhu, H. Evaluation of a Laser Scanning Sensor in Detection of Complex-Shaped Targets for Variable-Rate Sprayer Development. *Trans. ASABE* **2016**, *59*, 1181–1192. [CrossRef]
37. Gil, E.; Escolà, A.; Rosell-Polo, J.R.; Planas, S.; Val, L. Variable rate application of plant protection products in vineyard using ultrasonic sensors. *Crop. Prot.* **2007**, *26*, 1287–1297. [CrossRef]
38. Maghsoudi, H.; Minaei, S.; Ghobadian, B.; Masoudi, H. Ultrasonic sensing of pistachio canopy for low-volume precision spraying. *Comput. Electron. Agric.* **2015**, *112*, 149–160. [CrossRef]
39. Tao, Q.; Xin, S.; Yan, L.; Xu, H. Cavitation Process and Flow Characteristics inside Diesel Injector Nozzle. *Trans. Chin. Soc. Agric. Mach.* **2016**, *47*, 359–365. [CrossRef]
40. Sanz-Cortiella, R.; Llorens-Calveras, J.; Escolà, A.; Arnó-Satorra, J.; Ribes-Dasi, M.; Masip-Vilalta, J.; Camp, F.; Gràcia-Aguilá, F.; Solanelles-Batlle, F.; Planas-DeMartí, S.; et al. Innovative LIDAR 3D dynamic measurement system to estimate fruit-tree leaf area. *Sensors* **2011**, *11*, 5769–5791. [CrossRef] [PubMed]
41. Sanz, R.; Rosell, J.; Llorens, J.; Gil, E.; Planas, S. Relationship between tree row LIDAR-volume and leaf area density for fruit orchards and vineyards obtained with a LIDAR 3D Dynamic Measurement System. *Agric. For. Meteorol.* **2013**, *171–172*, 153–162. [CrossRef]

agronomy

MDPI

Article

Numerical Experiment and Optimized Design of Pipeline Spraying On-Line Pesticide Mixing Apparatus Based on CFD Orthogonal Experiment

Daozong Sun [1,2], Weikang Liu [1], Zhi Li [1], Xurui Zhan [1], Qiufang Dai [1,3], Xiuyun Xue [1,4] and Shuran Song [1,5,*]

[1] College of Electronic Engineering (College of Artificial Intelligence), South China Agricultural University, Guangzhou 510642, China; sundaozong@scau.edu.cn (D.S.); liuweikang@stu.scau.edu.cn (W.L.); lizhi@stu.scau.edu.cn (Z.L.); langzizhan@stu.scau.edu.cn (X.Z.); daiqiufang@scau.edu.cn (Q.D.); xuexiuyun@scau.edu.cn (X.X.)
[2] Guangdong Engineering Research Center for Monitoring Agricultural Information, Guangzhou 510642, China
[3] Division of Citrus Mechanization, China Agriculture Research System, Guangzhou 510642, China
[4] Guangdong Engineering Technology Research Center for Creative Hilly Orchard Machinery, Guangzhou 510642, China
[5] Guangdong Modern Agricultural Science and Technology Innovation Center for Intelligent Orchard, Guangzhou 510642, China
* Correspondence: songshuran@scau.edu.cn; Tel.: +86-136-6896-6908

Abstract: Pipeline spraying can be adopted for greatly improving spraying efficiency in hillside orchard spraying operations. However, residual phytosanitary product still remains in the pipeline after the completion of pipeline spraying operations. Currently, residual phytosanitary liquid is handled according to the following general method: pipeline flushing with fresh water. The method can easily lead to pesticide waste and environment pollution. On-line pesticide mixing technology can be adopted for reducing pesticide waste and environmental pollution. However, on-line pesticide mixing technology is not applied in pipeline spraying operations. Therefore, the mixing principle of jet-mixing apparatus is adopted as a reference in the paper for designing the basic structure of on-line pesticide mixing apparatus based on pipeline spraying. The structure is mainly composed of a constricted tube, suction chamber, Venturi, and diffusion tube. An analysis method based on the CFD orthogonal experiment is adopted for studying the influence of the changes of four key structure parameters on on-line pesticide mixing apparatus, pesticide dissolution performance, and pesticide mixing performance; the four parameters include constricted tube falloff angle, diffusion tube divergence angle, Venturi diameter, and Venturi length. Since there may be interaction among them, three experiment evaluation indexes of lifting height, turbulent kinetic energy, and pressure recovery distance are set for judgment. The change of three evaluation indexes with change of constricted tube falloff angle, diffusion tube divergence angle, Venturi diameter, and Venturi length, respectively, is revealed through single-index variance analysis; the three indexes include lifting height, turbulent kinetic energy, and pressure recovery distance. The primary and secondary sequences of respective influences of all structure parameters and their interaction on all evaluation indexes are obtained. Analysis results of all evaluation indexes are comprehensively considered in order to finally discover the comprehensive optimal pesticide mixing apparatus structure parameters, namely: constricted tube falloff angle is 22°, diffusion tube divergence angle is 9°, Venturi diameter is 2 mm, Venturi length is 6 mm, and pesticide mixing apparatus structure parameters are optimized. Theoretical reference is provided in the paper for on-line pesticide mixing apparatus prototype production on the basis of pipeline spraying.

Keywords: on-line pesticide mixing; pipeline spraying; pesticide mixing apparatus; CFD; orthogonal experiment; optimization

Citation: Sun, D.; Liu, W.; Li, Z.; Zhan, X.; Dai, Q.; Xue, X.; Song, S. Numerical Experiment and Optimized Design of Pipeline Spraying On-Line Pesticide Mixing Apparatus Based on CFD Orthogonal Experiment. *Agronomy* **2022**, *12*, 1059. https://doi.org/10.3390/agronomy12051059

Academic Editor: Roberto Marani

Received: 9 April 2022
Accepted: 27 April 2022
Published: 28 April 2022

1. Introduction

At present, most areas of China still use traditional premixed pesticide mixture for plant protection machinery spraying operations. The chosen pesticide and water–in a certain proportion–are combined in the tank to mix well before spraying. The operator is directly in contact with the pesticide under the pesticide mixing mode, and thus the operator is prone to pesticide positioning. The remaining pesticide liquid can lead to pesticide waste after mixing. If the pesticide liquid is not handled well, environment pollution also can occur [1]. However, the dosing tank is separated from the water tank according to on-line pesticide mixing mode, compared with the pre-mixed pesticide mixing mode. Sprayer pipeline system internal water flow and sprayer pipeline extra power are utilized for on-line mixing of pesticide and water, and the pesticide can be applied regularly, professionally, and accurately. This is based on environmental protection and operator safety, which is in line with the requirement of agricultural sustainable development [2].

On-line pesticide mixing technology is an important development direction in the plant protection machinery field, which is a topic of concern for experts and scholars all over the world, and has been widely researched [3–5]. The research on on-line pesticide mixing technology in China is mainly based on jet-mixing apparatus; extra power is not required for the pesticide mixing apparatus, the structure is compact, and the operation is convenient. A momentum exchange energy transfer mode is utilized. Pesticide and water online mixing is completed through the hydraulic power of the sprayer pipeline system. He et al. [6–8] engaged in numerical simulation calculations on jet-mixing apparatus mixing tubes, and the distribution law of radial velocity and axial velocity in the mixing tube was concluded. Li et al. [9] studied double-stage jet-mixing apparatus under the condition of different main-auxiliary jet flow nozzle outlet diameters and nozzle tube pitches through experiments; the influence of all parameters of jet-mixing apparatus on pressure loss and pesticide mixing ratio was revealed. Qiu et al. [10,11] studied the influence area ratio, nozzle inclination, and other parameters of jet-mixing apparatus flowing properties and the influence of area ratio on pesticide mixing uniformity through the CFD numerical simulation method. Chen et al. [12] regarded LabVIEW as a development environment for designing an on-line jet-mixing control system in order to solve the problems of jet-mixing apparatus, and realize accurate and intelligent control of jet-mixing apparatus; the problems include narrow pesticide mixing ratio adjustable scope and low control precision. The on-line jet-mixing control system was set for obtaining a suitable pesticide mixing ratio value, thereby significantly increasing the adjustment width of the phytosanitary mixing ratio. Zhou et al. [13] studied the influence of jet-mixing apparatus structure parameters on its performance through CFD numerical analog simulation analysis method and the combination of experiment data. The jet-mixing apparatus optimal area ratio and optimal nozzle tube pitch scope were finally determined. Ou et al. [14] adopted the CFD numerical value analysis method for studying the flow field characteristics of the jet-mixing apparatus under variable working condition, and obtained the following conclusion: the pesticide mixing ratio is linearly reduced with the increase of outlet static pressure; STC12C5A60S2 is regarded as the control core. Zhang et al. [15] established an on-line jet-mixing ratio control experiment platform, and an on-line jet-mixing ratio control experiment was carried out on the basis of the experiment platform. Experiment results showed that the experiment platform can control the pesticide mixing ratio according to different phytosanitary liquid concentrations; the adjustable scope of pesticide mixing ratio was expanded to improve real-time automatic pesticide mixing precision.

The research on on-line pesticide mixing technology in developed countries such as Europe and America is based on a pesticide direct injection system, which is different from the research focus of scholars in China. In the pesticide mixing mode, an extra power sprayer pipeline system is utilized for accurate metering of the pesticide, and for the mixing of pesticide and water in the pipeline. The compressed air pesticide direct injection system developed by Ghate and Phatak [16] can be used for adjusting the air pressure in the dosing tank and water tank, respectively, according to mechanical equipment driving speed, water, and pesticide flow out of the water tank and dosing tank according to certain proportions, thereby achieving pesticide and water online mixing, and meeting pesticide liquid concentration requirements. The peristaltic pump is utilized for quantitatively sucking pesticide in the dosing tank and feeding it into the main water pipeline; it is mixed with water, and control valves can be used for automatically adjusting the pesticide delivery dose according to the driving speed of the mechanical equipment in the Mid-West Technology CCI-2000 variable volume pump direct injection system, developed in America. This ensures that the mixed pesticide liquid concentration is in line with requirements [17]. There is also research that combines flow capacity control technology and pesticide direct injection: Koo and Summer [18] developed a pesticide direct injection system capable of real-time adjustment of phytosanitary liquid and water flow capacity with sprayer driving speed, and that maintained constant pesticide mixing concentration and dose in the unit. Slaughter et al. [3] applied the pesticide direct injection system in a lateral herbicide sprayer with the research development of pesticide direct injection technology, and its function of accurate targeted pesticide application was realized. Steward and Humburg [19] studied a control sub-system mathematics model aiming at the Raven SCS-700 pesticide direct injection system; the related parameters of flow capacity and valve are analyzed, and valve adjustment response and stability are determined for improving phytosanitary mixing performance and phytosanitary mixing precision. Gillis et al. [4] studied a pesticide direct injection system based on machine vision, and preliminary research results were obtained. Aissaoui et al. [20] developed an optical sensor with low power consumption, where the performance of pesticide direct injection system was dynamically measured; test results showed that the measurement data can be used for modifying the performance of the pesticide direct injection system. Luck et al. [21] adopted a dye/glycerinum mixture as a tracer agent for studying the working conditions of the pesticide direct injection system, and they proposed the solution that the dye/glycerinum mixture calibration curve can be utilized for simulating the pesticide mixing process. Felizardo et al. [22] constructed a variable speed spraying system model based on pesticide direct injection technology; a predictive control method is adopted for control, the carrier chemical mixing process and chemical flow rate error can be lower than 5% of the allowable level, and the effectiveness of the model and its control strategy is verified by experiment results.

Document literature consulted at present shows that pesticide direct injection systems studied in developed countries such as Europe and America, jet-mixing apparatus studied in China, and other on-line pesticide mixing apparatus, are mainly applied in medium- and large-scale mobile plant protection machinery equipment and on pipeline spraying system in hillside orchards, as shown in Figure 1. The CFD theory-based numerical simulation analysis method can be applied to better reveal various pesticide mixing apparatus internal flow field states in the aspect of the research method [14,23–27], and it is obvious that the orthogonal experiment method has prominent advantages in reducing experiment frequency and optimizing structure parameters [28,29].

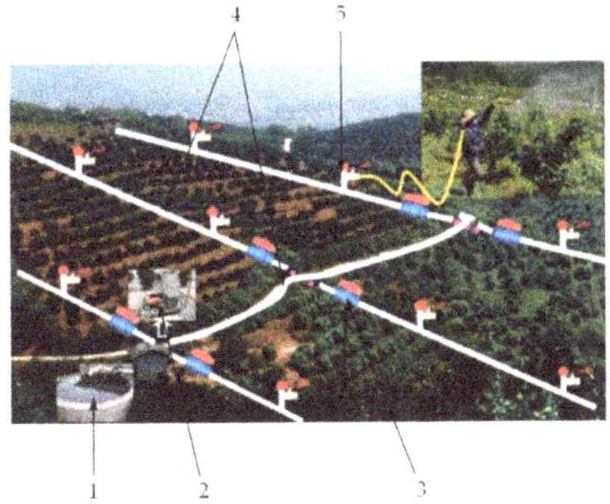

Figure 1. Schematic diagram of hillside orchard pipeline spraying. 1. Water pool. 2. Mixer. 3. Water valves. 4. Pipes. 5. Spray valves.

Therefore, the objective of the study was to analyze the internal flow field state of the mixer based on CFD orthogonal tests in order to obtain the optimum mixer construction parameters, which mainly include constricted tube falloff angle α, diffusion tube divergence angle β, Venturi diameter d, and Venturi length L. This would save on experiment costs and reduce the amount of testing needed.

2. Materials and Method

The poor road conditions in mountain orchards make it difficult for large- and medium-sized plant protection sprayers to access them for spraying operations. Pipeline spraying is adopted to greatly improve spraying efficiency, however residual pesticide liquid still remains in the pipeline after the pipeline spraying operation is complete. Currently, the pipeline is generally cleaned with fresh water as a common method, which can easily lead to pesticide waste and environment pollution. On-line pesticide mixing technology can be adopted for reducing pesticide waste and environmental pollution. On-line pesticide mixing technology will be applied for pipeline spraying in the paper in order to design the basic structure of on-line pesticide mixing apparatus based on pipeline spraying. An analysis method based on CFD orthogonal experiment is adopted for analyzing the pesticide mixing apparatus internal flow field state, thereby optimizing the pesticide mixing apparatus structure parameters, and finally discovering comprehensive optimal pesticide mixing apparatus structure parameters.

2.1. Pesticide Mixing Apparatus Structure and CFD Model

2.1.1. Design of Pesticide Mixing Apparatus Basic Structure

Jet-mixing apparatus is characterized by no requirement of extra power, compact structure, convenient operation, and low cost. Compared with other pesticide mixing apparatus, jet-mixing apparatus has the advantages of strong mixing, reliable working performance, quick mixing, and so on [1,2]. Therefore, the mixing principle of jet-mixing apparatus is adopted as reference in the paper; the actual dimension of hose diameter in pipeline spraying is combined to design the basic structure of on-line pesticide mixing apparatus based on pipeline spraying, as shown in Figure 2. Here, the diameter of water inlet and mixture outlet is 8.5 mm, the diameter of the pesticide mother liquid inlet is 2 mm, and the diameter is matched with the dimension of the spray gun hose in practical application at a flow rate of 10 L/min in the corresponding size; the length of the constricted tube L1 is 15 mm, the length of the diffusion tube L2 is 42 mm.

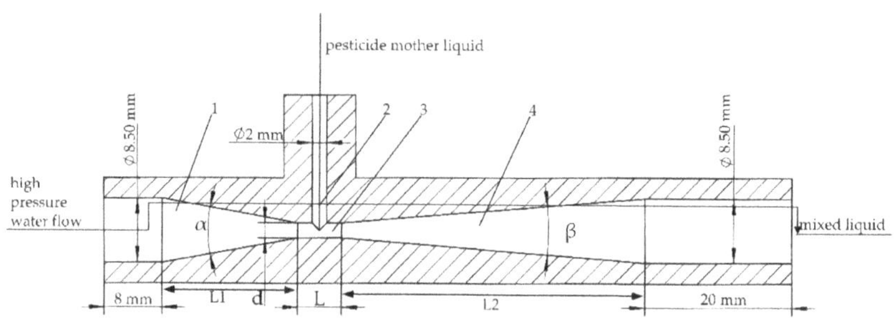

Figure 2. Structure of the jet mixer. 1. Constricted tube. 2. Suction chamber. 3. Venturi. 4. Diffusion tube.

The on-line pesticide mixing apparatus based on pipeline spraying is mainly composed of a constricted tube, suction chamber, Venturi, and diffusion tube, where the constricted tube falloff angle α, diffusion tube divergence angle β, Venturi diameter d, and Venturi length L are four key structure parameters of the on-line pesticide mixing apparatus; these parameters have an important influence on pesticide mixing performance [13,30,31]. Figure 2 shows that the velocity of high-pressure water flow penetrating through the constricted tube with a gradually reduced cross section area is gradually increased. Kinetic energy is increased, and pressure is gradually decreased, which is injected from the constricted tube outlet into the Venturi. Negative pressure is formed in the Venturi; the pesticide mother liquid in the dosing tank is entrained from the suction chamber to the pesticide mixing apparatus under the effect of barometric pressure, and it can exchange energy with the high-pressure water flow in the Venturi, where they are gradually and evenly mixed. Then, mixed liquid enters into the diffusion tube with a gradually increased cross section area. The velocity is gradually reduced, pressure is gradually increased, and the liquid flows out of the pesticide mixing apparatus outlet, thereby realizing online mixing of the pesticide mother liquid and water [6,7,32].

2.1.2. Establishment of Pesticide mixing Apparatus CFD Model

Before computational fluid dynamics (CFD) software Fluent is utilized for the numerical simulation calculation analysis of the on-line pesticide mixing apparatus internal flow field state on the basis of pipeline spraying, it is necessary to establish a CFD model for numerical simulation calculation. The pesticide mixing apparatus internal flow channel undergoes mesh generation through the integrated computer engineering and manufacturing code for computational fluid dynamics (ICEMCFD) mesh generation software in the paper, thereby generating a CFD model in line with requirements.

The pesticide mixing apparatus internal flow channel has a regular geometrical shape. Structure-based meshes can be adopted for division, thereby reducing mesh quantity and improving the numerical simulation calculation speed. Meshes vertical to the flowing direction can be easily generated by structure-based hexahedral mesh generation technology, thereby reducing pseudo diffusion, improving CFD model mesh quality, and ensuring computational accuracy. In the paper, structure-based hexahedral mesh is adopted for mesh generation in the pesticide mixing apparatus internal flow channel computational domain [33–35]. Since all structures of the pesticide mixing apparatus internal flow channel are basically circular tube sections, advanced O-shaped mesh generation technology of ICEMCFD is adopted, and blocks for generating structure-based hexahedral mesh are divided into an O-shape, thereby facilitating the generation of a CFD model with higher mesh quality, and improving the computational accuracy.

Normal velocity has very large gradients in areas close to the pesticide mixing apparatus internal flow channel wall, flow field conditions are complex, and standard wall-function method can be adopted for correct solution of the simulation value model wall condition according to CFD software Fluent. Six boundary layers are arranged in the pesticide mixing apparatus internal flow channel wall area for mesh encryption processing, thereby meeting numerical calculation requirements. The pesticide mixing apparatus internal flow channel suction chamber, Venturi, and areas with more complicated diffusion tube flow field states undergo local mesh encryption processing in order to evaluate pesticide mixing apparatus pesticide dissolution performance and pesticide mixing performance more accurately. In this process, 160 nodes are arranged on each longitudinal block line in suction chamber, 40 nodes are arranged on each transverse block line of each Venturi, and 150 nodes are arranged on each oblique block line of diffusion tube, thereby improving the computational accuracy, and better reflecting the pesticide mixing apparatus internal flow field state. A CFD model in line with numerical calculation requirements can be finally generated, as shown in Figure 3, and the mesh quantity is 416,160.

Figure 3. On-line pesticide mixing apparatus CFD model based on pipeline spraying.

2.2. *Design of Orthogonal Experiment*

2.2.1. Design of Orthogonal Experiment Plan

When medium and large plant protection machinery for field crop are used for spraying operations, better spraying effect can be achieved when the spraying pressure is 1 to 2 MPa. On the contrary, if the spraying object is a fruit tree hillside orchard, higher spraying pressure is required during pipeline spraying operations, which should be higher than 0.8MPa generally; this allows a pronounced spraying effect to be achieved [36]. Many field experiment studies show that excellent spraying effect can be achieved when the pipeline spraying operation spraying pressure is 1 MPa. Document literature is consulted, and it is obvious that pressure loss of jet-mixing apparatus is higher than or equal to 50% [6,10,13,14,31]. Therefore, the analysis method based on CFD orthogonal experiment is adopted in the paper, high pressure water flow inlet pressure of pipeline spraying on-line pesticide mixing apparatus is set as 2 MPa, pesticide mother liquid inlet pressure is set as 0 MPa (barometric pressure), mixed liquid outlet pressure is set as 1 MPa.

The use of the double dilution method for the concentration formulation of pharmaceutical solutions improves the dispersion and suspension of the active ingredients and allows for more accurate dosing. In order to make it easier and more intuitive for the subsequent operator to use the on-line mixing device based on pipeline spraying, the article uses a multiplicative representation for the preparation of a certain concentration of liquid. By consulting relevant information on pesticides on websites such as China Pesticide First and China Pesticide Network, it was determined that the mixing ratio of most pesticide stock solutions to water was between 1:200 and 1:4000, i.e., the dilution times were between 200 and 4000 times. In the whole process of preparing the concentration of the liquid, the mother liquor dilution method firstly dilutes the pesticide stock solution by 2 to 20 times to make the pesticide mother liquor, and then the pesticide mother liquid is mixed by the on-line mixing device, i.e., the pesticide is diluted by 100 to 200 times to spray. Ultimately, the pesticide stock solution can be diluted 200 to 4000 times to meet the needs of most pesticide dilutions. Therefore, the mixing ratio of the pesticide mother liquid and water in the online mixing device based on pipeline spraying designed in this paper is finally determined as 1:100~1:200.

The pesticide mixing apparatus internal flow field state is studied and analyzed to observe the influence of change of four key structure parameters on pesticide mixing apparatus pesticide dissolution performance and pesticide mixing performance; the four parameters include constricted tube falloff angle α, diffusion tube divergence angle β, Venturi diameter d, and Venturi length L, thereby finally discovering the comprehensive optimal pesticide mixing apparatus structure parameters.

Change in single structure parameters affects pesticide mixing apparatus pesticide dissolution performance and pesticide mixing performance. Therefore, some interaction thereof also can generate a certain influence on pesticide mixing apparatus pesticide dissolution performance and pesticide mixing performance according to the above four key structure parameters, and a main influence can be determined. Therefore, four principal factors and three interaction factors are finally selected as experiment factors based on CFD orthogonal experiment study in the paper. The four principal factors are the constricted tube falloff angle α (factor A), diffusion tube divergence angle β (factor B), Venturi diameter d (factor C), and Venturi length L (factor D); the three interaction factors are the interaction between constricted tube falloff angle α and diffusion tube divergence angle β (factor A*B), interaction between diffusion tube divergence angle β and Venturi diameter d (factor B*C), as well as the interaction between diffusion tube divergence angle β and Venturi length L (factor B*D).

2.2.2. Experiment Evaluation Indexes and Factor Level

Before the orthogonal experiment table is designed, suitable factor levels should be firstly selected for four principal factors in orthogonal experiment; the factor level is finally selected in the paper as Table 1 based a large number of literature [1,14,26,34] and previous pre-experiment study results.

Table 1. Factor level table.

Level	Factor			
	A α/ (°)	B β/ (°)	C d/mm	D L/mm
1	0.884	0.626	3.058	0.888
2	1.189	1.078	0.268	1.200
3	1.318	1.688	0.066	1.303

A = constricted tube falloff angle α; B = diffusion tube divergence angle β; C = Venturi diameter d; D = Venturi length L.

Three experiment evaluation indexes are selected for evaluation in the paper in order to correctly evaluate pesticide mixing apparatus pesticide dissolution performance and pesticide mixing performance, respectively, including lifting height H_1, turbulent kinetic energy k, and pressure recovery distance H_2, wherein lifting height evaluation indexes is used for evaluating pesticide mixing apparatus pesticide dissolution performance, and turbulent kinetic energy evaluation indexes and pressure recovery distance evaluation indexes are jointly used for evaluating pesticide mixing apparatus pesticide mixing performance. The influence of three experiment evaluation indexes on pesticide mixing apparatus pesticide dissolution performance and pesticide mixing performance is concretely described as follows:

1. Lifting heigh: the pesticide dissolution performance of the pesticide mixing apparatus is evaluated through observing the velocity vector diagram of the suction chamber traverse section, and recording the maximum height numerical value of high pressure water flow rising in the suction chamber. If the maximum height of high pressure water flow rising is lower, the obstruction of rising high pressure water flow to the sucked pesticide mother liquid is smaller, and the pesticide dissolution performance is better.
2. Turbulent kinetic energy: the numerical value of the maximum turbulent kinetic energy reached by the pesticide mixing apparatus internal flow field is observed and recorded for measuring the mixing ability of the pesticide mixing apparatus. If the maximum turbulent kinetic energy is higher, the pesticide mixing apparatus internal flow field turbulence development is better, the flow field state is more unstable, the water and pesticide mother liquid can be mixed better, and the mixing ability is stronger.
3. Pressure recovery distance: the distance for increasing the diffusion tube internal pressure from 0.1 MPa to 0.9 MPa is observed and recorded for measuring the pressure recovery ability of the pesticide mixing apparatus. If the distance is shorter, the pressure recovery is faster, the pressure recovery ability is stronger, and the pressure loss is smaller.

2.2.3. Design of Orthogonal Experiment Table

Since each principal factor contains three levels, the degree of freedom (DOF) of each column in the orthogonal experiment table is 2, the DOF of each interaction factor is 4, and each interaction should occupy two columns in the orthogonal experiment table. Therefore, the column number of factor A*B in the orthogonal experiment table is, respectively, AB1 and AB2, the column number of factor B*C is, respectively, BC1 and BC2, and the column number of factor B*D is, respectively, BD1 and BD2. Three error columns with column numbers of E1, E2, and E3, respectively, also should be finally reserved, and they are used for estimating random errors. An interaction orthogonal table with a table head of L27(313) is selected in the paper accordingly [28]. SPSS data statistical analysis software is utilized for generating the orthogonal experiment table in line with experiment requirements, as can be seen in Table 2.

Table 2. Orthogonal experiment table.

Experiment Number	Column Number												
	A	B	AB1	AB2	C	BD2	E1	BC1	D	E2	BC2	BD1	E3
1	2	2	3	2	3	2	3	1	1	3	1	1	3
2	1	2	1	3	3	3	3	2	3	2	2	1	1
3	2	1	3	3	1	3	1	3	2	2	3	1	3
4	3	2	2	1	3	1	3	3	2	1	3	1	2
5	3	2	1	3	2	1	1	3	1	3	2	2	3
6	2	2	2	1	2	2	1	1	3	2	3	2	1
7	2	3	2	3	1	1	3	2	2	3	1	2	1
8	1	1	3	3	3	1	2	1	3	3	3	2	2
9	3	1	1	1	3	2	2	2	2	2	1	2	3
10	1	3	3	1	1	2	3	3	1	2	2	2	2
11	3	3	1	2	1	3	3	1	3	1	3	2	3
12	2	2	1	3	1	2	2	1	2	1	2	3	2
13	2	3	1	2	3	1	1	2	1	2	3	3	2
14	1	2	2	1	1	3	2	2	1	3	3	3	3
15	1	1	2	2	2	1	3	1	2	2	2	3	3
16	2	1	1	1	2	3	3	3	3	3	1	3	2
17	3	1	3	3	2	2	3	2	1	1	3	3	1
18	3	3	2	3	2	3	2	1	1	2	1	1	2
19	1	3	2	3	3	2	1	3	3	1	1	3	3
20	2	3	3	1	2	1	2	2	3	1	2	1	3
21	1	2	3	2	2	3	1	2	2	1	1	2	2
22	3	2	3	2	1	1	2	3	3	2	1	3	1
23	2	1	2	2	3	3	2	3	1	1	2	2	1
24	1	3	1	2	2	2	2	3	2	3	3	1	1
25	1	1	1	1	1	1	1	1	1	1	1	1	1
26	3	1	2	2	1	2	1	2	3	3	2	1	2
27	3	3	3	1	3	3	1	1	2	3	2	3	1

A = constricted tube falloff angle α; B = diffusion tube divergence angle β; C = Venturi diameter d; D = Venturi length L; ABX = the column number of factor A*B in orthogonal experiment table (X = 1, 2); BCX = the column number of factor B*C in orthogonal experiment table (X = 1, 2); BDX = the column number of factor B*D in orthogonal experiment table (X = 1, 2); EX = error columns (X = 1, 2, 3).

2.3. Boundary Condition and Model Selection

Fluent software is adopted for computational fluid dynamics (CFD) numerical analog simulation research and analysis on pesticide mixing apparatus internal flow field state in the paper: high pressure water flow inlet boundary is set as pressure inlet, gauge pressure is 2 MPa; pesticide mother liquid inlet boundary is set as pressure inlet, gauge pressure is 0 MPa; mixed liquid outlet boundary is pressure outlet, gauge pressure is 1 MPa. Non-slipping wall boundary condition is adopted for the wall, standard wall-function method is adopted for treating the area near the wall, and operating environment belongs to standard barometric pressure [13,14,26,37].

The standard k-ε two-equation turbulence model belongs to a turbulence model, which is applied most widely in the engineering field at present. This can be better applied in a complex 3D flow field, and it is physically meshed in the aspect of flow field simulation with a lower calculated amount and rational precision. The model is economical and practical [1,10,13,26,31,33,38,39]. Therefore, a standard k-ε two-equation turbulence model is adopted for numerical simulation solution calculations on the CFD model in the paper.

Turbulent kinetic energy k equation:

$$\frac{\partial}{\partial_t}(\rho k) + div(\rho\mu k) = div\left[\left(\mu + \frac{\mu_t}{\sigma_k}\right) \times grad(k)\right] - \rho\varepsilon + \mu_t P_G \quad (1)$$

Turbulent dissipation rate ε equation:

$$\frac{\partial}{\partial_t}(\rho\varepsilon) + div(\rho\mu\varepsilon) = div\left[\left(\mu + \frac{\mu_t}{\sigma_z}\right) \times grad(\varepsilon)\right] - \rho C_2\frac{\varepsilon^2}{k} + \mu_t C_1\frac{\varepsilon}{k}P_G \tag{2}$$

In the above formula: k is the turbulent kinetic energy in $m^2 \cdot s^{-2}$, ε is the turbulent dissipation rate in $m^2 \cdot s^{-3}$, ρ is the fluid density in $kg \cdot m^{-3}$, μ is the dynamic coefficient of viscosity in $kg \cdot m^{-1} \cdot s^{-1}$, μ_t is the turbulence coefficient of viscosity in $kg \cdot m^{-1} \cdot s^{-1}$, and P_G is the turbulent kinetic energy generation item in s^{-2}.

SIMPLEC algorithm is adopted for pressure velocity coupling solution, PRESTO interpolating scheme is adopted for pressure equation solution in the aspect of solving method, other equation solutions are dispersed through Second Order Upwind method, residual error convergence standard is set as 1×10^{-4}, and Hybird Initialization is adopted for initialization [11,26,34,40].

Concrete steps of numerical simulation calculation are shown as follows:

1. Fluent solver selection; 3DFluent solver should be selected according to the pesticide mixing apparatus model.
2. Mesh importing; Menu bar File→Read→Mesh options are selected, and then msh files exported from ICEM software are selected.
3. Mesh checking; Menu bar Mesh→Check options are selected, and the operator should observe whether meshes suffer from negative volume. Meshes are divided again if there is negative volume.
4. Setting of computational domain dimensions; Since the basic unit of the pesticide mixing apparatus internal flow channel CFD model is mm, and the basic unit of Fluent software is m, it is necessary to select the menu bar Mesh→Scale options for setting the imported model as mm.
5. Selection of calculation model; Pressure-Based solver is selected for General parameter setting panel. Steady stable-state flow is selected for the panel Time option. Viscous-Standard *k-e* is used for standing the Models parameter setting panel, namely standard *k-ε* two-equation model.
6. Confirmation of fluid physical properties; Water-liquid is selected in the material depot equipped for the Fluent software; namely liquid water is regarded as the working liquid material.
7. Definition of operating environment; Menu bar Define→Operating Conditions option is selected, 101325 is the input into the Operating Pressure textbox, as the operating environment namely has standard barometric pressure.
8. Appointment of boundary condition; In the Boundary Conditions parameter settings panel, high pressure water flow inlet is set as pressure inlet, gauge pressure is 2 MPa, pesticide mother liquid inlet is set as pressure inlet, and gauge pressure is 0 MPa; mixed liquid outlet is set as pressure outlet, and gauge pressure is 1 MPa.
9. Solving method setting; Th pressure velocity coupling solution algorithm of SIMPLEC algorithm is set in the Solution Methods parameter setting panel, PRESTO interpolating scheme is adopted for pressure equation solution, and other equation solutions are dispersed by Second Order Upwind method.
10. Solution monitoring setting; Residual error convergence standard of 1×10^{-4} is set for Monitors parameter settings panel. One monitoring face is established for monitoring the change of mixture outlet mass flow rate.
11. Flow field initialization; Hybird Initialization is adopted for initializing flow field in the Solution initialization parameter settings panel.
12. Iterative solution; Iteration frequency of 4000 is set as the Run Calculation parameter settings panel to start the numerical simulation calculation.
13. Result checking.
14. Result saving and after-treatment.

3. Results and Discussion

3.1. Analysis of Chemical Mixer Internal Flow Field

Tables 1 and 2 in the paper, as well as setting in text 2.3, are used for the numerical calculation of the pesticide mixing apparatus internal flow field; calculation results occupy a large amount of written space. Experiment number 10 is only listed in the paper. The pesticide mixing apparatus internal flow field numerical calculation results for constricted tube falloff angle is 20°, diffusion tube divergence angle is 9°, Venturi diameter is 2 mm, and Venturi length is 6 mm, as shown in Figures 4–6.

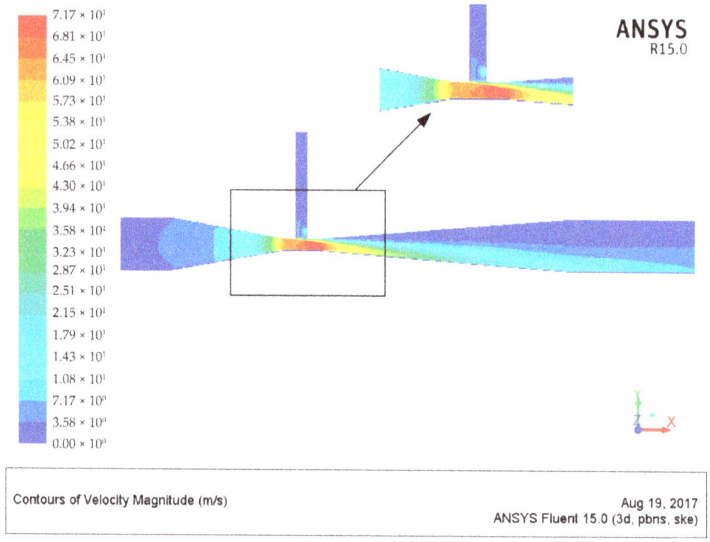

Figure 4. Cloud atlas of X-Y plane flow field velocity distribution. This is the image placed at the center of the pipe.

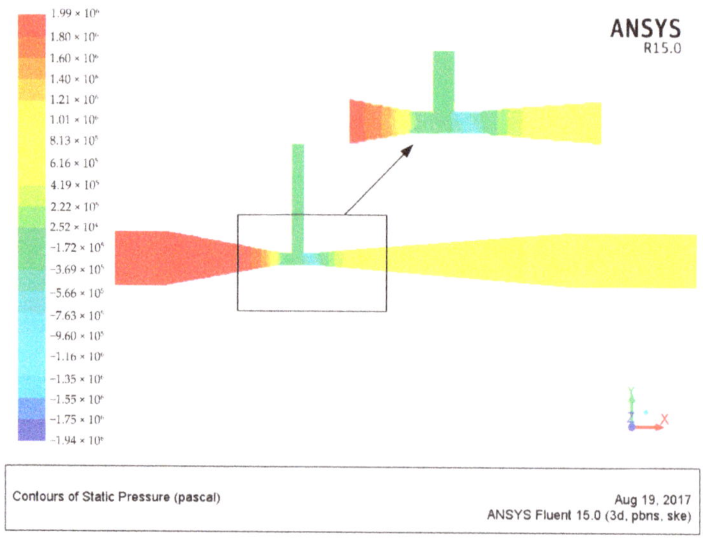

Figure 5. Cloud atlas of X-Y plane flow field pressure distribution. This is the image placed at the center of the pipe.

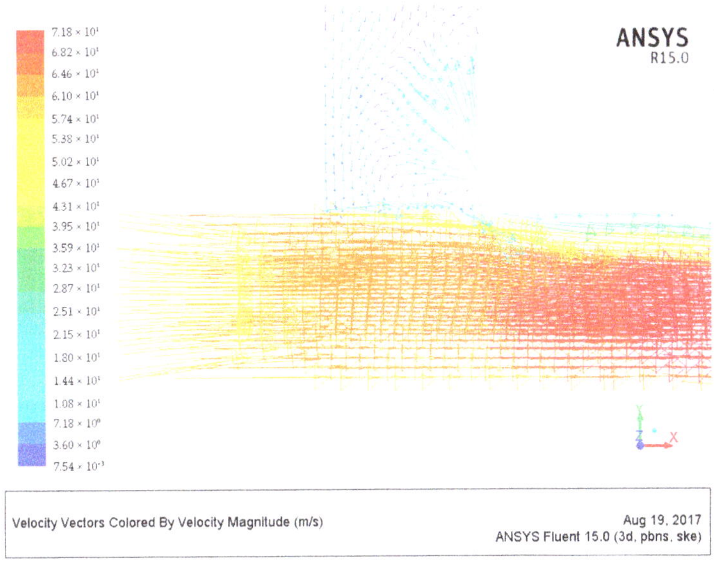

Figure 6. Velocity vector partial enlarged drawing at the joint of X-Y plane suction chamber and Venturi.

Figures 4 and 5 are combined. It is obvious that as low velocity high pressure water flow penetrates through the constricted tube with a gradually decreased cross section area, velocity is gradually increased, pressure is gradually decreased, and it is changed into a high-velocity low pressure water flow. Velocity reaches its maximum value in the Venturi, and pressure reaches the minimum level, thereby forming negative pressure; the pesticide mother liquid is sucked from the suction chamber into the pesticide mixing apparatus under the effect of the strong external barometric pressure and static pressure of the liquid. Two liquid flows with different velocities exchange energy in the Venturi, water flow rate is reduced, the flow rate of pesticide mother liquid is increased, and they are gradually and evenly mixed. Since the cross-section area of the diffusion tube is gradually increased, some kinetic energy is converted into pressure energy, mixed liquid flow rate is gradually decreased, and pressure is gradually increased, which is accompanied by certain velocity layering. Zhou et al. [13] have also concluded that the cross-sectional area can influence the suction performance of the mixing device, and that the best suction performance can be achieved with an area ratio between 1.0 and 1.6, but that the suction performance will gradually deteriorate when the area ratio exceeds 2.0.

Figure 6 shows that some velocity vectors of backflow suction chambers and local vortexes appear at the joint of the suction chamber and Venturi, which produces a certain influence on the phytosanitary dissolution performance of the chemical mixer. However, the velocity vectors are changed from a higher value to lower value with the increase of the suction chamber height, which is gradually attenuated, and finally changed into a completely downward velocity vector. It is obvious that the pesticide mother liquor can be smoothly sucked, since upward water flow has a smaller suction influence on the pesticide mother liquor.

3.2. Orthogonal Experiment Results and Analysis

3.2.1. Orthogonal Experiment Data

The CFD model of all experiment numbers in orthogonal experiment Table 2 undergoes numerical simulation calculation, and the performance evaluation indexes experiment data of the pesticide mixing apparatus under different experiment numbers is shown as Table 3.

Table 3. Orthogonal experiment data table.

Experiment Number	Evaluation Indexes		
	Lifting Height H_1/mm	Turbulent Kinetic Energy $k/m^2 \cdot s^{-2}$	Pressure Recovery Distance H_2/mm
1	0.07	302.45	24.97
2	0.06	279.09	24.81
3	1.88	239.12	17.62
4	0.05	292.47	24.68
5	0.24	266.51	20.57
6	0.20	239.33	20.58
7	4.84	287.50	15.02
8	0.05	316.23	25.72
9	0.05	315.28	25.63
10	3.58	281.30	15.21
11	5.27	285.64	14.83
12	3.05	266.88	16.16
13	0.10	299.32	24.70
14	2.41	265.10	16.30
15	0.17	228.03	22.20
16	0.13	270.18	21.96
17	0.17	290.23	21.94
18	0.44	272.16	19.61
19	0.08	249.28	24.43
20	0.38	242.90	19.56
21	0.26	232.11	20.84
22	3.36	265.35	15.98
23	0.05	330.57	25.81
24	0.42	244.15	19.87
25	0.93	233.11	17.71
26	2.20	240.78	17.47
27	0.08	267.00	24.31

3.2.2. Analysis on Orthogonal Experiment Variance

The orthogonal experiment in the paper contains three evaluation indexes, which belong to the multi-index orthogonal experiment. Each evaluation index undergoes single-index variance analysis, respectively, in order to discover the main influence factors and comprehensive optimal pesticide mixing apparatus structure parameters; the optimal level of each evaluation index is determined. Next, analysis results of all evaluation indexes are comprehensively considered and analyzed according to the experiment purpose; the comprehensive optimal level is finally determined, and comprehensive optimal pesticide mixing apparatus structure parameters are discovered. Results data of all variance analysis and all factor level data mean values calculated by SPSS data statistical analysis software are shown in Tables 4–10.

Table 4. Lifting height variance analysis.

Variance Source	Quadratic Sum	DOF	Mean Square Error	F	*p* Value
A	0.891	2	0.446	2.549	0.158
B	5.115	2	2.557	14.628	0.005
C	50.335	2	25.168	143.952	0.000
D	0.843	2	0.421	2.410	0.171
A*B	0.060	4	0.015	0.085	0.984
B*C	7.611	4	1.903	10.883	0.006
B*D	0.048	4	0.012	0.068	0.989
error	1.049	6	0.175		
Total	65.951	26			

Table 5. All principal factor level data mean values of lifting height.

Level	Factor H_1/mm			
	A	B	C	D
1	0.884	0.626	3.058	0.888
2	1.189	1.078	0.268	1.200
3	1.318	1.688	0.066	1.303

A = constricted tube falloff angle α; B = diffusion tube divergence angle β; C = Venturi diameter d; D = Venturi length L.

Table 6. Variance analysis of turbulent kinetic energy.

Variance Source	Quadratic Sum	DOF	Mean Square Error	F	*p* Value
A	1875.760	2	937.880	8.731	0.017
B	167.241	2	83.620	0.778	0.501
C	8244.767	2	4122.383	38.376	0.000
D	1913.082	2	956.541	8.905	0.016
A*B	107.812	4	26.953	0.251	0.899
B*C	7260.973	4	1815.243	16.898	0.002
B*D	443.888	4	110.972	1.033	0.462
error	644.527	6	107.421		
Total	20,658.050	26			

Table 7. Data mean values matched with interaction factor B*C of turbulent kinetic energy, respectively.

Factor C Level	Factor B Level $k/m^2{\cdot}s^{-2}$		
	1	2	3
1	237.670	265.777	284.813
2	262.813	245.983	253.070
3	320.693	291.337	271.867

Table 8. All principal factor level data mean values of turbulent kinetic energy.

Level	Factor $k/\ m^2{\cdot}s^{-2}$			
	A	B	C	D
1	258.711	273.726	262.753	282.306
2	275.361	267.699	253.956	263.616
3	277.269	269.917	294.632	265.420

A = constricted tube falloff angle α; B = diffusion tube divergence angle β; C = Venturi diameter d; D = Venturi length L.

Table 9. Analysis of pressure recovery distance variance.

Variance Source	Quadratic Sum	DOF	Mean Square Error	F	*p* Value
A	0.246	2	0.123	16.311	0.004
B	19.325	2	9.663	1282.019	0.000
C	344.774	2	172.387	22,872.018	0.000
D	0.126	2	0.063	8.380	0.018
A*B	0.004	4	0.001	0.122	0.969
B*C	1.558	4	0.389	51.676	0.000
B*D	0.014	4	0.004	0.474	0.755
error	0.045	6	0.008		
Total	366.093	26			

Table 10. All principal factor level data mean values of pressure recovery distance.

Level Levels	Factor H_2/mm			
	A	B	C	D
1	20.788	21.784	16.256	20.758
2	20.709	20.543	20.792	20.703
3	20.558	19.727	25.007	20.593

A = constricted tube falloff angle α; B = diffusion tube divergence angle β; C = Venturi diameter d; D = Venturi length L.

Variance analysis F test in Table 4 shows that *p* value of factor B, factor C, and interaction factor B*C is smaller than 0.05, and *p* value of factor A, factor D, interaction factor A*B, and interaction factor B*D is larger than 0.05. Therefore, factor B, factor C, and interaction factor B*C have significant influence on lifting height evaluation indexes at significance level $\alpha = 0.05$, and factor A, factor D, interaction factor A*B, and interaction factor B*D have insignificant influence difference on lifting height evaluation indexes. The primary and secondary influence sequences of all factors on lifting height can be discovered, namely C > B > B*C > A > D > A*B > B*D. Table 5 shows that the optimal level of lifting height evaluation indexes is $A_1B_1C_3D_1$, namely, the constricted tube falloff angle is 20°, diffusion tube divergence angle is 7°, Venturi diameter is 3 mm, and Venturi length is 6 mm.

The variance analysis F test in Table 6 shows that *p* value of factor A, factor C, factor D, and interaction factor B*C is smaller than 0.05, and the *p* value of factor B, interaction factor A*B, and interaction factor B*D is larger than 0.05. Therefore, factor A, factor C, factor D, and interaction factor B*C have significant influence on turbulent kinetic energy evaluation indexes at the significance level $\alpha = 0.05$, and factor B, interaction factor A*B, and interaction factor B*D have insignificant influence difference on turbulent kinetic energy evaluation indexes. The primary and secondary influence sequences of all factors on turbulent kinetic energy can be discovered, namely C > B*C > D>A > B*D > B>A*B. Since the diffusion tube divergence angle β is matched with Venturi diameter d, the interaction thereof has significantly greater influence on turbulent kinetic energy evaluation indexes than that on principal factor diffusion tube divergence angle β. Therefore, the optimal level of two structure parameters of diffusion tube divergence angle β and Venturi diameter d should be determined by their matched interaction optimal level. Tables 7 and 8 show that the optimal level of turbulent kinetic energy evaluation indexes is $A_3B_1C_3D_1$, namely, the constricted tube falloff angle is 22°, diffusion tube divergence angle is 7°, Venturi diameter is 3 mm, and Venturi length is 6 mm.

Variance analysis F test of Table 9 shows that the *p* value of factor A, factor B, factor C, factor D, and interaction factor B*C is smaller than 0.05, and the *p* value of interaction factor A*B and interaction factor B*D is larger than 0.05. Therefore, factor A, factor B, factor C, factor D, and interaction factor B*C have significant influence on pressure recovery distance evaluation indexes at significance level $\alpha = 0.05$, and interaction factor A*B and interaction factor B*D have insignificant influence difference on pressure recovery distance evaluation indexes. The primary and secondary sequences of all factors on pressure recovery distance is discovered, namely C > B > B*C > A>D > B*D > A*B. Table 10 shows that optimal level of pressure recovery distance evaluation indexes is $A_3B_3C_1D_3$, namely, constricted tube falloff angle is 22°, diffusion tube divergence angle is 9°, Venturi diameter is 2 mm, and Venturi length is 10 mm.

3.2.3. Determination of Comprehensive Optimal Level of Chemical Mixer Structure Parameters

When the influence of factor on indexes is analyzed, if single-index is available only under general conditions, interaction among factors is not considered, the optimal levels of all factors are combined, and it is regarded as the optimal optimization grouping. However, three evaluation indexes are considered for orthogonal experiment in the paper, and the experiment is thus a multi-index orthogonal experiment. The experiment results of the three evaluation indexes should be considered, respectively, and the analysis results of all evaluation indexes also should be considered and analyzed comprehensively according to experiment objectives, so that the comprehensive optimal level is finally determined. The orthogonal experiment in the paper aims at improving the pesticide mixing performance as much as possible, whilst guaranteeing that the pesticide dissolution performance is acceptable. Therefore, two evaluation indexes of turbulent kinetic energy and pressure recovery distance should be comprehensively considered first; it should be ensured that lifting height evaluation indexes are not too high, and that they are acceptable. The following are set: lifting height evaluation indexes account for 20%, turbulent kinetic energy evaluation indexes account for 40%, and pressure recovery distance evaluation indexes account for 40%. Finally, the determined comprehensive optimal level is $A_3B_3C_1D_1$, and through comprehensive consideration and analysis according to weight, the constricted tube falloff angle is 22°, diffusion tube divergence angle is 9°, Venturi diameter is 2 mm, and Venturi length is 6 mm.

3.2.4. Verification of Chemical Mixer Structure Parameter Comprehensive Optimal Level

Finally, the determined comprehensive optimal level combination is arranged in the orthogonal experiment of the paper; the combination should undergo a verification experiment, thereby proving that it is the comprehensive optimal level of the orthogonal experiment in the paper. The pesticide mixing apparatus internal flow channel simulation value model of combination structure parameters undergoes numerical simulation calculation analysis, and it is concluded that the lifting height is 3.79 mm, turbulent kinetic energy is 299.42 $m^2{\cdot}s^{-2}$, and pressure recovery distance is 14.93 mm; the experiment number is recorded as experiment 28. In order to prove that the determined comprehensive optimal level combination adheres to the comprehensive optimal pesticide mixing apparatus structure parameters, it is necessary to comprehensively score and calculate 27 groups of experiment results and the experiment results of experiment No. 28 in Table 3, according to the determined weight. When the comprehensive score of experiment No. 28 is the highest, it is obvious that the determined comprehensive optimal level combination adheres to the comprehensive optimal pesticide mixing apparatus structure parameters.

Since all evaluation indexes have great magnitude order difference, it is necessary to first normalize primary data of all evaluation indexes; the result values can be mapped in the interval [0, 1]. Min-max dispersed standardized method is adopted in the paper for normalizing primary data of all evaluation indexes, and the computational formula is shown in Formulas (3)–(5).

$$x^* = \frac{x - \min_x}{\max_x - \min_x} \tag{3}$$

In the formula: x^* is the value of lifting height evaluation index primary data after normalization, x is the lifting height evaluation index primary data in mm, $\min_x$ is the minimum value of lifting height evaluation indexes data in mm, $\max_x$ is the maximum value of lifting height evaluation indexes data in mm.

$$y^* = \frac{y - \min_y}{\max_y - \min_y} \tag{4}$$

In the formula: y^* is the value of turbulent kinetic energy evaluation index primary data after normalization, y is the turbulent kinetic energy evaluation index primary data in $m^2{\cdot}s^{-2}$, $\min_y$ is the minimum value of turbulent kinetic energy evaluation indexes data in $m^2{\cdot}s^{-2}$, $\max_y$ is the maximum value of turbulent kinetic energy evaluation indexes data in $m^2{\cdot}s^{-2}$.

$$z^* = \frac{z - \min_z}{\max_z - \min_z} \tag{5}$$

In the formula: z^* is the value of pressure recovery distance evaluation index primary data after normalization, z is the pressure recovery distance evaluation index primary data in mm, $\min_z$ is the minimum value of pressure recovery distance evaluation indexes data in mm, max_z is the maximum value of pressure recovery distance evaluation indexes data in mm.

Since the results of two evaluation indexes of lifting height and pressure recovery distance are smaller and better, results of turbulent kinetic energy evaluation indexes are higher and better. In order to facilitate subsequent correct and comprehensive scoring calculation and comparison, number 1 is respectively subtracted by the value of two evaluation indexes of lifting height or pressure recovery distance after primary data normalization, and these values are respectively regarded as the two evaluation index final normalization results; the computational formula is shown in Formulas (6) and (7). Formula (4) is still adopted for turbulent kinetic energy evaluation index final normalization results.

$$x' = 1 - x^* \tag{6}$$

In the formula: x' is the lifting height evaluation index final normalization results, x^* is the normalization result of lifting height evaluation indexes calculated through Formula (3).

$$z' = 1 - z^* \tag{7}$$

In the formula: z' is the pressure recovery distance evaluation index final normalization results, z^* is the normalization result of pressure recovery distance evaluation indexes calculated through Formula (5).

The comprehensive score concrete computational formula can be obtained as shown in Formula (8) according to the above-mentioned proportion of all evaluation indexes:

$$cs = 0.2x' + 0.4y^* + 0.4z' \tag{8}$$

In the formula: cs is the comprehensive score value of the calculation, x' is the final normalization result of the lifting height evaluation index, y^* is the final normalization result of the turbulent kinetic energy evaluation index, z' is the index final normalization result of the pressure recovery distance evaluation.

In Formulas (3)–(8): primary data of all evaluation indexes are normalized, and the comprehensive score is finally calculated. The obtained evaluation index final normalization results and comprehensive score results of the pesticide mixing apparatus under different experiment numbers are shown in Table 11, and the combination with the highest comprehensive score is the comprehensive optimal combination.

Table 11 shows that the comprehensive score of experiment No. 28 is prominently higher than the comprehensive scores of the remaining 27 groups of experiment combinations in Table 2. Therefore, it can be determined that it belongs to the comprehensive optimal level of orthogonal experiment in the paper.

According to the research, the best value of the contraction angle of the classical venturi structure is in the range of 20°~22° [41,42], and the optimal parameter of the contraction angle determined in the paper is 22°. With the increase of the contraction angle, the effective length of the contraction tube gets shorter, which can improve the mixing uniformity. The best value of the diffusion angle is in the range of 7°~14° [41], and the diffusion angle of 9° in the paper can achieve the uniformity of the pesticide mixing. Combined with related studies [43,44], it is found that the optimal structural parameters determined in the thesis are in line with objective laws, and the number of tests can be significantly reduced through orthogonal tests.

Table 11. Normalization results and comprehensive score results.

Experiment Number	Evaluation Indexes			Comprehensive Score
	Lifting Height H_1/mm	Turbulent Kinetic Energy $k/m^2 \cdot s^{-2}$	Pressure Recovery Distance H_2/mm	
1	0.996	0.726	0.077	0.520
2	0.998	0.498	0.091	0.435
3	0.649	0.108	0.746	0.472
4	1.000	0.628	0.103	0.493
5	0.964	0.375	0.477	0.534
6	0.971	0.110	0.476	0.429
7	0.082	0.580	0.983	0.642
8	1.000	0.860	0.008	0.547
9	1.000	0.851	0.016	0.547
10	0.324	0.520	0.965	0.659
11	0.000	0.562	1.000	0.625
12	0.425	0.379	0.879	0.588
13	0.990	0.695	0.101	0.517
14	0.548	0.362	0.866	0.601
15	0.977	0.000	0.329	0.327
16	0.985	0.411	0.351	0.502
17	0.977	0.607	0.352	0.579
18	0.925	0.430	0.565	0.583
19	0.994	0.207	0.126	0.332
20	0.937	0.145	0.569	0.473
21	0.960	0.040	0.453	0.389
22	0.366	0.364	0.895	0.577
23	1.000	1.000	0.000	0.600
24	0.929	0.157	0.541	0.465
25	0.831	0.050	0.738	0.481
26	0.588	0.124	0.760	0.471
27	0.994	0.380	0.137	0.406
28	0.284	0.696	0.991	0.732

4. Conclusions

An analysis method based on CFD orthogonal experiment is adopted in the paper, thereby optimizing pesticide mixing apparatus structure parameters, effectively saving experiment costs, and reducing the need for experiment frequency. The following conclusions are made through numerical simulation calculation and orthogonal experiment variance analysis:

1. Single-index variance analysis results show that diffusion tube divergence angle β and Venturi diameter d are the two structure parameters with the most important influence among the three experiment evaluation indexes, and the constricted tube falloff angle α and Venturi length L are two structure parameters with secondary influence. When it is only considered that turbulent kinetic energy is the evaluation index, interaction influence of diffusion tube divergence angle β and Venturi diameter d plays the decisive role, and the influence of interactions under other circumstances are from respective principal factors.
2. Four structure parameters of pesticide mixing apparatus have different optimal values under different evaluation indexes. Optimal pesticide mixing apparatus structure parameters include the following when aiming at lifting height evaluation indexes: constricted tube falloff angle is 20°, diffusion tube divergence angle is 7°, Venturi diameter is 3 mm, and Venturi length is 6 mm; optimal pesticide mixing apparatus structure parameters include the following when aiming at turbulent kinetic energy evaluation indexes: constricted tube falloff angle is 22°, diffusion tube divergence angle is 7°, Venturi diameter is 3 mm, Venturi length is 6 mm; optimal pesticide mixing apparatus structure parameters include the following when aiming at pressure

recovery distance evaluation indexes: constricted tube falloff angle is 22°, diffusion tube divergence angle is 9°, Venturi diameter is 2 mm, and Venturi length is 10 mm.

3. Analysis results of all evaluation indexes are comprehensively considered to finally conclude the following comprehensive optimal pesticide mixing apparatus structure parameters: constricted tube falloff angle is 22°, diffusion tube divergence angle is 9°, Venturi diameter is 2 mm, and Venturi length is 6 mm.

Author Contributions: Conceptualization, D.S., Z.L. and S.S.; methodology, D.S. and W.L.; software, X.Z.; validation, D.S., W.L. and Z.L; formal analysis, W.L. and Z.L.; investigation, D.S. and W.L.; resources, D.S., Q.D., X.X. and S.S.; data curation, W.L. and Z.L.; writing—original draft preparation, D.S., W.L. and Z.L.; writing—review and editing, D.S., W.L. and X.Z.; visualization, D.S.; supervision, D.S., Q.D., X.X. and S.S.; project administration, S.S.; funding acquisition, S.S. All authors have read and agreed to the published version of the manuscript.

Funding: This research was funded by National Natural Science Foundation of China, grant No. 31671591. It was also partly supported by the Guangdong Provincial Special Fund For Modern Agriculture Industry Technology Innovation Teams, grant No. 2019KJ108, Special Fund for 2018 Rural Revitalization Strategy of Agricultural Department of Guangdong Provincial (Yue Cai Nong [2018] No.125), Guangdong Provincial Science and Technology Plan Project, grant No.2017A020208049, China Agriculture Research System (No.CARS-27), and Guangzhou Science and Technology Plan Project, grant No.202002030245.

Data Availability Statement: The data can be requested from the corresponding authors.

Acknowledgments: We appreciate all the authors for their technical assistance in this study. We are thankful to the teachers of South China Agricultural University for the supply of the server.

Conflicts of Interest: The authors declare no conflict of interest.

References

1. Xu, Y.; Wang, X.; Zheng, J.; Zhou, F. Simulation and experiment on agricultural chemical mixing process for direct injection system based on CFD. *Trans. Chin. Soc. Agric. Eng.* **2010**, *26*, 148–152.
2. Li, J.; Jia, W.; Wei, X. On-line mixing pesticide device based on flow control valve and neural network. *Trans. Chin. Soc. Agric. Mach.* **2014**, *45*, 98–103.
3. Slaughter, D.; Giles, D.; Tauzer, C. Precision offset spray system for roadway shoulder weed control. *J. Transp. Eng.* **1999**, *125*, 364–371. [CrossRef]
4. Gillis, K.; Giles, D.; Slaughter, D.; Downey, D. Injection mixing system for boomless, target–activated herbicide spraying. *Trans. ASAE* **2003**, *46*, 997–1008. [CrossRef]
5. Liu, Z.; Xu, H.; Hong, T.; Zhang, W.; Zhu, Y.; Zhang, K. Key technology of variable-rate spraying system of online mixing pesticide. *Trans. Chin. Soc. Agric. Mach.* **2009**, *40*, 93–129.
6. He, P.; Wu, C.; Chen, C.; Li, X.; Wang, G. Experimental Investigation on Safe Mixing Apparatus. *China Saf. Sci. J.* **2001**, *4*, 38–42.
7. He, P.; Wu, C.; Chen, C.; Wang, G.; Li, X. Performance of a New Type Spraying mixing Apparatus. *Trans. Chin. Soc. Agric. Mach.* **2001**, *3*, 44–47.
8. Peijie, H.; Cuiying, C.; Chundu, W.; Xuehui, L. Numerical computation of mixing pipe flow-field in jet mixing apparatus pesticide mixture. *J. Jiangsu Univ. Nat. Sci. Ed.* **2002**, *23*, 13–16.
9. Li, Y.; Wu, C.; Fu, X. Experimental study on two-stage jet mixing apparatus. *Trans. Chin. Soc. Agric. Eng.* **2008**, *1*, 172–174.
10. Qiu, B.; Xu, X.; Deng, B.; Yang, N.; Jiang, G.; Wu, C. Effect of area ratio on mixing homogeneity in jet-mixing apparatus. *Trans. Chin. Soc. Agric. Mach.* **2011**, *42*, 95–100.
11. Qiu, B.; Xu, X.; Yang, N.; Deng, B.; Jiang, G.; Wu, C. Simulation analysis of structure parameters of jet-mixing apparatus on jet-mixing performance. *Trans. Chin. Soc. Agric. Mach.* **2011**, *42*, 70–69.
12. Chen, Z.; Zhu, S.; Qiu, B. Online jet mixing control system of pesticide concentration. *J. Drain. Irrig. Mach. Eng.* **2012**, *30*, 463–468.
13. Zhou, L.; Fu, X.; Xue, X. Design and experiment of jet mixing apparatus based on CFD. *Trans. Chin. Soc. Agric. Mach.* **2013**, *44*, 107–112.
14. Ou, M.; Jia, W.; Qiu, B.; Guan, X.; Sheng, Y. Experiment and numerical analysis of flow field in jet mixing device under variable working conditions. *Trans. Chin. Soc. Agric. Mach.* **2015**, *46*, 107–111.
15. Zhang, J.F.; Pang, X.B.; Liu, H.; Zhang, T.J.; Guo, X.; Zeng, F.C. Study on Proportional Control Experimental Platform of Online Jet Mixing Drug. *Hubei Agric. Sci.* **2016**, *55*, 2645–2648.
16. Ghate, S.R.; Phatak, S. A compressed air direct injection pesticide sprayer. *Appl. Eng. Agric.* **1991**, *7*, 158–162. [CrossRef]
17. Lin, M.; Zhao, G. Analysis for the security technology of chemical application in the plant protection mechinery from aborad. *Trans. Chin. Soc. Agric. Mach.* **1996**, *S1*, 153–158.

18. Koo, Y.M.; Sumner, H. Total flow control for a direct injection sprayer. *Appl. Eng. Agric.* **1998**, *14*, 363–367. [CrossRef]
19. Steward, B.L.; Humburg, D. Modeling the raven scs-700 chemical injection system with carrier control with sprayer simulation. *Trans. ASAE* **2000**, *43*, 231–245. [CrossRef]
20. Aissaoui, A.E.; Lebeau, F.; Destain, M.F. Development of an optical sensor to measure direct injection spraying system performance. *VDI-Ber.* **2007**, *2001*, 243–252.
21. Luck, J.D.; Shearer, S.A.; Luck, B.D.; Payne, F.A. Evaluation of a rhodamine-WT dye/glycerin mixture as a tracer for testing direct injection systems for agricultural sprayers. *Appl. Eng. Agric.* **2012**, *28*, 643–646. [CrossRef]
22. Felizardo, K.R.; Mercaldi, H.V.; Oliveira, V.A.; Cruvinel, P.E. Modeling and Predictive Control of a Variable-Rate Spraying System. In Proceedings of the 2013 8th EUROSIM Congress on Modelling and Simulation, Cardiff, UK, 10–13 September 2013; pp. 202–207.
23. Endalew, A.M.; Debaer, C.; Rutten, N.; Vercammen, J.; Delele, M.A.; Ramon, H.; Nicolaï, B.; Verboven, P. A new integrated CFD modelling approach towards air-assisted orchard spraying. Part I. Model development and effect of wind speed and direction on sprayer airflow. *Comput. Electron. Agric.* **2010**, *71*, 128–136. [CrossRef]
24. Wu, B. CFD simulation of gas mixing in anaerobic digesters. *Comput. Electron. Agric.* **2014**, *109*, 278–286. [CrossRef]
25. Salcedo, R.; Granell, R.; Palau, G.; Vallet, A.; Garcerá, C.; Chueca, P.; Moltó, E. Design and validation of a 2D CFD model of the airflow produced by an airblast sprayer during pesticide treatments of citrus. *Comput. Electron. Agric.* **2015**, *116*, 150–161. [CrossRef]
26. Zhou, L.; Zhou, L.; Xue, X.; Kong, W. Numerical analysis and test on cavitation of jet mixing apparatus. *Trans. Chin. Soc. Agric. Eng.* **2015**, *31*, 60–65.
27. Devarrewaere, W.; Heimbach, U.; Foqué, D.; Nicolai, B.; Nuyttens, D.; Verboven, P. Wind tunnel and CFD study of dust dispersion from pesticide-treated maize seed. *Comput. Electron. Agric.* **2016**, *128*, 27–33. [CrossRef]
28. Wang, H.; Shi, W.; Lu, W.; Zhou, L.; Wang, C. Optimization design of deep well pump based on latin square test. *Trans. Chin. Soc. Agric. Mach.* **2010**, *41*, 56–63.
29. Gao, X.; Shi, W.; Zhang, D.; Zhang, Q.; Fang, B. Optimization design and test of vortex pump based on CFD orthogonal test. *Trans. Chin. Soc. Agric. Mach.* **2014**, *45*, 101–106.
30. Ma, Y. Determination of Jet-type Pesticide Mixture Device. *J. Yangling Vocat. Tech. Coll.* **2014**, *13*, 31–33.
31. Sheng, Y.; Qiu, B.; Chen, J. Mixing Characteristic Experiment of Different Area Ratio of Jet-mixing Apparatus. *J. Agric. Mech. Res.* **2016**, *4*, 134–140.
32. Zhang, T.; Liang, J.; Xue, X. Numerical Simulation and Experimental Study of Internal Flow Field of the Jet Pesticide Mixer. *Chin. Agric. Mech.* **2011**, *03*, 63–67.
33. Lei, X.; Liao, Y.; Liao, Q. Simulation of seed motion in seed feeding device with DEM-CFD coupling approach for rapeseed and wheat. *Comput. Electron. Agric.* **2016**, *131*, 29–39. [CrossRef]
34. Song, H.; Xu, Y.; Zheng, J.; Zhu, H. Swirling jet mixture mechanism of fat-soluble pesticides and numerical simulation of mixer field. *Trans. Chin. Soc. Agric. Mach.* **2016**, *47*, 79–84.
35. Li, H.; Rong, L.; Zong, C.; Zhang, G. Assessing response surface methodology for modelling air distribution in an experimental pig room to improve air inlet design based on computational fluid dynamics. *Comput. Electron. Agric.* **2017**, *141*, 292–301. [CrossRef]
36. Dai, Q.; Hong, T.; Song, S.; Li, Z.; Chen, J. Influence of pressure and pore diameter on droplet parameters of hollow cone nozzle in pipeline spray. *Trans. Chin. Soc. Agric. Eng.* **2016**, *32*, 97–103.
37. Lee, I.-B.; Bitog, J.P.P.; Hong, S.-W.; Seo, I.-H.; Kwon, K.-S.; Bartzanas, T.; Kacira, M. The past, present and future of CFD for agro environmental applications. *Comput. Electron. Agric.* **2013**, *93*, 168–183. [CrossRef]
38. Norton, T.; Grant, J.; Fallon, R.; Sun, D.-W. Improving the representation of thermal boundary conditions of livestock during CFD modelling of the indoor environment. *Comput. Electron. Agric.* **2010**, *73*, 17–36. [CrossRef]
39. Li, H.; Rong, L.; Zhang, G. Study on convective heat transfer from pig models by CFD in a virtual wind tunnel. *Comput. Electron. Agric.* **2016**, *123*, 203–210. [CrossRef]
40. Qiu, B.; Xu, X. Contrast and analysis between 2D and 3D flow field of jet-mixing apparatus. *J. Drain. Irrig. Mach. Eng.* **2011**, *5*, 441–445.
41. Lu, H. *Theory and Application of Spraying Technology*; Wuhan University Press: Wuhan, China, 2004; ISBN 978-7-307-04074-8.
42. Sun, H.; Wang, J. *Design Manual for Flow Measurement Throttling Devices*, 2nd ed.; Chemical Industry Press: Beijing, China, 2006; Volume 21.
43. Song, H.; Xu, Y.; Zheng, J.; Dai, X. Structural parameters optimization of swirling jet mixer based on reducing effective length. *Trans. Chin. Soc. Agric. Eng.* **2018**, *34*, 62–69.
44. Song, H.; Xu, Y.; Zheng, J.; Wang, X.; Zhang, M. Structural analysis and mixing uniformity experiments of swirling jet mixer for applying fat-soluble pesticides. *Trans. Chin. Soc. Agric. Eng.* **2016**, *32*, 86–92.

agronomy

MDPI

Article

Development and Application of an Intelligent Plant Protection Monitoring System

Shubo Wang [1,2,3], Peng Qi [1,2,3], Wei Zhang [4] and Xiongkui He [1,2,3,*]

1 Centre for Chemicals Application Technology, China Agricultural University, Beijing 100193, China; wangshubo@cau.edu.cn (S.W.); qp@cau.edu.cn (P.Q.)
2 College of Agricultural Unmanned System, China Agricultural University, Beijing 100193, China
3 College of Science, China Agricultural University, Beijing 100193, China
4 Anhui Zhongke Intelligent Perception Industrial Technology Research Institute Co., Ltd., Wuhu 241000, China; zhangw@zkzngz.com
* Correspondence: xiongkui@cau.edu.cn

Abstract: Facing the need of modern agriculture to accurately grasp the information of farmland diseases and pests, this paper proposes an intelligent plant protection system. The system is composed of a wireless lens, temperature and humidity sensor, intelligent information terminal, and probe rod to realize the collection of plant images and meteorological information. At the same time, a software based on the mobile terminal and the computer terminal was developed. The plant images and meteorological data are transmitted to the server through Wi-Fi transmission. Combined with the expert knowledge model, a solution is generated, and the user can identify the current diseases and pests and obtain solutions at any time. The system can remotely and automatically monitor and warn of mainstream diseases and pests of field crops such as rice and wheat and provide support for fine plant protection management.

Keywords: farmland information; disease and pest identification; temperature and humidity acquisition; mobile monitoring

Citation: Wang, S.; Qi, P.; Zhang, W.; He, X. Development and Application of an Intelligent Plant Protection Monitoring System. *Agronomy* **2022**, *12*, 1046. https://doi.org/10.3390/agronomy12051046

Academic Editor: Roberto Marani

Received: 12 April 2022
Accepted: 26 April 2022
Published: 27 April 2022

1. Introduction

Diseases, pests, and weeds are the great enemies of crop growth. They occur in the entire growth period of crops, which can cause a large reduction in crop production [1,2]. The application of chemical pesticides can greatly reduce agricultural losses, but the "three causes" caused by the use of chemical pesticides (referring to the mutagenic, carcinogenic, and teratogenic effects of pesticides on higher animals), pesticide residues, environmental pollution, and other negative effects are becoming increasingly prominent. In addition to experienced experts, for non-professional researchers, especially farmers, identifying crop diseases through picture comparison or text description often leads to human judgment errors [3,4]. In this way, it is difficult to accurately and timely apply pesticides, resulting in a large reduction in crop production. More and more signs also show that the contradiction between the increasing demand for pest identification and the relatively few plant protection experts has become increasingly serious [5]. It is of great significance to find new ways to solve the problem of diseases and pests identification.

With the rapid development of digital image processing, the increasingly wide application of computer vision, and the maturity of various pattern recognition technologies, the purpose of intelligent recognition can be achieved by processing the images of crop diseases, pests, and weeds and extracting characteristic parameters. Many scholars have studied the intelligent perception of diseases, pests, and weeds. In terms of methodology, machine learning models for agricultural monitoring are widely developed and used at this stage [6,7]. Nagasubramanian et al. [8] developed an analytical statistics model for plant growth and disease patterns based on a convolutional neural network (CNN) and support vector machine (SVM). This framework supported irrigation, nutrition planning, and environmental compliance. Albanese et al. [9] presented an embedded system enhanced with ML functionalities, ensuring the continuous detection of pest infestation inside fruit orchards. In their work, three different machine learning (ML) algorithms had been trained and deployed. Mishra et al. [10] proposed a multi-node plant disease-monitoring strategy, collected plant images through nodes, and used the proposed sine cosine algorithm based on the rider neural network to identify diseases. The large-scale monitoring of farmland could be realized by using a multi-node method. Liu et al. [11] proposed a novel, deep learning-based automatic approach using hybrid and local activated features for pest monitoring. The authors tested their algorithm in seven years of large-scale pest data set containing 88,600 images (16 types of pests) with 582,100 pests and achieved good results.

In terms of hardware, there are ground-monitoring platforms and air-monitoring platforms [12–15]. For ground-monitoring platforms, Trilles et al. [16] developed a low-cost sensorized platform, capable of monitoring meteorological phenomena following the Internet of Things paradigm, with the goal to apply an alert disease model on the cultivation of the vine. However, the platform could only carry out meteorological monitoring, and it was difficult to support the implementation of intelligent agriculture due to insufficient information. Methun et al. [17] proposed an efficient carrot disease identification and classification method using a deep learning approach, especially VGG16, VGG19, MobileNet, and Inception v3. The accuracy achieved by Inception v3 was 97.4%. Udutalapally et al. [18] developed a disease prediction system disease prediction by trained a convolutional neural network (CNN) model. Considering the power problem of field environment, the solar sensor node was used for power supply. Khattab et al. [19] collected the environmental and soil information from a wireless sensor network installed in the planted area. It also provided an expert system that sent warning messages to the users before the outbreak of a disease. Mekala et al. [20] proposed a cloud-enabled CLAY-MIST measurement (CMM) index based on temperature and relative humidity to assess the comfort levels of a crop.

For an air-monitoring platform, He et al. [21] used multiangle remote sensing and hyperspectral remote sensing to monitor wheat powdery mildew (WPM). Dong et al. [22] developed an automatic system based on the Web GIS platform. At the same time, the author used it to predict wheat stripe rust and East Asian migratory locust and achieved good accuracy. Kim et al. [23] proposed an onion field monitoring system which composed of a Pan Tilt Zoom (PTZ) camera, a motor system, wireless translator, and image-logging module. Through pixel-level classification and location, six categories of disease recognition were realized.

In addition to ground systems and air systems, it is also a trend to integrate them for collaborative monitoring. Gao et al. [24] realized the joint detection of environmental information by ground sensors and UAVs. IoT and UAV could monitor the incidence of crop diseases and pests from the ground micro and air macro perspectives, respectively. Zhang et al. [25] proposed to use high-resolution remote sensing data to identify infected trees for controlling pine wilt disease (PWD). By processing the captured spectral, temporal, and spatial features, this paper effectively distinguished the withered pine from other easily confused objects.

In view of the above research status, as shown in the Figure 1, this paper has developed a variety of crop diseases and pests monitoring systems, taking into account meteorological conditions, plant images, and expert systems. The lower half of Figure 1 shows the acquisition of images and meteorological data. The wireless lens collects the crop images, and the Bluetooth temperature and humidity sensor collects the environmental information in real time. The upper half of Figure 1 shows the data upload and discrimination results. The obtained images are uploaded to the cloud expert system to realize the classification of diseases and pests. Additionally, the expert system can release prevention and control suggestions for users. The main contributions of this work are:

(1) A complete set of monitoring systems is constructed, including the software and hardware of the system.

(2) The system can collect meteorological data and plant images at the same time and realize the identification and counting of pests and diseases based on Yolo v3. A timely solution can be given through the expert system.

The rest of the paper is organized as follows. Section 2 introduces the diseases and pests identification. In Section 3, the hardware system is developed and applied. In Section 4, software systems based on terminal and computer are developed. Section 5 describes the detailed acquisition method and representative results. Section 6 concludes the paper.

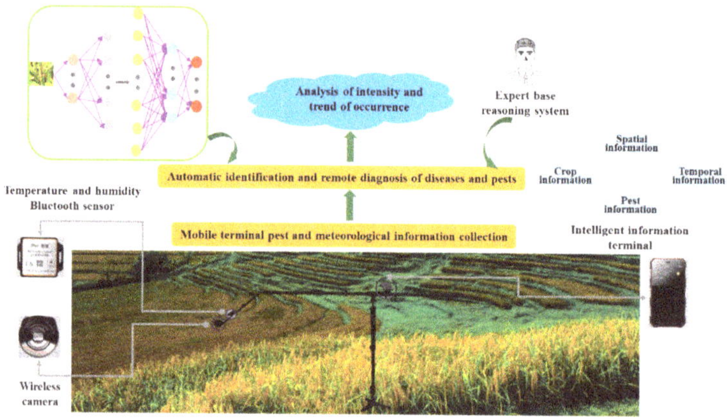

Figure 1. Overview of intelligent plant protection monitoring system.

2. Diseases and Pests Identification

2.1. Image Acquisition and Coding of Diseases and Pests in Macro Mode

As shown in Table 1, the names of diseases and pests are set to a number with a length of five. The combination of the first and second bits indicates the crop type; the third bit indicates diseases or pests; and the combination of the fourth and fifth bits indicates the specific kinds of diseases and pests.

Table 1. Coding table of diseases and pests with different crops.

Crop	Pests and Diseases	Specific Diseases and Pests	Coding
11: Wheat	1: Pest	11: Wheat spider	11111
		21: Aphid, wheat long tube aphid or wheat binary aphid (green)	11121
		22: Aphid, rhopalosiphus (black)	11122
		23: Aphids (red)	11123
		31: Slime worm	11131
	2: Disease	11: Normal wheat ears	11211
		12: Scab	11212
		21: Powdery mildew	11221
		31: Wheat leaf rust	11231
		41: Wheat stripe rust	11241

Table 1. *Cont.*

12: Rice	1: Pest	11: Rice planthopper	12111
		21: Normal rice leaves	12121
		22: Rice-leaf roller	12122
		31: Striped rice borer	12131
	2: Disease	11: Sheath blight	12211
		21: Flax spot	12221
		31: Rice-false smut	12231
		41: Rice blast (leaf)	12241
		42: Rice blast (panicle)	12242
		51: Bacterial streak	12251
13: Rape	1: Pest	11: Aphid	13111
	2: Disease	11: Sclerotinia (rhizome)	13211
		12: Sclerotinia (leaf)	13212
		21: Downy mildew	13221

2.2. Identification Based on Neural Network

This paper presents an integrated framework of diseases and pests identification based on YOLO V3 [26]. The YOLO algorithm has been optimized and iterated, and it is better than Single Shot MultiBox Detector (SSD), Faster RCNN, and other algorithms in detection performance. In terms of network structure, YOLO V3 integrates YOLO V2, Darknet-19, and other new residual networks, composed of continuous 3×3 and 1×1 convolutional layers. The YOLO V3 network preprocesses the remote sensing image, scales it to 416×416, and sends it to the convolutional neural network for inference. Then, the non-maximum suppression (NMS) and classification recognition of the detection results are carried out according to the confidence of the network model. The YOLO V3 network divides the input image into $S \times S$ grids. If there is a target object in the center of a grid, this grid is responsible for predicting the object. Each grid predicts the four offset coordinates of the bounding box of the detected object and the confidence score. The YOLO V3 network makes predictions on three different scales, namely 13×13, 26×26, and 52×52. Three bounding boxes are predicted for each scale, and the network structure is shown in the Figure 2.

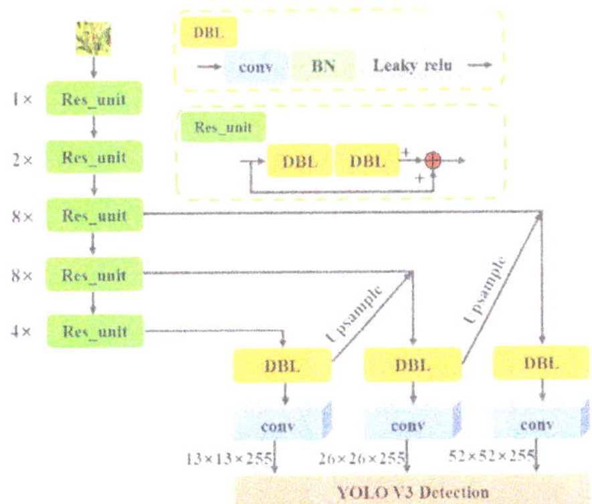

Figure 2. Structure diagram of Yolo V3. DBL: Darknetconv2d_BN_Leaky, BN: Batch Normalization, conv: convolution.

First, the input image is divided into 13 × 13 cells. Then, each cell will generate detection boxes, which are each composed of a five-dimensional prediction parameter, including the center point coordinate (x, y), width and height (w, h), and confidence score s_i. The confidence score is calculated as

$$s_i = P(C_i | O_{object}) \times P(O_{object}) \times IOU(k_T, k_p) \quad (1)$$

where $P(C_i | O_{object})$ represents the conditional probability of predicting class i in the cell when there are diseases and pests in the detection frame; and $P(O_{object})$ indicates the possibility of diseases and pests in the detection box of the current cell. If there are diseases and pests, the value is 1; otherwise, the value is 0. $IOU(k_T, k_p)$ is the intersection and union ratio of prediction detection frame and real annotation frame.

Then, the NMS algorithm is used to retain the target detection box with a high confidence score. The NMS algorithm formula is

$$s_i = \begin{cases} s_i & IOU(M, b_i) < N_i \\ 0 & IOU(M, b_i) \geq N_I \end{cases} \quad (2)$$

where M is the detection box with the highest confidence score in the current area; $IOU(M, b_i)$ is the intersection and union ratio of M and the adjacent overlapping frame b_i; N_I is the set overlap threshold.

In this paper, the loss function $CIOU$ is used to calculate the regression loss value of the prediction frame, and the formula is as follows:

$$Loss_{CIOU} = 1 - IOU + \frac{\rho^2(a, a^{gt})}{b^2} + \alpha v \quad (3)$$

$$v = \frac{4}{\pi^2}(\arctan\frac{w^{gt}}{h^{gt}} - \arctan\frac{w}{h}) \quad (4)$$

$$\alpha = \frac{v}{1 - IOU + v} \quad (5)$$

where $\alpha, a^{gt}, w, w^{gt}, h, h^{gt}$ are the center position coordinates, width, and height of the prediction box and the dimension box, respectively; ρ represents the Euclidean distance between the centers of the two frames; b represents the diagonal length of the smallest rectangle containing the prediction box and dimension box; αv represents the penalty item including the width and height of the prediction box and the dimension box; α is the weight parameter; and v represents the calculation item including the width and height of the prediction box and dimension box.

After using the loss function $\alpha, a^{gt}, w, w^{gt}, h.h^{gt}$, the optimization direction can still be given when the overlapping area of the prediction box and the annotation box is equal, disjoint, or included. This makes the prediction box close to the center of the annotation box and speeds up the convergence speed of network training.

Farmers would use mobile phones and other terminal devices to take pictures or videos and send them to the server through the Wi-Fi. After obtaining the pictures, the server uses the identification network proposed in this section combined with the expert database system to return the diseases and pests information and solutions to farmers.

2.3. Data Reporting and Location Information Acquisition

In this section, Geographic Information System (GIS) technology is further used to establish a geographic information system database to display the occurrence degree, transmission, and geographical distribution of crop diseases, pests, and weeds in the form of a map. Through the collected meteorological, crop varieties, pest information, control measures, and other information, combined with the mathematical model in the later stage, the temporal and spatial dynamics and law of pest and grass occurrence are analyzed. Finally, the occurrence trend of diseases and pests are predicted by evaluating the environmental and influencing factors.

3. Hardware System Development and Application

As shown in Table 2, the product type is selected according to the characteristics of crop objects. The macro mode is suitable for photographing diseases and pests with a very small size; The handheld probe mode is suitable for shooting close to the photographer; The support frame mode is suitable for shooting in a far and high range from the photographer.

Table 2. Hardware development and composition under different modes.

Form / Pattern	Wireless Lens	Probe Rod Kit	Intelligent Information Terminal	Temperature and Humidity Sensor	Support Frame	Macro Lens
Macro mode	√		√	√		√
Handheld probe mode	√	√	√	√		
Support frame mode	√	√	√	√	√	

Note: The probe rod kit consists of a hand-held probe rod, a vientiane folding rod, a lens connecting seat, a temperature and humidity sensor fixing seat, and an intelligent information terminal-fixing frame.

3.1. Perception Module Construction

As shown in Figure 3, the hardware system mainly includes temperature and humidity Bluetooth sensor, intelligent information terminal, wireless lens, and other firmware.

Figure 3. Hardware system module.

The temperature and humidity sensor measures the temperature and humidity information, connects with the intelligent information terminal through Bluetooth, and transmits information to the terminal to realize the real-time environmental information recording of farmland. The system adopts a waterproof and dustproof grade not lower than IP65 and has good drop resistance. The temperature measurement range is −20 °C to 50 °C, and the humidity is 0–100%. The detection accuracy of the temperature and humidity sensors are: humidity ±0.5% and temperature ±0.5 °C.

The intelligent information terminal includes a microprocessor, data memory, satellite positioning module, Wi-Fi communication module, Bluetooth communication module, power module, and display module. The Android version is not less than 6.0, and the rear camera is not less than 10 million pixels. It supports 4G wireless communication network communication, and the storage memory is 16G. The required image of plant diseases and pests can be clearly collected through the camera and supporting wireless camera carried by the intelligent information terminal and can be transmitted to the network service background together with GPS information through a communication protocol for further processing. The image shooting supports real-time shooting and real-time viewing.

The wireless lens realizes the front-end HD image acquisition and transmits it to the intelligent information terminal for data processing through Wi Fi, which can adapt to the harsh working environment of high temperature and humidity. The camera is 10 million pixels, supports single autofocus, and can clearly shoot single diseases and pests with a size of 1 mm–10 cm.

The telescopic probe rod can be extended to a maximum length of 2.5 m to ensure that it can collect pest images and climate information within 2.5 m away from the operator. At the same time, the long probe rod can ensure that the collection range of pests cannot be disturbed by human beings, and the collection point can represent the characteristics of typical farmland.

3.2. Positioning and Communication Module

The system positioning information includes the time, longitude, latitude, and other positioning status information corresponding to the collected image. The hardware developed in this paper can store the positioning information in the terminal and upload it to the monitoring center through wireless communication.

System communication modes include communication between camera, sensor, and intelligent terminal and communication between intelligent terminal and back-end service platform. The camera and the intelligent information terminal can be connected together to transmit the collected images; the environmental sensor can be connected with the intelligent information terminal to transmit the collected environmental data; and the intelligent information terminal and the back-end service platform transmit data using a wireless network. The camera communicates with the intelligent information terminal through a Wi-Fi network; the environmental sensor communicates with the intelligent information terminal through a Bluetooth network; and the intelligent information terminal communicates with the network service platform through 3G and 4G networks. The terminal sends pictures, environmental data, and other relevant data to the network service platform for analysis and processing through TCP/IP protocol.

4. Software System Development

According to the different needs of users, the software system developments of two platforms are carried out in this work. The software can be obtained through the website (www.zkzngz.com, accessed on 20 April 2022). At this stage, it is only available in Chinese. The team is developing versions in English and other languages to provide users in different regions.

4.1. Development of Mobile Terminal Software System

The terminal adopts a Wi-Fi connection with a wireless lens and sensor, and the system interface form is shown in Figure 4a. By clicking the environmental sensor and camera that need to be linked, the link between the mobile phone and device can be realized so as to establish wireless transmission. In the system design, the intelligent identification of plant diseases and pests is mainly divided into four steps: wireless shooting, image uploading, intelligent identification, and system storage. In the process of wireless shooting, the plant phenotype can be obtained at a macro distance through the connected wireless lens and the operation of retracting and stretching the lens, as shown in Figure 4b. After shooting, the obtained plant image is recognized by jumping the network link. In the recognition process, one or more pictures can be recognized continuously, and crop type and growth period need to be selected at the same time. The software system can also record the history of identification. At the same time, the recognition result of the collected image will be sent back to the user, and the solution of this kind of pest will be given.

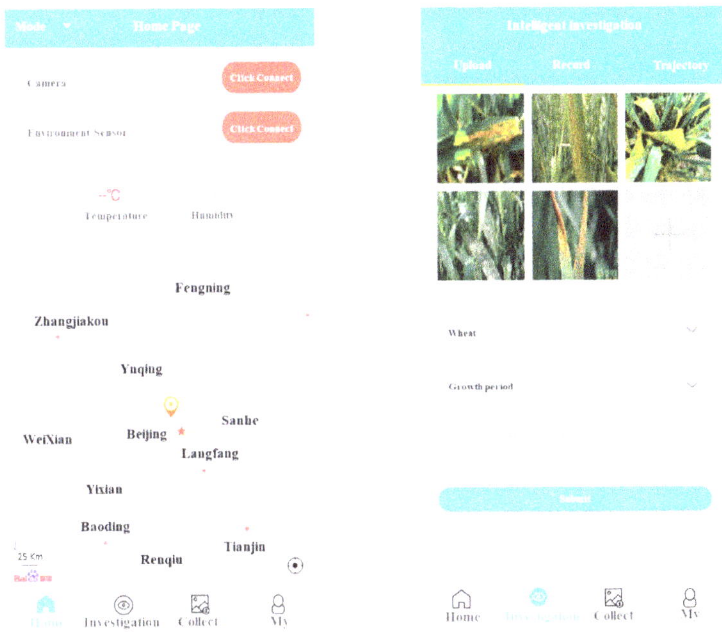

(a) (b)

Figure 4. Mobile terminal interface. (**a**) The main interface displays geographical location and meteorological data. (**b**) Image upload and selection of crops and growth period.

4.2. Development of Computer Terminal Software System

In order to record and analyze various diseases and pests in the field, it is necessary to set up multiple mobile monitoring systems in different regions and record and count the disease and pest information and environmental data obtained by the mobile monitoring system. The project has further developed the computer software system, which can accept and record the data of multiple mobile monitoring systems and display the regional location, intensity, and trend of diseases and pests.

5. Experiment and Discussion

5.1. Image Collection of Diseases and Pests

When using the mobile intelligent device for image acquisition, the user should see a relatively clear picture on the screen (i.e., the lens focusing is completed) and then click the shooting button to avoid dithering and fuzzy image data.

In the test experiments in this section, there were six kinds of common diseases and pests: rice planthopper, rice-leaf roller, wheat spider, wheat aphid, wheat scab, and sheath blight. For the area where each pest of rice or wheat gathers on the plant, the area where the disease occurs on the plant, and the significant characteristic area of different seedling conditions, the camera shall aim at the characteristic area for multi-angle shooting (such as side, top, middle, local, etc.), and the specific shooting standards are shown in Table 3.

Table 3. Collection mode under different crops with diseases and pests.

Type	Name	Viewfinder Range	Angle and Distance
Pest	Rice planthopper	The side is facing the aggregation area of rice planthopper above the rice root. The lens vertically penetrates into the cluster and overlooks the aggregation area of rice planthopper above the root.	Angle: 0~30°. Distance: 5~10 cm, 10~15 cm, mainly in a clear state with a distance of 5~10 cm.
	Rice-leaf roller	The viewfinder shall mainly focus on the side images of rice leaves, which can be distinguished by the naked eye.	Angle: 30~60°. Distance: 10~20 cm, 20~30 cm, mainly in a clear state with a distance of 20~30 cm.
	Wheat spider	The lens is close to the ground in the winter and perpendicular to the ground in the spring.	Angle: 0~90°. Distance: 5~10 cm, 10~15 cm, mainly in a clear state with a distance of 5~10 cm.
	Wheat aphid	The whole wheat plant shall be photographed before jointing, and the middle and upper parts of the wheat plant shall be photographed after booting.	Angle: 0~30°. Distance: 5~10 cm, 10~15 cm, mainly in a clear state with a distance of 5~10 cm.
	Others	Other pests, such as striped borer, leaf cicada, rice borer sandfly, and armyworm, can be photographed according to the actual situation of the investigation.	According to the survey characteristics, refer to the above form.
Disease	Sheath blight	The middle and lower part of rice is photographed (one side is used).	Angle: 0~30°. Distance: 10~20 cm, 20~30 cm, mainly in a clear state with a distance of 10~20 cm.
	Leaf rust	The viewfinder shall mainly focus on the side image of wheat ear (which can be distinguished by naked eyes).	Angle: 30~60°. Distance: 10~20 cm, 20~30 cm, mainly in a clear state with a distance of 20~30 cm.
	Others	Other diseases, such as bacterial stripe disease, rice false smut, and ear neck blast, can be photographed according to the actual situation of the investigation and with reference to the above types.	According to the survey characteristics, refer to the above form.
Seedling growth		According to the significant characteristic areas of different growth periods, the viewfinder should mainly focus on the side image of crop seedling.	Angle: 30~60°. Distance: 10~20 cm, 20~30 cm, mainly in a clear state with a distance of 20~30 cm.

5.2. Analysis of Acquisition Results

The six most common diseases and pests (rice planthopper, rice-leaf roller, rice sheath bright, wheat spider, wheat aphid, and wheat leaf rust) were selected for statistical analysis by different methods. The sample size of each kind of diseases and pests was 200, and the rice images were collected in Liyang, Jiangsu, China (119.67° E, 31.49° N); Huaiyuan, Anhui, China (117.04° E, 32.97° N); and Changsha, Hunan, China (113.33° E, 28.05° N). The wheat images were collected in Xuchang, Henan, China (113.77° E, 34.05° N); Baoding, HeBei, China (115.28° E, 38.89° N); and Linyi, Shandong, China (118.259° E, 35.19° N). The collection method is shown in Table 3. In the experiment, the average precision (*AP*) was used to evaluate the effects of different methods. The results are shown in Table 4. The *AP* is defined as

$$AP = \int_0^1 P(R)dR \tag{6}$$

$$R = \frac{TP}{TP + FN} \quad (7)$$

$$P = \frac{TP}{TP + FP} \quad (8)$$

where AP is the average detection accuracy, R is the recall rate, P is the accuracy, TP represents the positive samples predicted as positive, FN represents the positive samples predicted as negative, and FP represents the negative samples predicted as positive.

Table 4. Average precision (AP) of different methods on different diseases and pests.

	Rice Planthopper	Rice-Leaf Roller	Rice Sheath Blight	Wheat Spider	Wheat Aphid	Wheat Leaf Rust
Ours (Yolo V3)	83.32%	85.34%	82.03%	86.32%	85.53%	87.01%
Yolo V2 [27]	78.13%	80.78%	78.89%	81.32%	80.56%	83.65%
Faster RCNN [28]	75.52%	77.45%	76.76%	75.89%	73.20%	79.32%
SSD [29]	72.32%	75.32%	77.27%	74.12%	71.32%	74.78%

As shown in Table 4, the Yolo V3 method used in this paper had a higher accuracy than other detection methods for different diseases and pests. Yolo V3 had an average improvement of 10% compared with SSD in the detection of all diseases and pests; Yolo V3 had an average improvement of 5% compared with Yolo V2. The experimental statistical results show that Yolo V3 had a better perception and learning ability for diseases and pests images. The statistical methods and identification effects for different diseases and pests are as follows.

5.2.1. Rice Planthopper

The monitoring of rice planthopper should be conducted once every five days after the transplanting field returns to green. The direct seeding field starts from 30 days after rice sowing until the end of rice yellow maturity. During the experiment, two rows were chosen on each side of each field for a total of eight rows, one meter per row, and a total of eight meters. The front end of the camera of the mobile acquisition terminal went deep into the rice bush. The lenses were 5 cm–10 cm away from the rice base and aimed at the middle and lower part of the rice. One picture was taken at an interval of 20 cm, five pictures were taken in each line, and forty pictures were taken in each field. The number of shots per line can also be adjusted according to the actual situation. In principle, there is no remake or missed shot, or one shot per cluster. As shown in Figure 5, the server intelligently identifies the rice planthopper and counts it automatically.

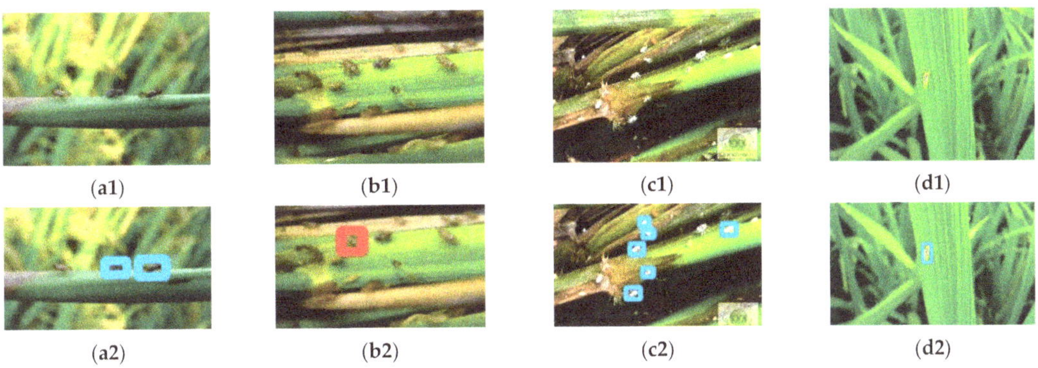

Figure 5. Rice planthopper. (**a1**), (**b1**), (**c1**), and (**d1**) are the original images; (**a2**), (**b2**), (**c2**), and (**d2**) are the recognition results.

5.2.2. Rice-Leaf Roller

The monitoring of the rice-leaf roller begins with adults in the field, driving moths, and taking photos day by day. Three representative paddy fields were selected and investigated. The researchers caught moths every morning, checked 10 m along the edge of each field, walked against the wind, slowly moved the rice plant with a pole, and counted the number of rice-leaf roller moths flying at the same time. The specific experimental operations were as follows: (1) Pest collection: one person drives moths, and the other takes photos immediately after them. When the person in front drives every square meter of rice, and the adults fly up, the person behind takes photos immediately. (2) Collection of leaf curl rate: five points are randomly sampled from each field; more than one third of the rice leaves are photographed at each point; and the damage status of the rice-leaf roller is photographed. The identification results are shown in Figure 6.

Figure 6. Rice-leaf roller. (**a1**), (**b1**), (**c1**), and (**d1**) are the original images; (**a2**), (**b2**), (**c2**), and (**d2**) are the recognition results.

5.2.3. Rice Sheath Blight

The monitoring of sheath blight began at the tillering stage of rice and was investigated every five days until the end of the milk ripening of the rice. Three representative types of fields were selected for the experiment. The four sides of each field were 2 m away from the side of the field. Two rows were taken along the ridge and rice-transplanting row. Each length was 1 m, and a total of 8 m was checked. The camera of the mobile acquisition terminal went deep into the rice bush; the lower end of the camera was 10–20 cm away from the rice base; and the camera was aimed at the middle and lower parts of the rice (one-side shooting was adopted). One photo was taken every 20 cm; five photos were taken in each line; and forty photos were taken in each field. The recognition results are shown in Figure 7.

Figure 7. Rice sheath blight. (**a1**), (**b1**), (**c1**), and (**d1**) are the original images; (**a2**), (**b2**), (**c2**), and (**d2**) are the recognition results.

5.2.4. Wheat Spider

Wheat spiders need to be monitored in the winter and spring. The collection was carried out every day from 2:00 p.m. to 3:00 p.m. in the winter and from 3:00 p.m. to 4:00 p.m. in the spring, once every five days. Three representative wheat fields of different types were selected, and the area of each field was not be less than 667 m^2. Two persons were required to cooperate in each data acquisition, one taking photos and one investigating data. Each field was photographed along a row of wheat in the middle of the field. The first photographing point was 2 m away from the field head. Later, 1 photo was taken every 2 m, and a total of 20 photos was taken. When taking pictures, the camera clung to the right ridge and took pictures to the left ridge. The lens was attached to the ground in the winter and 5 cm–10 cm away from the ground in the spring (the lens faces the wheat). When taking photos with mobile intelligent acquisition equipment, the lens was placed vertically. The collected pictures and recognition results are shown in Figure 8.

Figure 8. Wheat spider. (**a1**), (**b1**), (**c1**), and (**d1**) are the original images; (**a2**), (**b2**), (**c2**), and (**d2**) are the recognition results.

5.2.5. Wheat Aphid

The investigation was to be conducted once every seven days from the green jointing stage to the milk ripening stage of wheat. When the number of aphids rose sharply and the daily increase of aphids exceeded 300, the investigation was to be conducted once every five days. More than 10 representative wheat fields were selected according to sowing date, variety, growth, and other conditions. Samples were taken at five diagonal points of each field; 50 plants were investigated at each point before jointing; and 20 plants were investigated at each point after booting. The lenses were 5 cm–10 cm away from the wheat plant, close to the right ridge, and took photos to the left ridge. The whole wheat plant was to be photographed before jointing, and the middle and upper parts of the wheat plant were to be photographed after booting. The 5 photos were to be photographed for each field, and more than 50 photos were to be photographed for each survey. The collected pictures and recognition results are shown in Figure 9.

5.2.6. Wheat Leaf Rust

The investigation was conducted every five days from the jointing stage of wheat to the end of milk maturity of wheat. Each field was sampled in parallel at five points, and two pictures were taken at each point with an interval of 1 m for a total of 10 m. The lens of the equipment was aimed at more than one third of the wheat for photographing. The collected pictures and recognition results are shown in Figure 10.

Figure 9. Wheat aphid. (**a1**), (**b1**), (**c1**), and (**d1**) are the original images; (**a2**), (**b2**), (**c2**), and (**d2**) are the recognition results.

Figure 10. Wheat leaf rust. (**a1**), (**b1**), (**c1**), and (**d1**) are the original images; (**a2**), (**b2**), (**c2**), and (**d2**) are the recognition results.

6. Conclusions

The intelligent pest detection kit serves plant protection management through "sensing + artificial intelligence" technology to realize the accurate collection and mobile monitoring of plant information and data. The equipment has been loaded with more than 500 users in Qinghai, Jiangxi, Anhui, Zhejiang, Shanxi, and other places in China. The suite provides users with tools and means for plant protection image and microenvironment acquisition to help users implement more-refined plant protection management. Through artificial intelligence to identify diseases and pests of crops, the plant protection operation is easier, more accurate, and more efficient, so anyone can have a plant protection expert with him. At this stage, this project only identifies the main diseases and pests of wheat, corn, rice, and other major field crops and continues to increase the sample types to realize more crop disease and pest monitoring, which is the further research direction of intelligent agricultural monitoring.

Author Contributions: Conceptualization, S.W. and P.Q.; methodology, S.W.; software, S.W.; validation, S.W., P.Q., and X.H.; formal analysis, S.W.; investigation, S.W.; resources, W.Z.; data curation, W.Z.; writing—original draft preparation, S.W.; writing—review and editing, S.W.; visualization, S.W.; supervision, S.W.; project administration, X.H.; funding acquisition, X.H. All authors have read and agreed to the published version of the manuscript.

Funding: This research was funded by the Professor workstation of intelligent plant protection machinery and precision pesticide application technology, grant number 202005511011020; the

National Modern Agricultural Industrial Technology System of China (No. CARS-28-20); Deutsche Forschungsgemeinschaft (DFG, German Research Foundation)-328017493/GRK 2366.

Institutional Review Board Statement: Not applicable.

Informed Consent Statement: Not applicable.

Data Availability Statement: Not applicable.

Acknowledgments: We would like to thank Lijun Zhou and Jian Wang for their helpful work on their preliminary basic work.

Conflicts of Interest: The authors declare no conflict of interest.

References

1. Kaivosoja, J.; Hautsalo, J.; Heikkinen, J.; Hiltunen, L.; Ruuttunen, P.; Näsi, R.; Niemeläinen, O.; Lemsalu, M.; Honkavaara, E.; Salonen, J. Reference Measurements in Developing UAV Systems for Detecting Pests, Weeds, and Diseases. *Remote Sens.* **2021**, *13*, 1238. [CrossRef]
2. Duke, S.O. Interaction of chemical pesticides and their formulation ingredients with microbes associated with plants and plant pests. *J. Agric. Food Chem.* **2018**, *66*, 7553–7561. [CrossRef]
3. Kim, K.H.; Kabir, E.; Jahan, S.A. Exposure to pesticides and the associated human health effects. *Sci. Total Environ.* **2017**, *575*, 525–535. [CrossRef]
4. Manavalan, R. Automatic identification of diseases in grains crops through computational approaches: A review. *Comput. Electron. Agric.* **2020**, *178*, 105802. [CrossRef]
5. Li, L.; Zhang, S.; Wang, B. Plant disease detection and classification by deep learning—A review. *IEEE Access* **2021**, *9*, 56683–56698. [CrossRef]
6. Wang, S.; Han, Y.; Chen, J.; Zhang, K.; Zhang, Z.; Liu, X. Weed Density Extraction based on Few-shot Learning through UAV Remote Sensing RGB and Multi-spectral Images in Ecological Irrigation Area. *Front. Plant Sci.* **2022**, *12*, 735230. [CrossRef]
7. Waheed, A.; Goyal, M.; Gupta, D.; Khanna, A.; Hassanien, A.E.; Pandey, H.M. An optimized dense convolutional neural network model for disease recognition and classification in corn leaf. *Comput. Electron. Agric.* **2020**, *175*, 105456. [CrossRef]
8. Nagasubramanian, G.; Sakthivel, R.K.; Patan, R.; Sankayya, M.; Daneshmand, M.; Gandomi, A.H. Ensemble Classification and IoT Based Pattern Recognition for Crop Disease Monitoring System. *IEEE Internet Things J.* **2021**, *8*, 12847–12854. [CrossRef]
9. Albanese, A.; Nardello, M.; Brunelli, D. Automated Pest Detection with DNN on the Edge for Precision Agriculture. *IEEE J. Emerg. Sel. Top. Circuits Syst.* **2021**, *11*, 458–467. [CrossRef]
10. Mishra, M.; Choudhury, P.; Pati, B. Modified ride-NN optimizer for the IoT based plant disease detection. *J. Ambient. Intell. Humaniz. Comput.* **2021**, *12*, 691–703. [CrossRef]
11. Sudirman, S.; Liu, L.; Wang, R.; Xie, C.; Yang, P.; Wang, F.; Liu, W. Deep learning based automatic multi-class wild pest monitoring approach using hybrid global and local activated features. *IEEE Trans. Ind. Inform.* **2020**, *17*, 7589–7598.
12. Neupane, K.; Baysal-Gurel, F. Automatic identification and monitoring of plant diseases using unmanned aerial vehicles: A review. *Remote Sens.* **2021**, *13*, 3841. [CrossRef]
13. Wang, S.; Chen, J.; He, X. An adaptive composite disturbance rejection for attitude control of the agricultural quadrotor UAV. *ISA Trans.* **2022**. [CrossRef]
14. Sharma, A.; Jain, A.; Gupta, P.; Chowdary, V. Machine learning applications for precision agriculture: A comprehensive review. *IEEE Access* **2020**, *9*, 4843–4873. [CrossRef]
15. Wang, S.; Wang, L.; He, X.; Cao, Y. A Monocular Vision Obstacle Avoidance Method Applied to Indoor Tracking Robot. *Drones* **2021**, *5*, 105. [CrossRef]
16. Trilles, S.; Torres-Sospedra, J.; Belmonte, Ó.; Zarazaga-Soria, F.J.; González-Pérez, A.; Huerta, J. Development of an open sensorized platform in a smart agriculture context: A vineyard support system for monitoring mildew disease. *Sustain. Comput. Inform. Syst.* **2020**, *28*, 100309. [CrossRef]
17. Methun, N.R.; Yasmin, R.; Begum, N. Carrot Disease Recognition using Deep Learning Approach for Sustainable Agriculture. *Int. J. Adv. Comput. Sci. Appl.* **2021**, *12*, 732–741. [CrossRef]
18. Udutalapally, V.; Mohanty, S.P.; Pallagani, V.; Khandelwal, V. Scrop: A novel device for sustainable automatic disease prediction, crop selection, and irrigation in internet-of-agro-things for smart agriculture. *IEEE Sens. J.* **2020**, *21*, 17525–17538. [CrossRef]
19. Khattab, A.; Habib, S.E.; Ismail, H.; Zayan, S.; Fahmy, Y.; Khairy, M.M. An IoT-based cognitive monitoring system for early plant disease forecast. *Comput. Electron. Agric.* **2019**, *166*, 105028. [CrossRef]
20. Mekala, M.S.; Viswanathan, P. CLAY-MIST: IoT-cloud enabled CMM index for smart agriculture monitoring system. *Measurement* **2019**, *134*, 236–244. [CrossRef]
21. He, L.; Qi, S.-L.; Duan, J.-Z.; Guo, T.-C.; Feng, W.; He, D.-X. Monitoring of Wheat Powdery Mildew Disease Severity Using Multiangle Hyperspectral Remote Sensing. *IEEE Trans. Geosci. Remote Sens.* **2021**, *59*, 979–990. [CrossRef]
22. Dong, Y.; Xu, F.; Liu, L.; Du, X.; Ren, B.; Guo, A.; Geng, Y.; Ruan, C.; Ye, H.; Huang, W.; et al. Automatic System for Crop Pest and Disease Dynamic Monitoring and Early Forecasting. *IEEE J. Sel. Top. Appl. Earth Obs. Remote Sens.* **2020**, *13*, 4410–4418. [CrossRef]

23. Kim, W.S.; Lee, D.H.; Kim, Y.J. Machine vision-based automatic disease symptom detection of onion downy mildew. *Comput. Electron. Agric.* **2020**, *168*, 105099. [CrossRef]
24. Gao, D.; Sun, Q.; Hu, B.; Zhang, S. A framework for agricultural pest and disease monitoring based on internet-of-things and unmanned aerial vehicles. *Sensors* **2020**, *20*, 1487. [CrossRef]
25. Zhang, B.; Ye, H.; Lu, W.; Huang, W.; Wu, B.; Hao, Z.; Sun, H. A Spatiotemporal Change Detection Method for Monitoring Pine Wilt Disease in a Complex Landscape Using High-Resolution Remote Sensing Imagery. *Remote Sens.* **2021**, *13*, 2083. [CrossRef]
26. Redmon, J.; Farhadi, A. YOLOv3: An Incremental Improvement. *arXiv* **2018**, arXiv:1804.02767v1.
27. Redmon, J.; Farhadi, A. YOLO9000: Better, faster, stronger. In Proceedings of the IEEE Conference on Computer Vision and Pattern Recognition, Honolulu, HI, USA, 21–26 July 2017; pp. 7263–7271.
28. Ren, S.; He, K.; Girshick, R.; Sun, J. Faster R-CNN: Towards real-time object detection with region proposal networks. *Adv. Neural Inf. Process. Syst.* **2015**, 28. Available online: https://arxiv.org/abs/1506.01497 (accessed on 27 April 2021).
29. Liu, W.; Anguelov, D.; Erhan, D.; Szegedy, C.; Reed, S.; Fu, C.Y.; Berg, A.C. Ssd: Single shot multibox detector. In Proceedings of the European Conference on Computer Vision, Amsterdam, The Netherlands, 11–14 October 2016; Springer: Cham, Switzerland; pp. 21–37.

agronomy

MDPI

Article

Evaluation of the Effects of Airflow Distribution Patterns on Deposit Coverage and Spray Penetration in Multi-Unit Air-Assisted Sprayer

Tian Li [1,2], Peng Qi [1,2], Zhichong Wang [3], Shaoqing Xu [1,2], Zhan Huang [1,2], Leng Han [1,2] and Xiongkui He [1,2,*]

[1] Centre for Chemicals Application Technology, China Agricultural University, Beijing 100193, China; cault@cau.edu.cn (T.L.); qi-peng@139.com (P.Q.); b20213100722@cau.edu.cn (S.X.); derrick@cau.edu.cn (Z.H.); cauleng@163.com (L.H.)
[2] College of Science, China Agricultural University, Beijing 100193, China
[3] Tropics and Subtropics Group, Institute of Agricultural Engineering, University of Hohenheim, 70599 Stuttgart, Germany; wzcwyylk@foxmail.com
* Correspondence: xiongkui@cau.edu.cn

Abstract: Efficient utilization is a pre-requisite for pesticide reduction, and appropriate airflow distribution pattern plays a key role in enhancing the effectiveness of pesticide application by air-assisted orchard sprayers, yet the mechanism of this is unclear. In order to clarify the specific effects of airflow velocity and direction on spraying efficacy, a series of spray tests on pear and cherry and airflow distribution tests in open areas were conducted by a multi-unit air-assisted sprayer on ten different fan settings. Several deposit indicators were analyzed and contrasted with the air distribution. The results showed that an increase in airflow velocity inside the canopy improved the abaxial side deposit coverage of both pear (from 3.33% to 11.80% in the Top canopy and from 6.26% to 11.00% in the Upper canopy) and cherry leaves (from 3.61% to 10.87% in the Top canopy, from 1.36% to 9.04% in the Middle canopy, and from 3.40% to 9.04% in the Bottom canopy), but had no significant effect on the spray penetration. The correlation between deposit indicators and airflow velocities/directions was evaluated, and the results indicated that the enhanced airflow velocities, both in the forward and horizontal direction, improved the abaxial side deposit coverage (*CAB*) on the outside of pear canopy ($p < 0.001$), but for cherry, none of the airflow indicators had a significant impact on the *CAB* independently. On the other hand, the increased airflow direction angle in the cross-row plane for pear, as well as the increased airflow velocities in forward and vertical direction for cherry, both showed negative effects on the adaxial side deposit coverage ($p < 0.01$). The findings in this study might be helpful to improve the performance of pesticide application in orchards, especially for abaxial side deposition, and could provide a reference for the further investigations about the effect of airflow on spray canopy deposition.

Keywords: air-assisted sprayer; airflow distribution pattern; deposit coverage; spray penetration

Citation: Li, T.; Qi, P.; Wang, Z.; Xu, S.; Huang, Z.; Han, L.; He, X. Evaluation of the Effects of Airflow Distribution Patterns on Deposit Coverage and Spray Penetration in Multi-Unit Air-Assisted Sprayer. *Agronomy* **2022**, *12*, 944. https://doi.org/10.3390/agronomy12040944

Academic Editor: Karsten Schmidt

Received: 21 March 2022
Accepted: 12 April 2022
Published: 14 April 2022

1. Introduction

The reduction of pesticide application has been a consistent proposition of the Chinese government for many years, and it has also been included in the most important national development plan [1] as a significant goal of agricultural reform several times. Normally, the efficient utilization of pesticide, which means the active ingredient should be delivered to the target precisely, uniformly, and exclusively, was considered a key way to achieve pesticide reduction.

Pear and cherry are two cash crops grown on a large scale in several provinces in China, and their market demand has been increasing year by year. Recently, several investigations on pear and cherry orchards and their managers have been conducted by our team, and it was found that several kinds of diseases and pests occurred on the abaxial side of leaves, such as Pear psylla [2] and brown rot [3], which had a serious detrimental effect on the yield and quality of the fruits. Therefore, during the pesticide application in orchard, it is necessary to ensure sufficient deposit on the abaxial side of the leaves.

For decades, the air-assisted sprayer has been recognized as a high efficiency pesticide application equipment and is widely used for disease and pest control in the orchards. The main principle of air-assisted sprayers to improve the spraying effect is that the airflow generated by the fan sends droplets into and disturbs the tree canopy to increase spray deposition and enhance the uniformity of the droplet distribution [4–7]. In recent years, the axial fan sprayer [8,9] (AAS) and air tower sprayer [10,11] (ATS) have been the most widely used orchard application equipment, due to the stability of performance and convenience of operation. Various studies have focused on the spray effect of both the two types of sprayers, and there seems to be a general consensus that the spray deposition can directly or indirectly be affected by not only the canopy structure [12], but also the physicochemical properties of pesticides [13], the meteorological conditions [14] and operation techniques [15], among others. However, in most cases, the canopy structure of trees in the orchard is relatively stable with good management, where the application of pesticides mainly depends on the type and outbreak degree of pest and disease. Furthermore, there is no doubt that the artificial control of meteorological conditions is hard to achieve. All these factors make the adjustment of sprayer operation the most direct and effective way to improve the spray effect.

To date, the influence of the operating techniques on the spray effect, including travel speed, nozzle type, spray pressure, etc., have been explored by many researchers. Salyani, M. and Whitney, J.D. [16] characterized the effects of sprayer ground speed (1.6–6.4 km/h) on spray deposition at different locations within citrus tree canopies, and the results showed that travel speed did not have a significant effect on mean deposition in the tree canopy; however, variability of deposition increased as speed increased. The spray characteristics of the nozzle play an essential role in the pesticide application, especially for spray deposition. Hence, Nuyttens, D. et al. [17] tested the droplet size spectra of 32 nozzle-pressure combinations and expounded the effect of the nozzle type, size, and pressure on the droplet size and velocity spectra. It should be emphasized that droplet velocity is important with regard to canopy penetration and leaving surface retention. On the other hand, smaller droplets, especially those less than 150 [18] or 200 μm [19] in diameter, are more susceptible to airflow. According to applications based on an axial-fan air-blast orchard sprayer made by Derksen et al. [20], the smaller droplets can provide higher coverage than larger droplets on the axial side of leaves, and the authors believed that greater air turbulence was needed to improve this. However, considering that smaller droplets also have a higher risk of off target or drift, the balance between the advantages and the risk of off-target pollution should also be taken into account [21,22].

In addition to the above factors, the impact of spray distribution pattern on the pesticide application performance cannot be ignored. According to Gil et al. [23], an adequate adjustment of spray liquid distribution to the canopy structure can reduce spray drift up to 90% and reduce pesticide use up to 20%. Li et al. [24] conducted a comparative experiment on profile variable rate spray and conventional air assisted spray in orchard, and the results showed that a 23.2% reduction of spray drift and a 67.4% reduction of ground losses happened after adjusting the spray distribution pattern, moreover, a 45.7% saving of the solution was also obtained. It is now generally accepted that the spray distribution is directly determined by and dependent on the airflow pattern [4], including airflow velocity, direction, and their distribution in space. Farooq et al. [25] evaluated the spray penetration into the citrus canopy in two different airflow rates using air-carrier sprayers, results indicated that the higher airflow resulted in a better spray penetration and a significantly larger deposition in the far section of the canopy than the lower airflow. Holownicki et al. [26] compared three axial fan sprayers with different air discharge systems in the modern orchard, including conventional, cross-flow, and a high-volume directed air-jet (HVDAS) with the air spouts directed at 20 and 40° upwards, and the HVDAS with air spouts set

20° upwards produced a higher canopy deposit and a lower off-target loss than the other two sprayers. Another work [27], by investigating the effects of changing the fan speeds (1400, 2000 and 2500 rpm) and the direction of the air-jet (90 and 120° backward angle of the outlet side deflectors in relation to the treated row), studied the performance of a sprayer fitted with two vertical adjustable air outlets in vineyards, the highest deposit was observed with 1400 and 2000 rpm fan speeds and 120° backward angled air outlets, and the high 2500 rpm fan speed improved the degree and the uniformity of the spray coverage. Proper adjustment of the airflow velocity plays a key role in the efficiency of spray application [28,29]. However, excessive airflow will also reduce the canopy deposition [30] and lead to more spray loss and drift [31]. All these studies show that proper adjustment of airflow plays a key role in the efficiency of spray application. Therefore, as a prerequisite for determining the optimal adjustment, the specific effect of airflow on spray canopy deposition should be defined.

In order to clarify the correlations of airflow velocity distribution and direction with spray deposit in different parts of the canopy, we conducted a series of spray tests in the orchard and airflow distribution tests in an open field with different unit fan parameters, which were determined by the canopy characteristics of pear and cherry. In the end, the effect of the fan unit adjustment was evaluated by analyzing the correlation between the airflow velocity and direction with the canopy deposition. These were all based on a multi-unit air-assisted sprayer independently developed by our team.

2. Materials and Methods

2.1. Sprayer Characteristics

Figure 1 shows the multi-unit air-assisted sprayer (MAS). The length and width of the MAS were 3.2 m and 1.5 m, the rated power of the engine is 6.3 Kw. Three units were installed on both sides of the sprayer with a fan hood depth of 150 mm. Each of the units, equipped with 20-inch carbon-fiber fan blades driven by a brushless motor, were connected to the vertical support rod by a mechanical joint that ensured the manual adjustment of installation height and tilt angle could be achieved. The rotation rate of each brushless motor could be controlled independently through a specially made control box, and both of the brushless motors and the box were powered by a 24 V storage battery (Fengfan, Baoding, China) linked to a 3.5 kW range extender to provide stable working voltage. Four hollow cone nozzles HCI4001 (ARAG, Rubiera, Italy) were equidistantly installed at the air outlet of the unit. The spray angle of HCI4001 was 40°, the volume flow rate was 0.39 L/min, and the volume median diameter of the droplet was 121 μm at a spray pressure of 0.3 MPa. The droplet diameter was tested by Spraytech Particle size analyzers [32] (Malvern Panalytical, Worcestershire, UK) according to ISO 13320:2020 [33]

2.2. Spray Test Site

Figure 2a showed the spray test site in Haidian District, Beijing (116.19° E, 40.15° N), which was a plain area. The 15-years-old pear (*Pyrus bretschneideri*) orchard covered an area of 0.66 ha, trees were planted in a 6 m by 3 m row and tree spacing, and the average tree height was 3.1 m. The 4-years-old cherry (*Cerasus pseudocerasus*) orchard covered an area of 0.91 ha, the row and tree spacing were 4 m and 1 m, and the average tree height was 2.6 m. Tests were conducted in August 2021. During this period, the weather continued without rain, the temperature remained at 27.6–33.4 °C in the daytime, and the wind speed was below 1 m/s during the experiment—all of these provided a suitable meteorological condition for spraying.

Figure 1. The multi-unit air-assisted orchard sprayer used in this study: ① Top Unit, ② Electronic Speed Control, ③ Middle Unit, ④ Bottom Unit, ⑤ Speed control box, ⑥ Proportional control valve with pressure gauge, ⑦ Tank.

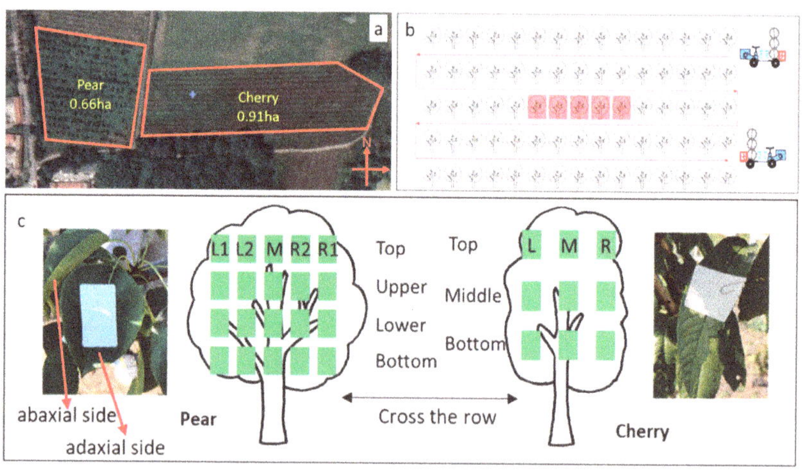

Figure 2. (**a**) Test fields for spray test (**b**) Layout of spray test: the red line showed the application route, trees with red background were sampled; (**c**) Sample site: green marks represented samples. The indicators L, M, R in the cherry canopy and the L1, L2, M, R2, R1 in the pear canopy represent the sample positions in per layer.

2.3. Experimental Design

2.3.1. Spray Test

Five rows in the middle of the orchard were selected for application (Figure 2b). Five consecutive trees were sampled in the middle row, and these trees were not less than 10 m from the beginning and end of the row to make sure the MAS was working stably when passing through. There were 4 layers set on each of the sample trees of pear and

5 sample points were set in each layer (Figure 2c). There were 3 layers for cherry trees and 3 sample points in each layer. All of the layer and sample parameters, including layer height, layer width, and the distance between sample and trunk, were measured and shown in Table 1. A well-grown leaf was chosen within the area 15 cm from the sample point and 2 white art papers (60 mm × 40 mm) were attached on the adaxial side (ADS) and abaxial side (ABS) of it. The application volume was 225 L/ha. Ponceau 4R, a dye usually used as a food coloring with no risk of environmental pollution and human and animal poisoning, was added as a tracer with a concentration of 5 g/L. During each test, the time when the MAS entered and left the test area was recorded to ensure that the deviation of the actual application volume from the set value did not exceed 5%. All tests were repeated 3 times (Table 1).

Table 1. Layers and samples setting parameters of pear and cherry.

Plant	Sample Layer	Layer Height/m	Width of Layer/m	Distance between Sample and Trunk/m				
				L		M	R	
				L1	L2		R2	R1
Pear	Top	2.8	2.8–3.2	1.3–1.6	0.5–0.9	0	0.4–0.7	1.0–1.2
	Upper	2.3	2.9–3.4	1.4–1.7	0.5–0.8	0	0.5–0.8	1.1–1.3
	Lower	1.8	2.6–3.1	1.3–1.5	0.6–0.9	0	0.5–0.8	1.2–1.4
	Bottom	1.3	2.0–2.6	1.0–1.3	0.4–0.6	0	0.4–0.7	1.0–1.2
Cherry	Top	2.5	1.2–1.4	0.5–0.6		0	0.5–0.7	
	Middle	1.9	1.3–1.5	0.6–0.8		0	0.6–0.8	
	Bottom	1.3	0.7–0.8	0.2–0.4		0	0.3–0.5	

According to the canopy characteristics of pear and cherry, 10 parameters were settled within the adjustable range of the fan (Table 2). OSC1–OSC6 were used on cherry, the fan speed was different between OSC1–OSC3, the installation height of the middle unit between OSC4 and OSC5 was different by 100 mm, and the vertical installation angle of the middle unit between OSC5 and OSC6 was different by 20°.

Table 2. The test parameters of MAS used in spray tests.

Parameters	Spray Units	Fan Speed (rpm)	Installation Height (m)	Tilt Angle (°)
OSC1	Top	2275	2.95	−40
	Middle	2275	1.95	0
	Bottom	2275	1.00	20
OSC2	Top	2635	2.95	−40
	Middle	2635	1.95	0
	Bottom	2635	1.00	20
OSC3	Top	2635	2.95	−40
	Middle	3090	1.95	0
	Bottom	2635	1.00	20
OSC4	Top	2635	2.95	−40
	Middle	3090	1.75	20
	Bottom	2635	1.00	20
OSC5	Top	2635	2.95	−40
	Middle	3090	1.85	20
	Bottom	2635	1.00	20
OSC6	Top	2635	2.95	−40
	Middle	3090	1.85	0
	Bottom	2635	1.00	20

Table 2. *Cont.*

Parameters	Spray Units	Fan Speed (rpm)	Installation Height (m)	Tilt Angle (°)
OSP1	Top	2635	2.95	−15
	Middle	2635	1.85	10
	Bottom	2635	1.00	20
OSP2	Top	2635	2.95	−15
	Middle	3090	1.85	10
	Bottom	2635	1.00	20
OSP3	Top	3090	2.95	−15
	Middle	3090	1.85	10
	Bottom	3090	1.00	20
OSP4	Top	2635	2.95	−15
	Middle	3090	1.75	30
	Bottom	2635	1.00	20

Note: Tilt angle < 0 means tilt down, Tilt angle > 0 means tilt up.

2.3.2. Airflow Pattern Test

The airflow pattern test of MAS was conducted in an open field outdoors in a static way according to ISO 9898:2000 [34]. Figure 3 showed the test layout, a group of 3 Axis wind sensors [35] (WindMaster 1590-PK-020, Gill Instruments Ltd., Hampshire, UK) was installed from 0.6 m away from the ground to 3.2 m according to the vertical interval of 0.2 m. All parameters listed in Table 2 were tested at 0.5 m, 1.0 m, 1.5 m, and 2.0 m away from the MAS, respectively. Fans on both sides of the sprayer opened simultaneously to keep the working condition consistent with the orchard application. All sensors were connected to the computer for the signal transmission, and the output data is received by the software at a frequency of 10 Hz and further calculated in Cartesian coordinates (X, Y, Z), including the airflow velocity and direction on the YZ plane and the air velocity on the X-axis. For each measuring point and parameter, data were recorded at least 15 s.

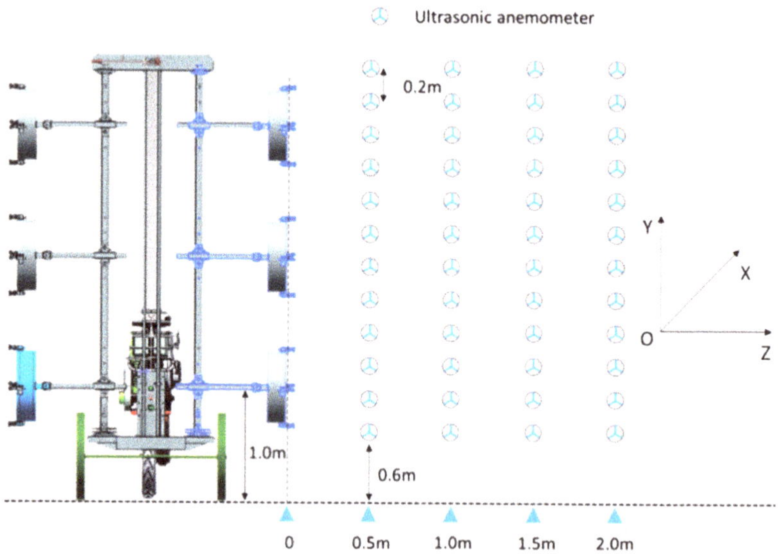

Figure 3. Layout of airflow pattern test of MAS in an open field. The X-axis represented the forward direction, the Y-axis represented the vertical direction, and the Z-axis represented the cross-row direction.

2.4. Data Analyses

All samples were scanned with a scanner (DS-1610, Epson (China), Beijing) at 400 dpi to obtain images. An image processing program ImageJ [36] (National Institutes of Health, United States) was used for the analysis to get spray coverage [37]. *CAD* and *CAB* represent deposit coverage on ADS and ABS of the leaf (%), and the deposit coverage of whole leaf (*CW*, %) was calculated as follows:

$$CW = CAD + CAB \tag{1}$$

Where the ratio of deposit coverage on ABS and whole leaf (*RBW*) was calculated to account for droplet distribution uniformity on leaves:

$$RBW = \frac{CAB}{CW} \tag{2}$$

Where the ratio of deposit coverage on the center sample and the total sample in the same layer (*RCT*) was calculated to account for the penetration effect of droplets:

$$RCT = \frac{CW_{M}}{CW_{L} + CW_{M} + CW_{R}} \tag{3}$$

where L, M, and R are sample sites in cherry canopy. As for pear, Left means L1 and L2, and R means R1 and R2 (Figure 2b).

The 3D airflow velocity and direction corresponding to all sample points are obtained through interpolation for 2D gridded data in meshgrid format [38] by Matlab 2021a (Mathworks, Natick, MA, USA). The interpolated value at a sample point is based on a cubic interpolation of the values at neighboring grid points in each respective dimension. The interpolation is based on a cubic spline using not-a-knot end conditions and 6 indicators were obtained: X-axis, *Y*-axis, and *Z*-axis airflow velocities, ZY plane (0° on Z-axis), ZX plane (0° on *Z*-axis), and YX (0° on *Y*-axis) plane airflow directions (*X-v, Y-v, Z-v, ZY-d, ZX-d, YX-d*). For assessing the impact of these indicators on canopy deposition, the method combining partial least square (PLS) and the variable importance in projection (VIP) scores (also called the PLS-VIP method [39]) have been introduced to quantify the contribution.

All statistical analyses were performed using IBM SPSS Statistics (IBM Corp., Armonk, NY, USA). The mean values of *CAD, CAB, CW, RBW,* and *RCT* in different test parameters were compared by using one-way analysis of variance (ANOVA) via the Duncan test (α = 0.05). The bivariate correlation analysis was adopted to evaluate the correlation between airflow indicators and *CAD, CAB, CW,* and *RBW* of all test parameters by using Pearson test and two-tailed significance test.

3. Results

3.1. Deposit Coverage Distribution in Layers

Table 3 shows mean values of all deposit indicators, including *CAD, CAB, CW, RBW,* and *RCT*, calculated from all samples in each layer of OSP1–OSP4 (results of all sample trees and replicates are involved). Intercomparison of OSP1–OSP3 was made on account of sharing the same installation height and tilt angle but different in fan speed. For the Top layer, the *CAD* and *CW* of OSP3 were significantly lower than those of OSP1 and had no significant difference with OSP2, and the *CAB* and *RBW* of OSP3 were significantly higher than those of OSP2 and had no significant difference with OSP1. For the Upper layer, the *CAD* of OSP3 was significantly lower than that of OSP1 and OSP2, the *CAB* of OSP3 was significantly higher than OSP1 and had no significant difference with OSP2, and the *RBW* of OSP3 was significantly higher than OSP1 and OSP2. The *RCT* of OSP1-OSP3 in all layers had no significant difference with each other. In the comparison between OSP2 and OSP4, the difference in deposition results was mainly due to the installation height and tilt angle of the Middle unit, the *CAD* and *RCT* of OSP4 in the Top layer were significantly lower than OSP2 but the *CAB* and *RBW* were higher. Meanwhile, in the Upper layer, the *CAD* of

OSP4 was significantly lower than OSP2, and in the Lower layer the *CAB* of OSP4 was also lower than OSP2.

Table 3. Results of spray tests on pear: *CAD, CAB, CW, RBW,* and *RCT* of OSP1-OSP4.

Sample Layer	Test Parameters	*CAD*		*CAB*		*CW*		*RBW*		*RCT*	
		Mean/%	SE	Mean/%	SE	Mean/%	SE	Mean	SE	Mean	SE
Top	OSP1	25.58 ^{ab}	3.19	6.91 ^{bc}	1.20	32.50 ^{abc}	2.96	0.28 ^{cd}	0.04	0.13 ^{bcde}	0.01
	OSP2	22.54 ^{abc}	2.54	3.33 ^{d}	0.44	25.87 ^{bcdef}	2.55	0.19 ^{c}	0.02	0.16 ^{abcde}	0.02
	OSP3	14.14 ^{cdef}	2.03	7.75 ^{abc}	1.55	21.89 ^{def}	2.30	0.37 ^{bc}	0.04	0.12 ^{abcdef}	0.03
	OSP4	13.83 ^{def}	1.81	11.80 ^{abc}	2.05	25.62 ^{bcdef}	2.54	0.41 ^{abc}	0.04	0.08 ^{f}	0.01
Upper	OSP1	13.93 ^{ef}	1.27	6.26 ^{c}	0.93	20.18 ^{ef}	1.39	0.30 ^{c}	0.03	0.19 ^{a}	0.02
	OSP2	16.26 ^{bcde}	1.69	7.67 ^{bc}	0.95	23.93 ^{cdef}	1.87	0.35 ^{bc}	0.03	0.17 ^{abc}	0.02
	OSP3	7.81 ^{g}	0.93	11.00 ^{ab}	1.37	18.81 ^{f}	1.48	0.53 ^{a}	0.04	0.17 ^{ab}	0.02
	OSP4	8.98 ^{fg}	1.15	13.25 ^{ab}	2.23	22.23 ^{def}	2.07	0.49 ^{ab}	0.05	0.11 ^{bcdef}	0.02
Lower	OSP1	29.97 ^{a}	3.11	14.40 ^{a}	2.10	44.37 ^{a}	3.17	0.33 ^{bc}	0.04	0.10 ^{cdef}	0.02
	OSP2	22.14 ^{abcd}	2.51	13.88 ^{a}	1.88	36.02 ^{ab}	2.84	0.36 ^{bc}	0.04	0.10 ^{cdef}	0.02
	OSP3	18.25 ^{bcde}	2.08	9.17 ^{abc}	1.62	27.42 ^{bcde}	2.38	0.33 ^{bc}	0.04	0.10 ^{ef}	0.02
	OSP4	14.32 ^{cdef}	1.70	5.81 ^{cd}	1.27	20.13 ^{def}	2.01	0.30 ^{ab}	0.04	0.16 ^{abcdef}	0.04
Bottom	OSP1	17.49 ^{bcde}	1.54	11.03 ^{abc}	2.45	28.52 ^{bcd}	2.56	0.27 ^{cd}	0.04	0.15 ^{abcde}	0.02
	OSP2	18.07 ^{bcde}	2.88	8.13 ^{abc}	1.82	26.20 ^{bcdef}	3.07	0.26 ^{cd}	0.04	0.15 ^{abcd}	0.01
	OSP3	12.81 ^{efg}	1.65	7.93 ^{abc}	1.87	20.74 ^{def}	1.97	0.32 ^{bc}	0.05	0.11 ^{def}	0.02
	OSP4	15.94 ^{bcde}	2.25	6.90 ^{abc}	1.95	22.84 ^{cdef}	2.65	0.24 ^{cd}	0.04	0.10 ^{cdef}	0.02

Note: SE means standard error. Different letters in the same column indicate significant differences at $p < 0.05$ level.

The only difference between OSC1-OSC3 is the fan speed setting, which was similar to OSP1-OSP3. There was no significant difference for *CAD* and *RCT* in all layers of OSC1-OSC3 as shown in Table 4, however in the Middle layer, the *CAB* of OSC2 and OSC3 were significantly higher than OSC1. In the Bottom layer, OSC3 also showed a higher *CAB* than OSC1. There was no significant difference in the deposition effect between OSC3 and OSC6, nor between OSC4 and OSC5, however, the *CAB* of OSC5 in the Top layer was significantly higher than that of OSC6.

Table 4. Results of spray tests on cherry: *CAD, CAB, CW, RBW,* and *RCT* of OSC1-OSC6.

Sample Layers	Test Parameters	*CAD*		*CAB*		*CW*		*RBW*		*RCT*	
		Mean/%	SE	Mean/%	SE	Mean/%	SE	Mean	SE	Mean	SE
Top	OSC1	20.69 ^{ab}	2.65	4.00 ^{bc}	0.84	24.70 ^{abc}	2.78	0.22 ^{bcd}	0.04	0.27 ^{a}	0.04
	OSC2	20.38 ^{ab}	3.33	8.53 ^{abc}	2.00	28.91 ^{abc}	3.78	0.31 ^{abcd}	0.05	0.23 ^{ab}	0.04
	OSC3	20.54 ^{ab}	2.82	6.36 ^{abc}	1.50	26.90 ^{abc}	2.70	0.25 ^{bcd}	0.05	0.22 ^{ab}	0.03
	OSC4	20.99 ^{ab}	3.17	7.85 ^{abc}	1.68	28.84 ^{abc}	3.03	0.33 ^{abc}	0.06	0.25 ^{a}	0.04
	OSC5	19.71 ^{ab}	3.22	10.87 ^{a}	1.63	30.58 ^{ab}	3.34	0.46 ^{a}	0.06	0.27 ^{a}	0.05
	OSC6	14.90 ^{ab}	2.35	3.61 ^{bcd}	0.94	18.52 ^{c}	2.36	0.29 ^{abcd}	0.05	0.24 ^{abc}	0.06
Middle	OSC1	18.94 ^{ab}	3.20	1.36 ^{d}	0.48	20.30 ^{bc}	3.30	0.14 ^{d}	0.04	0.12 ^{bc}	0.03
	OSC2	20.78 ^{ab}	2.87	4.81 ^{bc}	0.80	25.59 ^{abc}	2.86	0.25 ^{bcd}	0.04	0.18 ^{abc}	0.04
	OSC3	21.67 ^{ab}	2.88	9.04 ^{ab}	2.00	30.71 ^{ab}	2.62	0.31 ^{abc}	0.05	0.23 ^{ab}	0.03
	OSC4	21.74 ^{ab}	3.72	8.60 ^{abc}	2.15	30.34 ^{abc}	3.83	0.37 ^{ab}	0.06	0.11 ^{c}	0.01
	OSC5	21.76 ^{ab}	3.57	8.98 ^{ab}	1.90	30.74 ^{ab}	3.22	0.36 ^{ab}	0.06	0.15 ^{abc}	0.03
	OSC6	27.65 ^{a}	3.40	7.43 ^{abc}	1.65	35.08 ^{a}	3.36	0.24 ^{bcd}	0.05	0.16 ^{abc}	0.03
Bottom	OSC1	19.09 ^{ab}	2.27	3.40 ^{cd}	0.63	22.48 ^{bc}	2.22	0.18 ^{cd}	0.03	0.20 ^{ab}	0.03
	OSC2	20.57 ^{ab}	2.81	9.04 ^{ab}	2.02	29.60 ^{ab}	2.85	0.32 ^{abcd}	0.06	0.18 ^{ab}	0.02
	OSC3	24.63 ^{ab}	2.75	6.90 ^{abc}	1.52	31.52 ^{ab}	2.52	0.23 ^{bcd}	0.05	0.27 ^{a}	0.04
	OSC4	25.54 ^{ab}	3.49	5.01 ^{bc}	0.99	30.54 ^{ab}	3.54	0.21 ^{bcd}	0.04	0.23 ^{ab}	0.04
	OSC5	22.18 ^{ab}	2.92	6.85 ^{abc}	1.39	29.02 ^{ab}	2.88	0.28 ^{abcd}	0.05	0.21 ^{ab}	0.04
	OSC6	26.00a ^{b}	3.38	7.59 ^{ab}	1.38	33.60 ^{a}	3.21	0.27 ^{abcd}	0.05	0.22 ^{ab}	0.03

Note: SE means standard error. Different letters in the same column indicate significant differences at $p < 0.05$ level.

3.2. Comparison between Airflow Velocity Distribution and Deposit Coverage Distribution

The distribution of both airflow and spray canopy deposit varied according to the fan setting. For this reason, a comparison was made between them. As can be seen in Figure 4, the airflow velocities of OSP1-OSP3 decreased with distance at Top, Lower, and Bottom heights, while at Upper height the airflow showed the opposite trend. In the meantime, the *CAB* of OSP3 improved, but the *CAD* decreased in the Top and Upper layers. Compared to

OSP2, the air velocity of OSP4 attenuated less at Top height but was lower at Lower height. As for the canopy deposit coverage, the *CAB* of OSP4 was higher in the Top but lower in the Lower layer, meanwhile the *CAD* was lower in both of the Top and Upper layers, which was consistent with the results shown in Section 3.1.

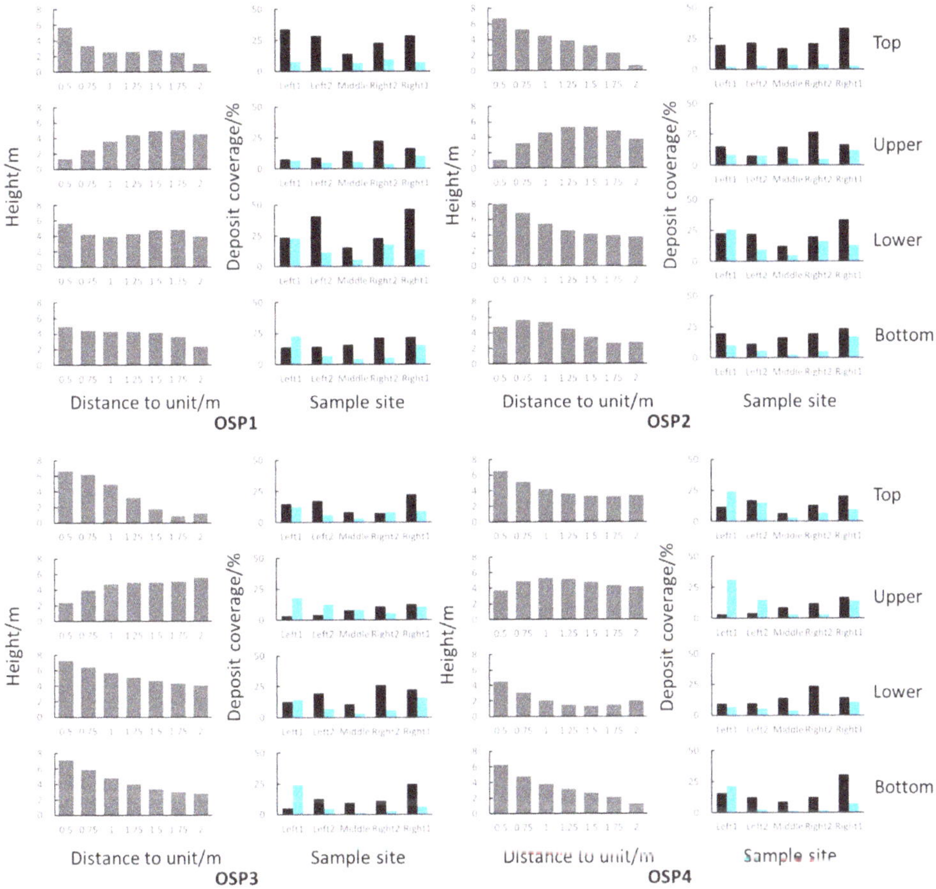

Figure 4. The airflow distribution at different distances from the MAS (left), *CAD* (black block), and *CAB* (blue block) of samples in all layers (right) of OSP1–OSP4.

OSC1–OSC3 shared a similar airflow velocity distribution pattern just like OSP1. OSP3, OSC2, and OSC3 have higher air velocities than OSC1 at Middle height, which made the *CAB* of OSC1 less than OSC2 and OSC3 at the same height. OSC3 has a higher airflow velocity at Middle height compared to OSC6, and OSC5 also has a higher airflow velocity at Top height compared to OSC4, however in the corresponding layers OSC6 has a similar CAB to OSC3 and OSC4 to OSC5. In addition, the OSC5 has a higher air velocity at Top height as well as a *CAB* compared to the OSC6 (see Figure 5).

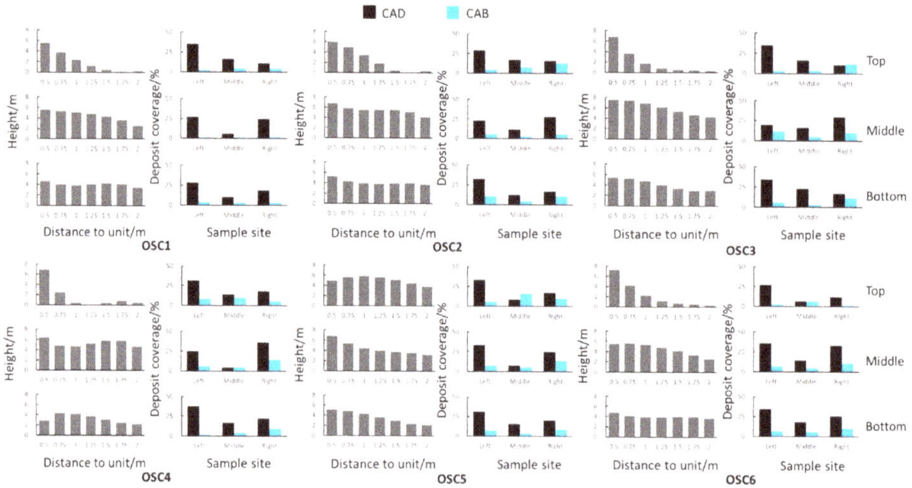

Figure 5. The airflow distribution at different distances from the MAS (left); *CAD* (black block) and *CAB* (blue block) of samples in all layers (right) of OSC1–OSC6.

3.3. PLS-VIP and Correlation Analysis between Airflow Velocity, Direction, and Deposit Indicators

3.3.1. Results of OSP1-OSP4

For OSP1-OSP4, a PLS-VIP analysis was carried out between deposit indicators (*CAD*, *CAB*, *CW*, and *RBW*) and airflow indicators (*X-v*, *Y-v*, *Z-v*, *ZY-d*, *ZX-d*, *YX-d*). Results exhibited in Figure 6 showed that *Z-v* represented a more significant influence on *CAB* and RBW than the other indicators, while *ZY-d* made a greater contribution to *CAD* and *CW*. In addition, the effect of *X-v* on *CAB* and *Z-v* on *CW* cannot be ignored as well.

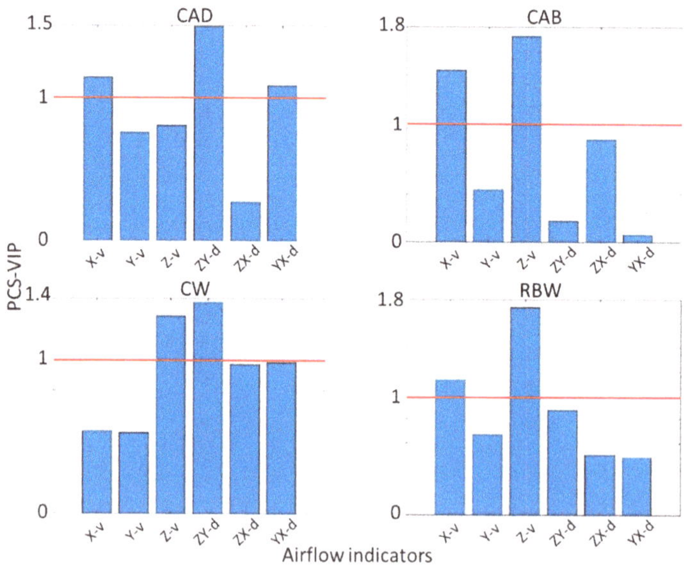

Figure 6. PLS-VIP analysis results between the airflow indicators and the deposit indicators of OSP1–OSP4. The red line indicates that the PLS-VIP = 1, PLS-VIP > 1 indicates a significant effect [40]. The red line marks the position where the PLS-VIP value is 1.

According to Table 5, *X-v* and *Z-v* showed extremely significant positive correlations ($p < 0.001$) with *CAB* and *RBW* in the total and outside canopy, but also negative correlations ($p < 0.05$) with *CAD* in the outside canopy, while *ZY-d* showed significant negative correlations ($p < 0.01$) with both *CAD* and *CW* in the total and outside canopy. All these were largely consistent with the results represented in Figure 7.

Table 5. Results of the correlation analysis between airflow and deposit indicators at different parts of pear canopy.

Deposit Indicators	Canopy Parts	*X-v*		*Y-v*		*Z-v*		*ZY-d*		*ZX-d*		*YX-d*	
		r	*p*	*r*	*p*	*r*	*p*	*r*	*p*	*r*	*p*	*r*	*p*
CAD	Total	−0.067	0.060	−0.027	0.454	−0.060	0.092	−0.102	0.004	−0.007	0.845	−0.066	0.063
	Inside	−0.047	0.554	−0.008	0.916	−0.016	0.844	−0.115	0.148	−0.105	0.185	−0.013	0.868
	Outside	−0.083	0.036	−0.028	0.479	−0.128	0.001	−0.119	0.002	−0.006	0.873	−0.060	0.128
CAB	Total	0.203	0.000	−0.037	0.292	0.205	0.000	−0.002	0.945	0.093	0.009	0.001	0.966
	Inside	0.118	0.136	0.163	0.039	−0.016	0.837	−0.074	0.351	−0.033	0.678	−0.079	0.319
	Outside	0.192	0.000	−0.048	0.228	0.180	0.000	−0.006	0.870	0.097	0.014	0.028	0.483
CW	Total	0.075	0.033	−0.048	0.177	0.083	0.019	−0.091	0.010	0.055	0.119	−0.057	0.109
	Inside	0.018	0.823	0.071	0.369	−0.021	0.793	−0.131	0.099	−0.103	0.194	−0.049	0.538
	Outside	0.056	0.159	−0.058	0.143	0.007	0.864	−0.112	0.004	0.061	0.125	−0.035	0.370
RBW	Total	0.136	0.000	0.061	0.084	0.172	0.000	0.081	0.022	0.062	0.082	0.038	0.281
	Inside	0.191	0.016	0.182	0.022	−0.064	0.424	0.024	0.765	0.063	0.427	−0.065	0.411
	Outside	0.124	0.002	0.052	0.188	0.174	0.000	0.090	0.023	0.050	0.208	0.070	0.076

Note: The *r* value represents the Pearson correlation coefficient, $p < 0.05$ indicates a correlation between the two indicators, $p < 0.01$ indicates a significant correlation, and $p < 0.001$ indicates an extremely significant correlation.

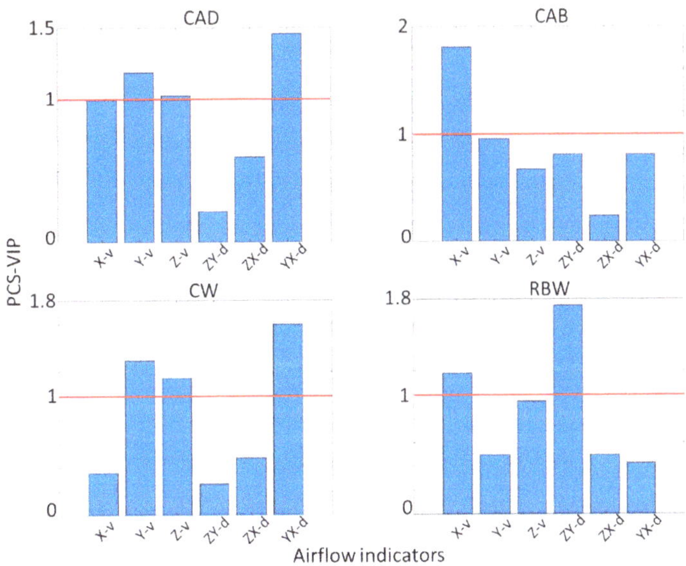

Figure 7. PLS-VIP analysis results between the airflow indicators and the deposit indicators of OSC1–OSC6. The red line marks the position where the PLS-VIP value is 1.

3.3.2. Results of OSC1-OSC6

According to Figure 7, for the *CAD* and *CW* of cherry, the most influential indicator was *YX-d*, followed by *Y-v*, then *X-v* showed a significant influence on *CAB*. Moreover, the most influential indicator on *RBW* was *ZY-d*.

As represented in Table 6, *X-v* had a relatively complex effect on canopy deposit coverage, showing positive correlations with *CAB* ($p < 0.05$) and *RBW* ($p < 0.001$) on the outside of the cherry canopy, but negative correlations with *CAB* and *RBW* ($p < 0.05$) on the inside of the canopy and *CAD* ($p < 0.01$) on the outside of the canopy. *Y-v* showed a negative correlation with *CAD* ($p < 0.01$) and *CW* ($p < 0.05$) inside the canopy, meanwhile *YX-d* showed the same negative correlation with *CAD* and *CW* of the total canopy ($p < 0.05$), in addition, *ZY-d* showed a significant positive correlation with *RBW* inside the canopy ($p < 0.001$). Although a few of the airflow indicators in Table 6 showed different contributions to those in Figure 7, this does not prevent the overall results from being consistent.

Table 6. Results of the correlation analysis between airflow and deposit indicators at different parts of cherry canopy.

Deposit Indicators	Canopy Parts	*X-v*		*Y-v*		*Z-v*		*ZY-d*		*ZX-d*		*YX-d*	
		r	*p*	*r*	*p*	*r*	*p*	*r*	*p*	*r*	*p*	*r*	*p*
CAD	Total	−0.068	0.112	−0.053	0.217	0.052	0.230	−0.007	0.879	−0.039	0.364	−0.085	0.048
	Inside	0.015	0.845	−0.202	0.007	−0.039	0.605	−0.156	0.037	0.011	0.884	0.116	0.121
	Outside	−0.157	0.003	−0.030	0.571	0.013	0.811	−0.060	0.259	−0.128	0.015	−0.034	0.518
CAB	Total	0.083	0.055	−0.004	0.923	0.028	0.521	0.034	0.435	0.011	0.805	−0.035	0.418
	Inside	−0.173	0.020	0.030	0.694	−0.146	0.051	0.147	0.049	−0.104	0.163	−0.113	0.131
	Outside	0.130	0.014	−0.013	0.809	0.056	0.287	−0.036	0.499	0.045	0.390	0.034	0.516
CW	Total	−0.027	0.539	−0.055	0.205	0.065	0.132	0.010	0.812	−0.033	0.439	−0.102	0.018
	Inside	−0.055	0.463	−0.157	0.036	−0.089	0.234	−0.073	0.332	−0.031	0.675	0.053	0.482
	Outside	−0.092	0.082	−0.040	0.450	0.047	0.372	−0.086	0.105	−0.110	0.036	−0.016	0.757
RBW	Total	0.068	0.112	0.036	0.403	−0.031	0.473	0.079	0.066	0.022	0.614	−0.019	0.658
	Inside	−0.174	0.020	0.190	0.011	−0.026	0.731	0.245	0.001	−0.086	0.253	−0.213	0.004
	Outside	0.172	0.001	−0.005	0.930	−0.009	0.863	0.035	0.505	0.106	0.044	0.035	0.510

4. Discussion

In most cases, the air adjustments of AAS and ATS are based on the rotational speed of the fan and on the position of air deflectors [41] which impact air velocity and air direction. For this reason, there are limited options for adjusting the airflow pattern and vertical spray distribution. In this research the MAS allowed for more adjustments by changing the installation height, tilt angle, and fan speed, which offered a multivariate scenario in this study. As mentioned before, in the absence of a tree canopy, the spray volume distribution was directly influenced by the airflow pattern [4]. However, in the case of pesticide applications, the effects of airflow velocity and direction (mainly determined by fan parameters) on canopy deposition were relatively complex. By the influence of airflow, droplets are forced into the canopy and attached to the leaves. In the meantime, leaves are swung and lifted, which can somehow increase the deposition on the leaves, especially on the ABS.

For the spray test conducted on pear in this study, the distance between spray units and the trunk was 2.25 m, which means the airflow between 1.5 m and 2 m from the fan entered the canopy and had a significant impact on droplet deposition, therefore the velocity of airflow inside the canopy (AI) was used as the main reference for evaluating pear canopy deposition. Compared to the airflow distribution patterns of OSP1, the AI velocity of OSP3 was higher at Upper and lower at Top (Figure 4), which makes the leaves in both the two layers more susceptible to be uplifted. Due to this, the exposed area of the ABS was increased, resulting in a higher *CAB* for OSP3 (11.00%) than OSP1 (6.26%) at Upper. On the other hand, considering that the distance between units and the cherry trunk is 1.25 m, the airflow between 1.0 m and 1.5 m from the fan should be defined as AI for the evaluation of the spray effect. For OSC1-OSC3 (Figure 5), the AI velocities of OSC2 and OSC3 were higher than OSC1 at Middle height, which made the twisting frequency of leaves and the exposure probability of the ABS increase, and thus both OSC2 (4.81%) and OSC3 (9.04%) had greater *CAB* than OSC1 (1.36) in the Middle layer.

Compared to OSC6, OSC3 had greater AI velocity at Middle height, however there was no significant difference in *CAD* and *CAB* at Middle layer between them, which did not seem to be consistent with the previous analysis. On the other hand, OSC6 got higher AI velocity at Bottom than OSC3, so it was presumed that under the combined effect of airflow at both Middle and Bottom, the twisting frequencies of their Middle layer leaves were close. A similar situation occurred between OSC4 and OSC5, where OSC5 got a higher AI velocity at Top height but a lower AI velocity at Middle height compared to OSC4, from this point the *CAD* and *CAB* of OSC5 in the Top layer were also not significantly different from OSC4.

In general, the enhanced velocity of airflow inside the canopy resulted in an increase of deposition coverage on the ABS of leaves, for both pear and cherry, however it did not provide a significant improvement to the spray penetration. The difference is that the smaller pear leaves, compared to the cherry leaves, oscillated mainly in the longitudinal direction by airflow, which made that an increased exposed area of the ABS which was always accompanied by a decrease exposure of the ADS. For this reason, the increase of *CAB* happened frequently in conjunction with the decrease of *CAD*. The cherry leaves were slender and mainly twisted in the axial direction, which lead to an increase of *CAB* without decreasing the *CAD*. Furthermore, in many cases the spray coverage in a particular layer (e.g., Top layer of OSP3, Middle layer of OSC3, Top layer of OSC5) was simultaneously influenced by the effect of both the airflow in this height and that above or below it, which can be referred to as a regional airflow contribution, but the rationale for this phenomenon is unclear and needs to be explored in further studies.

Over the years, many reports have focused on spray penetration [25,42], distribution [43,44], and the mass balance [8,45] in orchards, which helps to clarify the practical application effects of different spray technologies and fan operating parameters. However, from the point of disease and pest control, the deposit on the ADS and ABS of leaves should also be taken into account.

The results of the PLS-VIP and correlation analysis showed that the enhanced *X-v* and *Z-v* improved ($p < 0.001$) the *CAB* on the outside of pear canopy, but for cherry, none of the airflow indicators had a significant impact on the CAB. On the other hand, for the increased *ZY-d* ($p < 0.01$) for pear as well as the increased *X-v* and *Y-v* ($p < 0.01$) for cherry, both showed negative effects on the *CAD*. Obviously, pear and cherry are very different in canopy structure and leaf shape. By the influence of airflow, the pear leaves swung violently and the exposure of the ABS increased, which resulted in the improvement of the *CAB*. However, for the cherry leaves, the changes in airflow did not significantly increase the exposure of the ABS. Moreover, it is worth noting that the reduction of *CAD* is very detrimental to the efficacy of pesticide control and should be avoided in the application. For pear and cherry orchards, we should adopt different airflow patterns to improve the spraying effect. Based on the structure of the MAS, the adjustment of *X-v* can be achieved by changing the angle between the unit and the travel direction, *Z-v* is mainly dependent on the fan speed setting, while the *ZY-d* can be adjusted by changing the tilt angle.

In recent years, efforts have focused on improving the design of pesticide application equipment. Grella et al. [46] tested the effect of different fan settings, air-conveyor orientation, and nozzle configuration on both airflow pattern and vertical spray distribution, and the results showed that a forward air-conveyor orientation was selected as one of the most optimal settings for spray applications in a trellised vineyard. Interestingly, in this study a similar trend was observed (the enhanced *X-v* could be adjusted by changing the install angle between the unit and the travel direction) that improved the deposit coverage in the ABS. Therefore, to some extent we can conclude that the forward airflow has a positive contribution to the spray effect.

In the last decade, agricultural unmanned aerial vehicle (UAV) has become a widely used piece of equipment for orchard pesticide application, especially in East Asian countries [47]. The downwash airflow of the UAV plays an important role in the transportation of droplets to the target [48,49], which was similar to the horizontal and/or vertical airflow from air-assisted sprayers. However, the UAV had a much lower deposition on leaves compared to the AAS [28,50,51]. Zhang et al. [52] pointed out that, in order to improve the spraying efficacy of UAV sprayers and reduce off-target losses, it is essential to understand the distribution characteristics of downwash airflow inside and around the target tree canopy. From this point of view, we suggest that the velocity and directional distribution characteristics of the downwash airflow should be focused on, which is a key factor affecting the droplets deposit distribution in the canopy, as reported in the present study.

Clearly, the airflow velocity and direction had a significant effect on spray deposit, and the fundamental cause of this is that the behaviors of droplets [53] and leaves [54] are influenced by airflow, which will be the focus of our future research. Although there is a correlation between the spray indicators, considering that this study aims to make a preliminary exploration of the effects of airflow indicators on deposition indicators, a large number of field tests for different fan parameters are still needed before we can build a predictive model.

5. Conclusions

Spray tests and airflow distribution tests were conducted with ten different fan settings using a multi-unit air-assisted sprayer. The effects of various airflow distribution patterns on spray deposition coverage and penetration were compared, and the correlation between airflow indicators and deposition indicators was evaluated. The results showed that the increased AI improved the *CAB* of pear (from 3.33% to 11.80% in the Top canopy and from 6.26% to 11.00% in the Upper canopy) and cherry leaves (from 3.61% to 10.87% in the Top canopy, from 1.36% to 9.04% in the Middle canopy, and from 3.40% to 9.04% in the Bottom canopy) but had no significant effect on spray penetration. As a whole, spray deposition in one part of the canopy was influenced by the combined effects of airflow in a wider area. The enhanced *X-v* and *Z-v* improved the *CAB* on the outside of pear canopy ($p < 0.001$), but for cherry, none of the airflow indicators had a significant impact on the *CAB* independently. On the other hand, the increased *ZY-d* for pear as well as the increased *X-v* and *Y-v* for cherry both showed negative effects on the *CAD* ($p < 0.01$).

According to our findings, for fruits with small leaves, such as pears, apples, citrus, etc., it is recommended to increase the airflow velocity in the horizontal and forward direction to optimize the spraying effect. As for cherry and other fruits with larger leaves, such as mangoes, lychees, etc., an increase in the upward airflow and an appropriate reduction of airflow velocity are suggested. For years, we have been committed to the improvement of spray performance on the abaxial side of leaves in orchards by setting optimal application parameters and improving the spray system. More research is still needed to clarify the multiple correlations between airflow velocity/direction and spray deposit distribution. Furthermore, the dynamical behavior of droplets and leaves in the airflow, as a fundamental principle of droplet deposition in the canopy, will be further investigated to improve our understanding of air-assisted spraying.

Author Contributions: Conceptualization, X.H. and Z.W.; methodology, T.L., P.Q. and Z.W.; validation, X.H.; formal analysis, T.L. and Z.W.; investigation, T.L., P.Q., S.X., Z.H. and L.H.; resources, S.X., Z.H. and L.H.; data curation, T.L., S.X., Z.H. and L.H.; writing—original draft preparation, T.L.; writing—review and editing, X.H. and Z.W.; visualization, T.L., Z.W.; supervision, X.H.; funding acquisition, X.H. All authors have read and agreed to the published version of the manuscript.

Funding: This research was funded by the National Natural Science Foundation of China (No. 31761133019), the earmarked fund for China Agriculture Research System (CARS-28) and the Deutsche Forschungsgemeinschaft (DFG, German Research Foundation)—project number 391697993.

Institutional Review Board Statement: Not applicable.

Informed Consent Statement: Not applicable.

Data Availability Statement: The data presented in this study are available on request from the corresponding author.

Acknowledgments: The authors would like to give special thanks to Dong Xue for providing the test orchard and the sprayer storage.

Conflicts of Interest: The authors declare no conflict of interest.

References

1. The 14th Five-Year Plan for National Economic and Social Development of the People's Republic of China and Outline of the Vision for 2035. Available online: http://www.gov.cn/xinwen/2021-03/13/content_5592681.htm (accessed on 13 April 2022).

2. Dong, Z.X.; Wang, Y.W.; Liu, Q.Z.; Tian, B.L.; Liu, Z.L. Laboratory Screening of 26 Essential Oils Against Cacopsylla chinensis (Hemiptera: Psyllidae) and Field Confirmation of the Top Performer, Perilla frutescens (Lamiales: Lamiaceae). *J. Econ. Entomol.* **2019**, *112*, 1299–1305. [CrossRef]
3. De Curtis, F.; Ianiri, G.; Raiola, A.; Ritieni, A.; Succi, M.; Tremonte, P.; Castoria, R. Integration of biological and chemical control of brown rot of stone fruits to reduce disease incidence on fruits and minimize fungicide residues in juice. *Crop Prot.* **2019**, *119*, 158–165. [CrossRef]
4. Dekeyser, D.; Duga, A.T.; Verboven, P.; Endalew, A.M.; Hendrickx, N.; Nuyttens, D. Assessment of orchard sprayers using laboratory experiments and computational fluid dynamics modelling. *Biosyst. Eng.* **2013**, *114*, 157–169. [CrossRef]
5. Delele, M.A.; De Moor, A.; Sonck, B.; Ramon, H.; Nicolaï, B.M.; Verboven, P. Modelling and Validation of the Air Flow generated by a Cross Flow Air Sprayer as affected by Travel Speed and Fan Speed. *Biosyst. Eng.* **2005**, *92*, 165–174. [CrossRef]
6. Escolà, A.; Rosell-Polo, J.R.; Planas, S.; Gil, E.; Pomar, J.; Camp, F.; Llorens, J.; Solanelles, F. Variable rate sprayer. Part 1–Orchard prototype: Design, implementation and validation. *Comput. Electron. Agric.* **2013**, *95*, 122–135. [CrossRef]
7. Khot, L.R.; Ehsani, R.; Albrigo, G.; Larbi, P.A.; Landers, A.; Campoy, J.; Wellington, C. Air-assisted sprayer adapted for precision horticulture: Spray patterns and deposition assessments in small-sized citrus canopies. *Biosyst. Eng.* **2012**, *113*, 76–85. [CrossRef]
8. Garcerá, C.; Moltó, E.; Chueca, P. Spray pesticide applications in Mediterranean citrus orchards: Canopy deposition and off-target losses. *Sci. Total Environ.* **2017**, *599–600*, 1344–1362. [CrossRef]
9. Sinha, R.; Ranjan, R.; Khot, L.R.; Hoheisel, G.A.; Grieshop, M.J. Comparison of within canopy deposition for a solid set canopy delivery system (SSCDS) and an axial–fan airblast sprayer in a vineyard. *Crop Prot.* **2020**, *132*, 105124. [CrossRef]
10. Michael, C.; Gil, E.; Gallart, M.; Stavrinides, M.C. Influence of Spray Technology and Application Rate on Leaf Deposit and Ground Losses in Mountain Viticulture. *Agriculture* **2020**, *10*, 615. [CrossRef]
11. Blanco, M.N.; Fenske, R.A.; Kasner, E.J.; Yost, M.G.; Seto, E.; Austin, E. Real-Time Monitoring of Spray Drift from Three Different Orchard Sprayers. *Chemosphere* **2019**, *222*, 46–55. [CrossRef]
12. Rosell, J.R.; Sanz, R. A review of methods and applications of the geometric characterization of tree crops in agricultural activities. *Comput. Electron. Agric.* **2012**, *81*, 124–141. [CrossRef]
13. Vallet, A.; Tinet, C. Characteristics of droplets from single and twin jet air induction nozzles: A preliminary investigation. *Crop Prot.* **2013**, *48*, 63–68. [CrossRef]
14. Grella, M.; Gallart, M.; Marucco, P.; Balsari, P.; Gil, E. Ground Deposition and Airborne Spray Drift Assessment in Vineyard and Orchard: The Influence of Environmental Variables and Sprayer Settings. *Sustainability* **2017**, *9*, 728. [CrossRef]
15. Nuyttens, D.; Taylor, W.A.; De Schampheleire, M.; Verboven, P.; Dekeyser, D. Influence of nozzle type and size on drift potential by means of different wind tunnel evaluation methods. *Biosyst. Eng.* **2009**, *103*, 271–280. [CrossRef]
16. Salyani, M.; Whitney, J.D. GROUND SPEED EFFECT ON SPRAY DEPOSITION INSIDE CITRUS TREES. *Trans. ASAE* **1990**, *33*, 0361–0366. [CrossRef]
17. Nuyttens, D.; Baetens, K.; De Schampheleire, M.; Sonck, B. Effect of nozzle type, size and pressure on spray droplet characteristics. *Biosyst. Eng.* **2007**, *97*, 333–345. [CrossRef]
18. Combellack, J.H.; Westernt, N.M.; Richardson, R.G. A comparison of the drift potential of a novel twin fluid nozzle with conventional low volume flat fan nozzles when using a range of adjuvants. *Crop Prot.* **1996**, *15*, 147–152. [CrossRef]
19. Bouse, L.F.; Kirk, I.W.; Bode, L.E. Effect of spray mixture on droplet size. *Trans. ASAE* **1990**, *33*, 0783–0788. [CrossRef]
20. Derksen, R.C.; Fox, R.D.; Brazee, R.D.; Krause, C.R. Coverage and drift produced by air induction and conventional hydraulic nozzles used for orchard applications. *Am. Soc. Agric. Biol. Eng.* **2000**, *1*, 3131–3149. [CrossRef]
21. Cross, J.V.; Walklate, P.J.; Murray, R.A.; Richardson, G.M. Spray deposits and losses in different sized apple trees from an axial fan orchard sprayer: 2. Effects of spray quality. *Crop Prot.* **2001**, *20*, 333–343. [CrossRef]
22. Garcerá, C.; Román, C.; Moltó, E.; Abad, R.; Insa, J.A.; Torrent, X.; Planas, S.; Chueca, P. Comparison between standard and drift reducing nozzles for pesticide application in citrus: Part II. Effects on canopy spray distribution, control efficacy of Aonidiella aurantii (Maskell), beneficial parasitoids and pesticide residues on fruit. *Crop Prot.* **2017**, *94*, 83–96. [CrossRef]
23. Gil, E.; Landers, A.; Gallart, M.; Llorens, J. Development of two portable patternators to improve drift control and operator training in the operation of vineyard sprayers. *Span. J. Agric. Res.* **2013**, *11*, 615–625. [CrossRef]
24. Li, L.; He, X.; Song, J.; Liu, Y.; Wang, Z.; Li, J.; Jia, X.; Liu, Z. Comparative experiment on profile variable rate spray and conventional air assisted spray in orchards. *Nongye Gongcheng Xuebao/Trans. Chin. Soc. Agric. Eng.* **2017**, *33*, 56–63. [CrossRef]
25. Farooq, M.; Salyani, M.; Engineering, B.; Largo, K. Spray Penetration into the Citrus Tree Canopy from Two Air-Carrier Sprayers. In Proceedings of the 2002 ASAE Annual Meeting, Chicago, IL, USA, 28–31 July 2002; p. 1. [CrossRef]
26. Holownicki, R.; Doruchowski, G.; Godyn, A.; Swiechowski, W. Variation of spray deposit and loss with air-jet directions applied in orchards. *J. Agric. Eng. Res.* **2000**, *77*, 129–136. [CrossRef]
27. Pezzi, F.; Rondelli, V. The performance of an air-assisted sprayer operating in vines. *J. Agric. Eng. Res.* **2000**, *76*, 331–340. [CrossRef]
28. Miranda-Fuentes, A.; Rodríguez-Lizana, A.; Gil, E.; Agüera-Vega, J.; Gil-Ribes, J.A. Influence of liquid-volume and airflow rates on spray application quality and homogeneity in super-intensive olive tree canopies. *Sci. Total Environ.* **2015**, *537*, 250–259. [CrossRef]
29. Cross, J.V.; Walklate, P.J.; Murray, R.A.; Richardson, G.M. Spray deposits and losses in different sized apple trees from an axial fan orchard sprayer: 3. Effects of air volumetric flow rate. *Crop Prot.* **2003**, *22*, 381–394. [CrossRef]

30. Li, J.; Li, Z.; Ma, Y.; Cui, H.; Yang, Z.; Lu, H. Effects of leaf response velocity on spray deposition with an air-assisted orchard sprayer. *Int. J. Agric. Biol. Eng.* **2021**, *14*, 123–132. [CrossRef]
31. Dekeyser, D.; Foqué, D.; Duga, A.T.; Verboven, P.; Hendrickx, N.; Nuyttens, D. Spray deposition assessment using different application techniques in artificial orchard trees. *Crop Prot.* **2014**, *64*, 187–197. [CrossRef]
32. Wachter, S.; Jakobs, T.; Kolb, T. Effect of solid particles on droplet size applying the time-shift method for spray investigation. *Appl. Sci.* **2020**, *10*, 7615. [CrossRef]
33. ISO-ISO 13320:2020-Particle Size Analysis—Laser Diffraction Methods. Available online: https://www.iso.org/standard/69111.html (accessed on 3 March 2022).
34. ISO-ISO 9898:2000-Equipment for Crop Protection—Test Methods for Air-Assisted Sprayers for Bush and Tree Crops. Available online: https://www.iso.org/standard/17781.html (accessed on 4 March 2022).
35. Javier García-Ramos, F.; Vidal, M.; Boné, A.; Malón, H.; Aguirre, J. Analysis of the Air Flow Generated by an Air-Assisted Sprayer Equipped with Two Axial Fans Using a 3D Sonic Anemometer. *Sensors* **2012**, *12*, 7598. [CrossRef] [PubMed]
36. Zhu, H.; Salyani, M.; Fox, R.D. A portable scanning system for evaluation of spray deposit distribution. *Comput. Electron. Agric.* **2011**, *76*, 38–43. [CrossRef]
37. Owen-Smith, P.; Perry, R.; Wise, J.; Jamil, R.Z.R.; Gut, L.; Sundin, G.; Grieshop, M. Spray coverage and pest management efficacy of a solid set canopy delivery system in high density apples. *Pest Manag. Sci.* **2019**, *75*, 3050–3059. [CrossRef] [PubMed]
38. Interpolation for 2-D Gridded Data in Meshgrid Format. Available online: https://www.mathworks.com/help/releases/R2021a/matlab/ref/interp2.html#btyq8s0-1-method (accessed on 4 March 2021).
39. Mehmood, T.; Liland, K.H.; Snipen, L.; Sæbø, S. A review of variable selection methods in Partial Least Squares Regression. *Chemom. Intell. Lab. Syst.* **2012**, *118*, 62–69. [CrossRef]
40. Chong, I.G.; Jun, C.H. Performance of some variable selection methods when multicollinearity is present. *Chemom. Intell. Lab. Syst.* **2005**, *78*, 103–112. [CrossRef]
41. Marucco, P.; Tamagnone, M.; Balsari, P. Study of Air Velocity Adjustment to Maximise Spray Deposition in Peach Orchards. *Agr. Eng. Int. CIGR J.* **2008**, *X*, 1–13.
42. Cross, J.V.; Walklate, P.J.; Murray, R.A.; Richardson, G.M. Spray deposits and losses in different sized apple trees from an axial fan orchard sprayer: 1. Effects of spray liquid flow rate. *Crop Prot.* **2001**, *20*, 13–30. [CrossRef]
43. Musiu, E.M.; Qi, L.; Wu, Y. Spray deposition and distribution on the targets and losses to the ground as affected by application volume rate, airflow rate and target position. *Crop Prot.* **2019**, *116*, 170–180. [CrossRef]
44. Glotfelty, D.E.; Schomburg, C.J.; McChesney, M.M.; Sagebiel, J.C.; Seiber, J.N. Studies of the distribution, drift, and volatilization of diazinon resulting from spray application to a dormant peach orchard. *Chemosphere* **1990**, *21*, 1303–1314. [CrossRef]
45. Balsari, P.; Marucco, P.; Tamagnone, M. A system to assess the mass balance of spray applied to tree crops. *Trans. ASAE* **2005**, *48*, 1689–1694. [CrossRef]
46. Grella, M.; Marucco, P.; Zwertvaegher, I.; Gioelli, F.; Bozzer, C.; Biglia, A.; Manzone, M.; Caffini, A.; Fountas, S.; Nuyttens, D.; et al. The effect of fan setting, air-conveyor orientation and nozzle configuration on airblast sprayer efficiency: Insights relevant to trellised vineyards. *Crop Prot.* **2022**, *155*, 105921. [CrossRef]
47. He, X.K.; Bonds, J.; Herbst, A.; Langenakens, J. Recent development of unmanned aerial vehicle for plant protection in East Asia. *Int. J. Agric. Biol. Eng.* **2017**, *10*, 18–30. [CrossRef]
48. Zhang, H.; Qi, L.; Wu, Y.; Liu, W.; Cheng, Z.; Musiu, E. Spatio-temporal Distribution of Down-wash Airflow for Multi-rotor Plant Protection UAV Based on Porous Model. *Nongye Jixie Xuebao/Trans. Chin. Soc. Agric. Mach.* **2019**, *50*, 112–122. [CrossRef]
49. Zhang, H.; Qi, L.; Wu, Y.; Cheng, Z.; Liu, W.; Musiu, E.; Xiao, Y.; Yang, Z. Distribution characteristics of rotor downwash airflow field under spraying on orchard using unmanned aerial vehicle. *Nongye Gongcheng Xuebao/Trans. Chin. Soc. Agric. Eng.* **2019**, *35*, 44–54. [CrossRef]
50. Pan, Z.; Lie, D.; Qiang, L.; Shaolan, H.; Shilai, Y.; Yande, L.; Yongxu, Y.; Haiyang, P. Effects of citrus tree-shape and spraying height of small unmanned aerial vehicle on droplet distribution. *Int. J. Agric. Biol. Eng.* **2016**, *9*, 45–52. [CrossRef]
51. Zhang, H.; Qi, L.; Wan, J.; Musiu, E.M.; Zhou, J.; Lu, Z.; Wang, P. Numerical simulation of downwash airflow distribution inside tree canopies of an apple orchard from a multirotor unmanned aerial vehicle (UAV) sprayer. *Comput. Electron. Agric.* **2022**, *195*, 106817. [CrossRef]
52. Xue, S.; Xi, X.; Lan, Z.; Wen, R.; Ma, X. Longitudinal drift behaviors and spatial transport efficiency for spraying pesticide droplets. *Int. J. Heat Mass Transf.* **2021**, *177*, 121516. [CrossRef]
53. Shao, C.P.; Chen, Y.J.; Lin, J.Z. Wind induced deformation and vibration of a Platanus acerifolia leaf. *Acta Mech. Sin. 2012 283* **2012**, *28*, 583–594. [CrossRef]
54. Jiang, H.; Xin, D.; Zhang, H. Wind-tunnel study of the aerodynamic characteristics and mechanical response of the leaves of Betula platyphylla Sukaczev. *Biosyst. Eng.* **2021**, *207*, 162–176. [CrossRef]

MDPI AG
Grosspeteranlage 5
4052 Basel
Switzerland
Tel.: +41 61 683 77 34

Agronomy Editorial Office
E-mail: agronomy@mdpi.com
www.mdpi.com/journal/agronomy

www.ingramcontent.com/pod-product-compliance
Lightning Source LLC
Chambersburg PA
CBHW040759250926
51850CB00007BA/1036
* 9 7 8 3 7 2 5 8 1 5 8 3 8 *